Applied Biochemistry and Bioengineering

VOLUME 1

Advisory Board

Applied Biochemistry and Bioengineering

VOLUME 1
Immobilized Enzyme Principles

Edited by

Lemuel B. Wingard, Jr.

Department of Pharmacology, School of Medicine
University of Pittsburgh
Pittsburgh, Pennsylvania

Ephraim Katchalski-Katzir

Department of Biophysics
The Weizmann Institute of Science
Rehovot, Israel

Leon Goldstein

Department of Biochemistry
Tel Aviv University
Tel Aviv, Israel

ACADEMIC PRESS NEW YORK SAN FRANCISCO LONDON 1976
A Subsidiary of Harcourt Brace Jovanovich, Publishers

7226 - 2412

CHEMISTRY

ACADEMIC PRESS, INC.
111 Fifth Avenue, New York, New York 10003

United Kingdom Edition published by
ACADEMIC PRESS, INC. (LONDON) LTD.
24/28 Oval Road, London NW1

LIBRARY OF CONGRESS CATALOG CARD NUMBER: 76-9161

ISBN 0–12–041101–6

PRINTED IN THE UNITED STATES OF AMERICA

Contents

TP248
.3
I45
CHEM

List of Contributors

Numbers in parentheses indicate the pages on which the authors' contributions begin.

Ichiro Chibata (329), Research Laboratory of Applied Biochemistry, Tanabe Seiyaku Co. Ltd., Osaka, Japan

A. Constantinides (221), Department of Chemical and Biochemical Engineering, Rutgers University, New Brunswick, New Jersey

B. Davidson (221), Department of Chemical and Biochemical Engineering, Rutgers University, New Brunswick, New Jersey

Jean-Marc Engasser (127), Laboratoire des Sciences du Génie Chimique, E.N.S.I.C., Nancy, France

Leon Goldstein (1, 23), Department of Biochemistry, The George S. Wise Center for Life Sciences, Tel Aviv University, Tel Aviv, Israel

Csaba Horvath (127), Biochemical Engineering Group, Department of Engineering and Applied Science, Yale University, New Haven, Connecticut

Ephraim Katchalski-Katzir (1), Department of Biophysics, The Weizmann Institute of Science, Rehovot, Israel

Georg Manecke (23), Institut für Organische Chemie der Freien Universität Berlin and Fritz-Haber-Institut der Max-Planck-Gesellschaft, Berlin, Germany

Tetsuya Tosa (329), Research Laboratory of Applied Biochemistry, Tanabe Seiyaku Co. Ltd., Osaka, Japan

K. Venkatasubramanian (221), Department of Chemical and Biochemical Engineering, Rutgers University, New Brunswick, New Jersey

W. R. Vieth (221), Department of Chemical and Biochemical Engineering, Rutgers University, New Brunswick, New Jersey

Introduction to the Series

The biological sciences have made remarkable progress during the last two decades. A wealth of information has accumulated on the structure and function of the materials comprising the living organism, on the chemical and physical aspects of a great number of biological processes, on catalysis in biological systems, and on the relationship between structure and function in enzymes and other biospecific macromolecules. This work on fundamentals has been accompanied by salient achievements in the fields of microbial genetics, tissue culture, and related areas. Nonetheless, the communication gap between pure and applied science has still to be narrowed to make better use of the potential of some of the recent discoveries in biology. The cooperation and mutual esteem and understanding between basic scientists and engineers is thus needed to attain cross-fertilization between the diverse approaches and experiences of the two disciplines; hence it is the aim of this publication series to bring together comprehensive summaries of work being done in the overlapping areas of engineering and biology.

Several areas of interaction between biological scientists and engineers have already begun to emerge; the term bioengineering has been coined to cover this range of interactions. Chemical engineers and microbiologists have been working together in the industrial production of foods, beverages, and chemicals by fermentation. Mechanical engineers and chemical engineers have cooperated with physiologists and people in the medical sciences to develop artificial organs, special life-support machines, artificial materials, and prosthetic devices. Electrical engineers together with physiologists have begun to apply system approaches to the study of biological control mechanisms. More recently biochemists and biophysicists have interacted with chemical engineers to explore the utilization of enzymes as special catalysts for use in industrial processing, analytical chemistry, and medicine.

The basic understanding of biological phenomena appears rooted in events that occur at the molecular level. Since current biological research is heavily committed in this direction, it seems logical to stress the underlying common denominator that biochemistry can bring to the understanding of the many facets of biological systems.

Thus the title of this serial publication, *Applied Biochemistry and Bioengineering,* has been selected to emphasize the biochemical common denominator underlying the interaction of engineering practice and the biological sciences for technological development. It is hoped that the series will provide guidance in the application of these technological developments for the benefit of mankind.

THE EDITORS
February 1976

Preface

Utilization of immobilization techniques for the study and application of enzyme catalysts in a variety of potential end uses seems to be an especially appropriate subject for the first volume of this series. Both scientific and engineering inputs are required, with a strong reliance on basic biochemistry and biophysics. This volume places a major emphasis on the preparation of enzyme-support systems, on the effects caused by the concurrent phenomena of enzyme-catalyzed reaction kinetics and mass transfer resistances, and on how these are incorporated into the design of enzyme-catalyzed reactor systems. An additional chapter is included to show some examples of the practical application of immobilized enzymes.

Immobilized Enzymes—A Survey[1]

Leon Goldstein

*Department of Biochemistry, The George S. Wise Center for Life Sciences,
Tel Aviv University, Tel Aviv, Israel,*

and

Ephraim Katchalski-Katzir

*Department of Biophysics, The Weizmann Institute of Science,
Rehovot, Israel*

Until rather recently, immobilized enzymes were more of a curiosity, conceived in the initial stages of the rapprochement between biochemistry and polymer chemistry. The motivations underlying the early activities could be rationalized as an attempt to apply the accumulated experience in "making" macromolecules to the more exacting task of grafting a biocatalyst onto a polymeric structure designed to lead to a biologically active conjugate. From the biochemist's point of view, such conjugates could serve as water-insoluble highly specific reagents, easily removable from the reaction mixture at any predetermined stage of the reaction.

This approach contained the seeds of essentially all basic concepts and developments that materialized in the decade that followed the preparation of the first stable and reusable water-insoluble enzyme derivatives in the late 1950s. Hence, realization of the potential of immobilized enzymes as a new type of model system for the investigation of isolated aspects of complex biological phenomena on the one hand, and of their industrial potential as a new type of highly specific heterogeneous catalyst for continuous processes on the other, brought together chemical engineers, organic and physical chemists, biochemists, biologists, and microbiologists, each with his own expertise. This meeting of disciplines, within the loose framework of "enzyme engineering" has generated new concepts as well as new technologies.

[1] The authors dedicate this article to Georg Manecke on his sixtieth birthday.

Historically, the earliest reported cases of protein immobilization involved physical adsorption of the protein onto particles such as charcoal, kaolinite, red blood cell stroma, cellulose, and glass beads (Nelson and Hitchcocks, 1921; Langmuir and Schaefer, 1938, 1939). The first attempts to make use of such preparations were soon to follow, and already in the 1930s work on the application of adsorbed antigens for the isolation of specific antibodies could be found in the immunological literature (for review, see Isliker, 1957). The unpredictable behavior of these systems and the inability to obtain clean separations led the early investigators to the realization that fixation by forces stronger than adsorption was necessary. It is thus not surprising that the initial attempts at covalent fixation onto water-insoluble supports were carried out by the immunologists. Landsteiner and Van der Scheer (1936) described the coupling of diazotized haptens to blood cell stroma, and the utilization of the insoluble preparations for the isolation of the corresponding antibodies. Their work was followed, after the interruption of World War II, by the first experiments on covalent binding of a variety of proteins including enzymes as well as antigens to chemically well defined water-insoluble polymeric supports. The methodology of binding, however, was limited to the commercial polymers available at the time—derivatized celluloses and styrene polymers. In 1949, Micheel and Evers described the covalent binding of proteins to carboxymethyl cellulose azide. Campbell *et al.* (1951) reported on the coupling of ovalbumin to diazotized *p*-aminobenzyl cellulose, and the isolation of ovalbumin antibodies on the immunoadsorbent thus obtained.

These first steps were soon followed by other methods of coupling proteins to polymers. Isliker (1953) prepared immunoadsorbents with carboxychloride and sulfonylchloride derivatives of polystyrene; Manecke (Manecke and Gillert, 1955) utilized diazotized poly(*p*-aminostyrene) and later a poly(4-isocyanatostyrene) derivative (Manecke *et al.*, 1958) for the same purpose. In parallel the immobilization of enzymes by similar approaches was tried by Grubhofer and Schleith (1953, 1954), who coupled carboxypeptidase and amylase to diazotized poly(*p*-aminostyrene) and by Manecke (Manecke and Gillert, 1955; Manecke *et al.*, 1958; Manecke and Singer, 1960) and Brandenberger (1955, 1956, 1957), who used poly(*p*-aminostyrene) and poly(4-isocyanatostyrene) to bind enzymes. The amounts of bound protein and the enzymic activities retained in the immobilized preparations obtained by these methods were, however, relatively poor, presumably owing to the hydrophobicity of the supports. This early work was improved upon by Mitz and Summaria (1961), who coupled trypsin and chymotrypsin to *p*-aminobenzyl cellulose and carboxymethyl cellulose hydrazide preparations of known degrees of substitution, and by Katchalski (Bar-Eli and Katchalski, 1960, 1963; Cebra *et al.*, 1961; Katchalski, 1962), who prepared water-insoluble derivatives of trypsin and papain by coupling the enzymes to diazotized leucine–*p*-aminophenylalanine copolymers. In the case of trypsin, a polytyrosyl derivative of the enzyme was used to protect it from inactivation in the course of the coupling reaction. Concurrently a series of copolymers of methacrylic acid and methacrylic acid-3-fluoro-4,6-dinitroanilide of varying ratios of comonomers were prepared by Manecke (Manecke and Singer, 1960; Manecke, 1962). In these preparations the 3-fluoro-4,6-dinitroanilide group served as the reactive moiety, and the carboxylic groups as the component bestowing hydrophilicity.

This work was paramount in delineating the objectives as well as the problems facing the chemist aiming at the covalent immobilization of biologically active proteins. The main conclusions to be drawn were as follows: (1) Derivatized polymers with groups of different chemical specificities are needed for attaining biologically active immobilized preparations of different proteins. (2) The chemical nature of the support material may determine not only the amount of bound protein, but also the extent to which its biological activity is retained; more specifically, supports rich in hydrophobic groups give immobilized preparations of low stability while the presence of hydrophilic groups enhances the stability and may in some cases counteract the deleterious effects of a hydrophobic environment. (3) Protection of the enzyme by chemical modification prior to coupling may sometimes be necessary.

These ideas, summarized in several reviews (Manecke, 1962; Katchalski, 1962; Silman and Katchalski, 1966), led to more coherent attempts at designing polymers of predetermined characteristics in terms of their mechanical properties, their effect on the stability of the bound protein and the type of functional group through which they would attach to the protein. Thus the leucine-*p*-aminophenylalanine and methacrylic acid–methacrylic acid-3-fluoro-4,6-dinitroanilide copolymers were soon followed by ethylene-maleic anhydride copolymers (Levin *et al.*, 1964; Goldstein *et al.*, 1964), derivatized cellulose (Lilly *et al.*, 1965, 1966; Kay and Lilly, 1970), cyanogen bromide-activated Sephadex and Sepharose (Axén *et al.*, 1967; Porath *et al.*, 1967), and, somewhat later, derivatized acrylic polymers and copolymers (Inman and Dintzis, 1969; Mosbach, 1970; Barker *et al.*, 1970; Manecke *et al.*, 1970), derivatized porous glass (Weetall, 1969, 1970), and derivatized nylons (Hornby and Filippusson, 1970; Inman and Hornby, 1972; Goldstein *et al.*, 1974; Campbell *et al.*, 1975). The work on enzyme immobilization has been extensively reviewed (Goldman *et al.*, 1971b; Melrose, 1971; Smiley and Strandberg, 1972; Orth and Brümmer, 1972; Royer *et al.*, 1973; Zaborsky, 1973; Manecke, 1974; Goldstein and Manecke, this volume).

The work of Inman and Dintzis (1969) on derivatized polyacrylamide suggested a more general approach to the problems of enzyme immobilization, i.e., the use of "parent carrier polymers," which can by consecutive chemical manipulations be transformed into the chemical species suitable for a specific task.

Although the mainstream in the methodology of enzyme immobilization centered until recently on covalent linking, considerable effort was devoted throughout the years to developing methods for noncovalent fixation of enzymes. Such methods would be more general, and of particular significance in the case of enzymes sensitive to chemical modification. The methods that have eventually gained acceptance are: physical adsorption of enzymes or enzyme derivatives onto supports of superior adsorptive properties, mainly ion exchangers (Tosa *et al.*, 1966; Messing, 1975; Hofstee and Otillio, 1973; Stanley and Palter, 1973; Gladishev *et al.*, 1973; Solomon and Levin, 1974); occlusion into cross-linked polymer gels (Bernfeld and Wan, 1963; Mosbach and Mosbach, 1966; Bernfeld *et al.*, 1968; Mosbach, 1970); and recently encapsulation into microcapsules (Chang, 1964, 1972; Chang *et al.*, 1966), fibers (Dinelli, 1972; Marconi *et al.*, 1974), and liposomes (Gregoriadis *et al.*, 1971; Gregoriadis, 1974).

The idea that the specificity of biological macromolecules as reflected in their high binding constants for substrates, inhibitors, or effectors, can be used for separation and purification through formation of insoluble complexes had been in the air since the early work on immunoadsorbents (Campbell *et al.*, 1951; Lerman, 1953a,b). This concept applied by Schramm and co-workers to the purification of α-amylase by precipitating the enzyme–substrate complex out of solution (Schramm and Loyter, 1962, 1966; Levitzki *et al.*, 1964), gained in importance with the perfection of immobilization techniques.

Immobilized derivatives of trypsin and chymotrypsin were used for the selective adsorption of the pancreatic inhibitors of these enzymes from crude extracts. The inhibitors were subsequently eluted under conditions where binding was weakest (Fritz *et al.*, 1966, 1967, 1968, 1969). The reversal of the procedure, i.e., the use of the purified inhibitors in immobilized form for the isolation of pure enzymes, was a natural extension of the same basic concept. The immense importance of enzyme purification by selective adsorption was recognized soon thereafter in the report by Cuatrecasas, Wilchek, and Anfinsen (1968) on the *affinity chromatography* purification of staphylococcal nuclease, chymotrypsin, and carboxypeptidase A on columns containing synthetic low-molecular-weight inhibitors of these enzymes, covalently attached to a solid matrix. The method has found application in the purification of a wide variety of biological substances (for reviews, see Cuatrecasas and Anfinsen, 1971; Cuatrecasas, 1972; Wilchek, 1974; Dunlap, 1974; Jakoby and Wilchek, 1974; Wilchek and Hexter, 1976).

As was shown recently, immobilized analogs of cofactors, such as adenosine 5'-monophosphate (AMP), adenine nicotinamide dinucleotide (NAD$^+$) and pyridoxal 5'-phosphate, which have affinity for a broad spectrum of enzymes, could be used to adsorb an entire family of enzymes, individual members being then eluted by appropriate "specific elution" procedures (Mosbach, 1974; Mosbach *et al.*, 1971, 1972; Kaplan *et al.*, 1974). These advances in "general ligand" affinity chromatography cross-fertilized the field of immobilized enzymes, instigating methods for the fixation of enzymes via or together with immobilized cofactors or cofactor analogs. Some current work could illustrate these new trends: Enzymically active water-insoluble glycogen-phosphorylase *b* could be prepared by immobilizing the enzyme on an insoluble derivative of its effector, AMP[N^6-(6-aminohexyl)adenosine 5'-phosphate Sepharose] (Mosbach and Gestrelius, 1974). By a similar approach Fukui *et al.* (1975) immobilized tyrosinase and tryptophanase on water-insoluble derivatives of pyridoxal 5'-phosphate—an effector of these enzymes; their work furthermore demonstrated that, in the case of multisubunit enzymes, attachment to an insoluble matrix, containing immobilized effector, via site-directed binding to one subunit only, could be sufficient to attain immobilization (Fukui *et al.*, 1975).

Most of the initial work on the methodology of enzyme immobilization was done with hydrolases, in particular proteases, owing to their accessibility and relative simplicity. Insoluble derivatives of papain, trypsin, and chymotrypsin using different types of support materials, charged as well as electrically neutral, were thus among the first immobilized enzymes to be employed in the next phase of development, clarification of some of the more fundamental aspects of the kinetic behavior of immobilized enzymes and the first serious attempts to apply immobilized enzymes in the laboratory, as stable reusable and removable reagents. Work in these areas has been extensively reviewed and will be only briefly highlighted here (see Goldstein and

Katchalski, 1968; Goldstein, 1969, 1970; Stark, 1971; Laidler and Sundaram, 1971; Goldman *et al.*, 1971a,b; Katchalski *et al.*, 1971; Lilly and Dunnill, 1971, 1972; Vieth and Venkatasubramanian, 1973, 1974; Zaborsky, 1973; Bunting and Laidler, 1973; Weetall, 1975).

In 1964 Goldstein and co-workers showed that the pH-activity profiles of polyanionic derivatives of trypsin and chymotrypsin were displaced toward more alkaline pH values relative to the native enzymes; conversely, the pH-activity profiles of polycationic derivatives of the same enzymes were displaced toward more acidic pH values (Goldstein *et al.*, 1964; Goldstein and Katchalski, 1968; Pecht and Levin, 1972; Goldstein, 1972; Manecke, 1975). These effects could be interpreted in terms of changes in the local concentration of hydrogen and hydroxyl ions in the domain of the bound enzyme, i.e., by a modified microenvironment, due to Donnan-type partitioning of hydrogen ions between the bulk solution and the charged enzyme particles. The polyelectrolyte nature of these effects could be demonstrated by their cancellation at high ionic strength.

In the case of charged substrates (e.g., the systems, esters or amides of arginine, acted upon by polyanionic or polycationic derivatives of trypsin, bromelain, ficin, or papain), partitioning of substrate resulting from attractive or repulsive interactions with the polyelectrolyte support, i.e., higher or lower local substrate concentration, could account for the observed lowering or increase in the values of the apparent Michaelis constants (Goldstein *et al.*, 1964; Wharton *et al.*, 1968). Assuming a Boltzmann distribution for the charged low-molecular-weight species in solution, Goldstein *et al.* (1964) could relate the observed shifts in pH-activity curves (ΔpH) and Michaelis constants (ΔpK_m) to the electrostatic potential in the domain of a charged enzyme particle. These phenomena were later analyzed in greater detail (Wharton *et al.*, 1968; Hornby *et al.*, 1968; Shuler *et al.*, 1972; Sundaram *et al.*, 1970; Bunting and Laidler, 1973; Kobayashi and Laidler, 1973). It should be noted that some of the findings of Goldstein *et al.* (1964) were anticipated by McLaren (McLaren and Estermann, 1957; Esterman *et al.*, 1959; McLaren and Babcock, 1959; McLaren, 1960; McLaren and Packer, 1970), who reported alkaline shifts in the pH-activity curves of chymotrypsin adsorbed on kaolinite particles and ascribed the changes to differences in the surface pH of the particles. These authors were also the first to point out the biological implications of the observed phenomena. Their work, however, coming too early, and addressed to a different audience—the soil chemist interested in clays—remained relatively unknown until the mid-1960s.

The main conclusion to be drawn from these studies was that the magnitude of the perturbation of the apparent kinetic parameters of an immobilized enzyme could serve in principle as a measure of the effective concentrations of substrate, modifier, or inhibitor at the site of the enzymic reaction. Moreover, the microenvironment concept emphasized the uncertainties and limitations of the prevalent approach of reconstituting metabolic pathways in cells via solubilization of the multienzyme complexes and the study of individual enzymic reactions *in vitro* (see, e.g., Green and Silman, 1967; Brown, 1971; Katchalski *et al.*, 1971). These aspects gained in significance after the preparation of enzyme membranes (Goldman *et al.*, 1965, 1968a,b, 1971a,b; Selegny *et al.*, 1968; Broun *et al.*, 1969), enzyme columns (Bar-Eli and Katchalski, 1963; Lilly *et al.*, 1966; Lilly and Sharp, 1968; Hornby *et al.*, 1968), and immobilized multienzyme systems (Mosbach and Mattiasson, 1970; Mattiasson and Mossbach, 1971; Goldman and Katchalski, 1971; Broun *et al.*, 1972; Lecoq *et al.*, 1975).

In 1965 Goldman and co-workers found that a papain–collodion membrane acting on ester substrates displayed distorted pH-activity profiles (Goldman *et al.*, 1965, 1968a). The anomalies were attributed to the local accumulation of hydrogen ions, produced by the hydrolysis of the ester substrates within the porous membrane. This interpretation was supported by the finding that grinding the enzyme membrane into powder led to cancellation of the effect. Using coupled reaction-diffusion models, Goldman and others showed that substrate and product concentration gradients are established within an enzyme membrane owing to diffusional limitations on the translocation of substrate and product. Hence substrate depletion is reflected in an increase in the value of the experimentally determined Michaelis constant. Moreover, the full enzymic activity of the membrane could be realized only in the case of very poor substrate, viz., extremely slow reactions (Goldman *et al.*, 1968a,b; Sundaram *et al.*, 1970; Selegny *et al.*, 1971). These studies introduced the concept of a microenvironment generated by an enzymic reaction taking place in a sterically constrained system.

Extension of the experimental investigation of enzyme membranes to very fast enzymes, such as alkaline phosphatase or glucose oxidase (Goldman *et al.*, 1971a,b; Broun *et al.*, 1969; Selegny *et al.*, 1971), showed that theoretical models based solely on internal diffusional resistances within a porous support could not fully account for the highly perturbed values of the Michaelis constants. The experimental findings could, however, be explained if in addition the existence of concentration gradients across unstirred layers (the Nernst diffusion layers; Nernst, 1904) around the enzyme membranes were assumed (Goldman *et al.*, 1971a,b; Goldman, 1973).

In parallel, intensive work was being carried out on continuous-flow packed-bed and continuous-stirred tank-enzyme reactors; analytical expressions correlating the degree of conversion of substrate for systems obeying Michaelis–Menten kinetics with the rate of flow of solution through the column or the agitation rate in the case of stirred-tank reactors were developed (Lilly *et al.*, 1966, 1974; Lilly and Sharp, 1968; Lilly and Dunnill, 1972; Hornby *et al.*, 1968). Here again, the high values obtained for the Michaelis constants, which could not be accounted for by the simple kinetic models, led to the assumption that substrate concentration gradients across a stagnant, unstirred layer surrounding the immobilized enzyme particles were responsible for the anomalous kinetic behavior and hence to the extension of the theoretical models to include these effects (Lilly and Sharp, 1968; Hornby *et al.*, 1968).

The kinetic consequences of diffusional limitations in immobilized enzyme systems were further demonstrated in several experiments: Axén showed that, in the case of particulate chymotrypsin–Sepharose conjugates of highly perturbed Michaelis constants, the values of the latter dropped to essentially those of the native enzyme, after solubilization with dextranase (Axén *et al.*, 1970). Mosbach, who investigated the behavior of multienzyme systems, showed that, in the case of two enzymes that carry out consecutive reactions, the initial rate of appearance of the last product is enhanced when the enzymes are immobilized together; moreover, the lag usually observed in the appearance of the last product, with the soluble enzymes, was absent with the immobilized two-enzyme system. The observations suggested that owing to the spatial proximity of the two enzymes on the supporting matrix, and the diffusional resistances deriving from unstirred layers, higher local concentrations of the intermediate product could be attained in the immobilized two-enzyme system (Mosbach and Mattiasson, 1970; Mattiasson and Mosbach, 1971; Gestrelius *et al.*, 1972, 1973; Mosbach *et al.*, 1974a,b). A theoretical analysis based on these assumptions (Goldman and Katchalski, 1971) gave predictions in good agreement with the experimental observations.

The microenvironment and diffusional resistance concepts roughly outlined here in the sequence of their formulation have been applied to rather sophisticated model enzyme membranes and particulate immobilized-enzyme systems to study aspects of structure-modulated kinetics; these include the precise physical meaning of experimentally determined kinetic parameters (Engasser and Horvath, 1973, 1974a; Hamilton *et al.*, 1974a,b; Moo-Young and Kobayashi, 1972; Kobayashi and Laidler, 1973; Buchholz and Rüth, 1976), regulatory effects, ion-selective, facilitated and active transport (Mitz, 1971; Broun *et al.*, 1970, 1972; Selegny *et al.*, 1971; Lecoq *et al.*, 1975; Goldstein, 1972, 1973; Gestrelius *et al.*, 1972, 1973; Kasche and Bergwall, 1974; Johansson and Mosbach, 1974a,b; Engasser and Horvath, 1974b; Thomas *et al.*, 1974; Thomas and Broun, 1973; Hervagault *et al.*, 1975), as well as new concepts, such as asymmetrical behavior, hysteresis, and oscillations (Thomas *et al.*, 1972; Caplan *et al.*, 1973; Naparstek *et al.*, 1973, 1974; Thomas and Caplan, 1976). Moreover, serious attempts are being made to apply the experience accumulated in the study of model systems for the quantitative description of metabolic pathways and metabolic compartmentalization in intact cells (Blum and Jenden, 1957; Roughton, 1959; Connett and Blum, 1971, 1972; Connett *et al.*, 1972; Raugi *et al.*, 1973a,b, 1975; Liang *et al.*, 1973; Blum, 1974; Srere *et al.*, 1973; Srere and Mosbach, 1974). Most of these aspects are discussed in depth in the chapter by Engasser and Horvath in this volume.

The study of the engineering aspects of coupled mass transfer-reaction kinetics, initiated in the early work on enzyme columns (Lilly *et al.*, 1966; Lilly and Sharp, 1968), led through integration of the approaches of the physical chemist dealing with the fundamentals of diffusion and the chemical engineer well versed in mass-transfer and heterogeneous catalysis, to a high degree of sophistication in enzyme-reactor analysis and design (Wingard, 1972a,b; O'Neill, 1972; Lilly and Dunnill, 1972; Lilly *et al.*, 1972, 1974; Vieth and Venkatasubramanian, 1974). The reader is referred to the chapter by Vieth *et al.* in this volume for a comprehensive survey of current status of design and analysis of immobilized-enzyme flow reactors.

The advances in the study of the basic properties of immobilized enzymes were accompanied by venues into laboratory-scale application and in analysis.

The controlled degradation of complex biological macromolecules with immobilized enzymes was first described by Cebra and co-workers, who isolated F_{ab} and F_c fragments from short digests of rabbit γ-globulin with immobilized papain (Cebra *et al.*, 1961, 1962; Cebra, 1964). Along the same lines, Lowey and others used immobilized derivatives of trypsin and papain to obtain and characterize subfragments of myosin in their studies on the structure of muscle proteins (Lowey *et al.*, 1966, 1967, 1968; Slayter and Lowey, 1967; Wolodko and Kay, 1975). Similar work on other biological macromolecules has been summarized elsewhere (Goldman *et al.*, 1971a,b; Katchalski *et al.*, 1971).

The use of immobilized enzymes in conjunction with a sensing device, to give highly specific bioprobes, originated in the glucose-specific electrode constructed by forming a layer of polyacrylamide gel-entrapped glucose oxidase over a polarographic oxygen electrode (Updike and Hicks, 1967; see also Guilbault, 1970; Clark, 1972). Other types of sensors based on ion-selective electrodes using glass or liquid membranes or thermistor devices have been used in combination with immobilized enzymes in a variety of configurations. The developments in this field have reached a high degree of sophistication (Guilbault, 1970, 1974; Weibel, 1974; Cooney *et al.*, 1974; Rechnitz, 1975). Parallel developments in the medical and analytical fields have been oriented toward exploration of the potential of immobilized enzymes for clinical and diagnostic application. Devices for continuous analysis using enzyme columns or enzyme-coated tubes have been developed to the stage where they can be commercialized (Hornby and Filippusson, 1970; Inman and Hornby, 1972; Campbell *et al.*, 1975; Morris *et al.*, 1975). Concentrated efforts are now being made to bring extracorporeal shunts based on immobilized enzyme columns, membranes, or microcapsules to an operational stage. The work in these areas has been summarized in several reviews, books, and conference proceedings published recently (Chang, 1972, 1976; Wingard, 1972a,b; Zaborsky, 1973; Pye and Wingard, 1974; Olson and Cooney, 1974; Spencer, 1974; Vandegaar, 1974; Dunlap, 1974; Salmona *et al.*, 1974; Rechnitz, 1975; Weetall, 1975).

The use of immobilized enzymes for the production of large quantities of a substance has required careful assessment of the economic as well as of the engineering problems involved. Although early research workers held rather strong views about the industrial aspects of immobilized enzymes (Katchalski, 1962; Manecke, 1962; Silman and Katchalski, 1966; Lilly *et al.*, 1966), industry was far from ready to introduce changes in the traditional fermentation processes. It was not until biochemists and chemical engineers joined forces that this aspects of enzyme technology was finally initiated. The definition of aims and means that took place in the late 1960s with the first attempts at developing immobilized enzyme-based processes was accompanied by a burst of activity in product-oriented research in several directions (see Wingard, 1972a,b; Aiba *et al.*, 1973; Pye, 1974):
(1) Development of new types of carriers and methods for enzyme immobilization from the point of view of their end-use as components of biochemical reactors, with special emphasis on mechanical proper-

ties and on the operational stability of the enzyme adducts (see Goldstein and Manecke, this volume). (2) Improved methods for selection and growth of microorganisms, and for production and isolation of enzymes from bacterial sources, recently with emphasis on heat-stable enzymes (Perlman, 1969; Teru, 1972; Hédén, 1973). (3) The first serious attempts to evaluate the nonengineering parameters critical for the commercial feasibility of a process based on immobilized enzymes: cost, operational lifetime, and specific activity of the enzyme; cost of reagents, supports, and equipment peculiar to the enzymic reaction; purity and market price of end product.

The first process in which immobilized enzymes were used on an industrial scale was the preparation of L-amino acids by the resolution of racemic mixtures of amino acids, such as alanine, phenylalanine, or methionine, with immobilized aminoacylase (Tosa *et al.*, 1966, 1967; Chibata *et al.*, 1972, 1974a,b; Chibata and Tosa, this volume). In this process, used by the Tanabe Company in Japan, chemically synthesized α-N-acetyl-DL-amino acids are passed through a column of immobilized enzyme. The stereospecificity of L-aminoacylase ensures that only the acetyl-L-amino acid is cleaved to yield the free L-amino acid, readily isolated in pure form.

Immobilized glucose isomerase has been recently introduced for the large-scale production of fructose-enriched syrups from corn starch hydrolyzates. The benefits of using immobilized glucose isomerase stem primarily from the high cost of the enzyme and the simple continuous operation of the process (Weetall, 1973; Hamilton *et al.*, 1974c). Similar considerations, as well as the fact that the enzyme does not remain in the final product, have led to the introduction of immobilized penicillin amidase in the production of 6-aminopenicillanic acid (Lilly *et al.*, 1970, 1972; Kamogashira *et al.*, 1972a,b; Warburton *et al.*, 1973; Hueper *et al.*, 1973a,b; see also Weetall and Suzuki, 1975; Chibata and Tosa, this volume).

Another process that shows economic promise and might also be of help in solving ecological problems, is the treatment of milk and cheese whey with lactase (β-galactosidase) to hydrolyze milk sugar (Charles *et al.*, 1974; Hasselberger *et al.*, 1974; Weetall *et al.*, 1974; Morisi *et al.*, 1973; Husted *et al.*, 1973; Woychik and Wondolowski, 1973; Dahlquist *et al.*, 1973). Removal of lactose from milk and unfermented milk products will permit their consumption by lactose-intolerant children and adults; these comprise a considerable part of the world population. By reducing the concentration of lactose, whey could be used as a cheap additive to food products such as ice creams and protein supplements.

The eventual uses of several other immobilized enzymes are being evaluated at present: steroid modification with immobilized enzymes; immobilized α-amylases and amyloglucosidases for the production of glucose syrups from corn starch; immobilized α-galactosidase to remove raffinose from crude beet-sugar extracts; immobilized papain and other proteases in chill-proofing of beer; immobilized pectinases for the clarification of fruit juices and wines; immobilized invertase for the production of invert sugar from sucrose (see Pye and Wingard, 1974; Spencer, 1974; Olson and Cooney, 1974; Messing, 1975; Weetall, 1975; Weetall and Suzuki, 1975).

Processes based on immobilized whole microbial cells are recently gaining prominence (Tosa *et al.*, 1974). Despite several problems intrinsic to the use of intact microorganisms, like the need to suppress or reduce the extent of side reactions caused by unwanted enzymic activities present in the cell, work has been carried out on several such processes (see Chibata and Tosa, this volume). An industrial reactor based on immobilized *Escherichia coli* cells with aspartase activity, for the production of aspartic acid from fumaric acid and ammonia, has been in operation in Japan since 1973. For perspectives and further details the reader is referred to the chapter by Chibata and Tosa in this volume.

The first generation of immobilized enzymes to be used in industry served in effect to replace the corresponding native, soluble enzymes in the conventional technology, where enzymes carried out relatively simple hydrolytic reactions.

The versatility and specificity of enzymes capable of carrying out many different types of reactions, such as stereospecific reductions, oxidations, isomerizations, specific degradations, and complex syntheses, suggests possible industrial applications of enzymes other than hydrolases. Since such enzymes require as a rule expensive coenzymes or effectors, their use would depend on the availability of means for recycling and regeneration of these costly compounds. Considerable effort is being made to develop procedures to that end, with emphasis on the preparation of active, soluble and insoluble, high-molecular-weight cofactor analogs. Of particular interest in this connection is the work on the immobilized derivatives of NAD(H), NADP(H), and AMP (Mosbach *et al.*, 1971, 1972, 1974a,b; Weibel *et al.*, 1974; Chibata *et al.*, 1974a,b; Chambers *et al.*, 1974).

Enzyme systems carrying out complex synthetic reactions are being investigated in several laboratories. These attempts can be illustrated by the exploration of the enzymic synthesis of gramicidin S being carried out at the Massachusetts Institute of Technology. The work is part of a coordinated effort to demonstrate large-scale, cell-free enzymic synthesis of useful products with simultaneous regeneration of the ATP consumed in the biosynthetic reaction (Hamilton *et al.*, 1974d; Gardner *et al.*, 1974; Whitesides *et al.*, 1974; Marshall, 1974a,b). Since ATP is an essential component of many biosynthetic systems, methods for its large-scale regeneration are of crucial importance for any further progress in the technology of synthetic biochemical processes.

Living organisms have developed a number of organized molecular structures that have the capability of efficiently transforming energy from one form to another (see King and Klingsberg, 1971). Such structures, associated with mitochondria, chloroplasts, and other subcellular particles, are responsible for converting chemical or light energy supplied to the cells into intermediate forms of chemical energy or thermodynamic work. Although the detailed mechanism of electron flow in the enzyme-transfer processes involved remains unclear, enough is known concerning the high efficiency of biological energy-transfer reactions to suggest that such processes might be employed to design an efficient transducer for the conversion of chemical into electrical energy. As a matter of fact, work on biochemical fuel cells is beginning to take shape and concentrates mainly on the system based on hydrogen/hydrogenase, methanol/methanol oxidase, and glucose/glucose oxidase. Such cells should in principle operate via immobilized enzyme electrodes connected to an external circuit. Current research on biochemical fuel cells still centers around basic problems, such as selection, production, and purification of the suitable oxidizing enzymes and ground work on electrodes and performance of model systems. The major technical problems that have to be solved at the present stage of development are: (1) electron transfer from substrate–enzyme–cofactor complex to an external electrode, (2) mass transfer (concentration polarization) and electrode design, and (3) enzyme stability under operational conditions.

The advances of the last decade briefly outlined here have provided the basis for the wealth of research and development related to immobilized enzymes now under way both in academic institutions and in industry.

In concluding this survey we would like to present a very personal view of topics where research should be further pursued.

1. Effects of immobilization on the stability and catalytic behavior of an enzyme. This should include the following: (a) Examination of the effects of the chemical nature of the support material, and the microenvironment thus imposed, on the conformational and opera-

tional stability of an immobilized enzyme. (b) Study of the dependence of stability on the number and nature of covalent links that the enzyme forms with the support. (c) Exploration of the possibility of modifying the stability of an immobilized enzyme through the freezing of a given conformation by cross-linking with bi- and multifunctional reagents in the presence or in the absence of substrates, inhibitors, or modifiers (see, for example, Zaborsky, 1974). The use, in such studies, of enzymes labeled with reporter molecules and of enzymes derived from both thermophilic and mesophilic organisms could provide valuable information. (d) Development of polymeric reagents of high chemical specificity and selectivity and of methods for site-directed covalent binding; such approaches might enable the detection of regions in the protein molecule, crucial for its stability. (e) Along similar lines, investigation of the effects of the microenvironment imposed by the support on the catalytic behavior of an immobilized enzyme. Such studies should be aimed at delineating the limits within which a preselected microenvironment can affect the intrinsic catalytic efficiency of an enzyme as well as the partitioning of substrate.

2. Immobilization of multisubunit enzymes. The potential of the immobilization approach for the investigation of structure–function relationships in multisubunit enzymes has been elegantly demonstrated by Chan in his study of immobilized aldolase (Chan, 1970; Chan *et al.*, 1973) and more recently by Fukui, who immobilized tyrosinase via a water-insoluble derivative of an effector of the enzyme, pyridoxal 5′-phosphate (Fukui *et al.*, 1975). These studies could serve as pointers to strategies that can be adopted to investigate (a) Subunit interactions in enzymes composed of nonidentical subunits; a corollary to this problem is exploring the potential of immobilization techniques for the preparation of well characterized isoenzyme hybrids. (b) Stabilization of multisubunit enzyme complexes by intersubunit cross-linking. (c) Effects of the state of association of an enzyme on its kinetic behavior and operational stability. (d) The effects of low-molecular-weight modifiers on the stability and state of association of an enzyme. In this context the comprehensive list of multisubunit enzymes recently published by Darnall and Klotz (1975) should be of great value.

3. Immobilization of cofactors and modifiers. Some of the recent work on the use of immobilized cofactors and effectors in "general ligand" chromatography (Mosbach, 1974) and as anchors for site-directed binding of enzymes (Fukui *et al.*, 1975) was summarized in earlier sections. These studies call for further work along the following lines: (a) The possibility of obtaining enzymically active insoluble

conjugates of an enzyme requiring a cofactor for its activity by binding the enzyme onto an immobilized conjugate of the cofactor or cofactor analog (see Mosbach and Gestrelius, 1974). (b) Covalent fixation of cofactors, modifiers, or analogs thereof on their specific binding site on the protein. Such studies may lead to methods for the modulation of enzymic activity and to enzyme preparations of improved operational stability. (c) The feasibility of coupling enzymic oxidation–reduction reactions with electrode processes via pyridine nucleotide coenzymes covalently coupled to systems capable of conducting electrons.

4. Theoretical studies: (a) Extension of the studies on relatively simple enzyme membranes and particulate enzyme systems to aspects of structure-modulated kinetics, such as regulatory effects and feedback mechanisms, asymmetrical behavior, and oscillatory phenomena. (b) Adaptation of the techniques used in the investigation of artificial enzyme membranes for the quantitative description of metabolic pathways and metabolite compartmentalization in intact cells.

5. Applied research: (a) Exploration of enzyme systems capable of catalyzing reactions leading to new useful products, e.g., steroid-transforming enzymes, oxidoreductases and synthetases, and in particular enzymes capable of carrying out asymmetric syntheses. (b) Examination of the relative efficiency of the various immobilized-enzyme reactor types with respect to process engineering. Model studies should be carried out to determine, for any given enzyme reaction, the most adequate reactor configuration, viz., batch, packed-bed, fluidized-bed, or membrane. (c) Exploration of new, more efficient reactor configurations; recent work has indicated that, although in the case of enzymes immobilized on rigid particles or within gels the rates of substrate conversion are diffusion controlled and hence relatively slow, when substrate is applied under pressure to an enzyme membrane very high rates of conversion are observed (Gregor and Rauf, 1975). This suggests that pressure-driven enzyme-membrane reactors can be operated under conditions where diffusional restrictions do not obtain. Work in this area should benefit greatly from the vast experience accumulated in the study of the fundamental as well as engineering principles underlying the behavior of ultrafiltration membranes.

REFERENCES

Aiba, S., Humphrey, A. E., and Millis, N. F. (1973). "Biochemical Engineering," 2nd Ed. Univ. of Tokyo Press, Tokyo.
Axén, R., Porath, J., and Ernback, S. (1967). *Nature (London)* **214**, 1302.

Axén, R., Myrin, P. A., and Janson, J. C. (1970). *Biopolymers* **9**, 401.

Bar-Eli, A., and Katchalski, E. (1960). *Nature (London)* **188**, 856.

Bar-Eli, A., and Katchalski, E. (1963). *J. Biol. Chem.* **238**, 1690.

Barker, S. A., Somers, P. J., Epton, R., and McLaren, J. V. (1970). *Carbohydr. Res.* **14**, 287.

Bernfeld, P., and Wan, J. (1963). *Science* **142**, 678.

Bernfeld, P., Bieber, R. E., and McDonell, P. C. (1968). *Arch. Biochem. Biophys.* **127**, 779.

Blum, J. J. (1974). *J. Cell. Physiol.* **83**, 275.

Blum, J. J., and Jenden, D. J. (1957). *Arch. Biochem. Biophys.* **66**, 316.

Brandenberger, H. (1955). *Angew. Chem.* **67**, 661.

Brandenberger, H. (1956). *J. Polym. Sci.* **20**, 215.

Brandenberger, H. (1957). *Helv. Chim. Acta* **40**, 61.

Broun, G., Selegny, E., Avrameas, S., and Thomas, D. (1969). *Biochim. Biophys. Acta* **185**, 260.

Broun, G., Selegny, E., Tran Minh, C., and Thomas, D. (1970). *FEBS Lett.* **7**, 223.

Broun, G., Thomas, D., and Selegny, E. (1972). *J. Membr. Biol.* **8**, 313.

Brown, H. D., ed. (1971). "Chemistry of the Cell Interface," Parts A and B. Academic Press, New York.

Buchholz, K., and Rüth, W. (1976). *Biotechnol. Bioeng.* **18**, 95.

Bunting, P. S., and Laidler, K. J. (1973). "The Chemical Kinetics of Enzyme Action." Oxford Univ. Press (Clarendon), London and New York.

Campbell, D. H., Luescher, F., and Lerman, L. S. (1951). *Proc. Natl. Acad. Sci. U.S.A.* **37**, 575.

Campbell, J., Hornby, W. E., and Morris, D. L. (1975). *Biochim. Biophys. Acta* **384**, 307.

Caplan, S. R., Naparstek, A., and Zabusky, N. J. (1973). *Nature (London)* **245**, 364.

Cebra, J. J. (1964). *J. Immunol.* **92**, 977.

Cebra, J. J., Givol, D., Silman, H. I., and Katchalski, E. (1961). *J. Biol. Chem.* **236**, 1720.

Cebra, J. J., Givol, D., and Katchalski, E. (1962). *J. Biol. Chem.* **237**, 751.

Chambers, R. P., Ford, J. R., Allender, J. H., Baricos, W. H., and Cohen, W. (1974). *In* "Enzyme Engineering" (E. K. Pye and L. B. Wingard, eds.), Vol. 2, p. 195. Plenum, New York.

Chan, W. W. C. (1970). *Biochem. Biophys. Res. Commun.* **41**, 1198.

Chan, W. W. C., Schutt, H., and Brand, H. (1973). *Eur. J. Biochem.* **40**, 533.

Chang, T. M. S. (1964). *Science* **146**, 524.

Chang, T. M. S. (1972). "Artificial Cells." Thomas, Springfield, Illinois.

Chang, T. M. S., ed. (1976). "Biomedical Applications of Immobilized Enzymes and Proteins." Plenum, New York.

Chang, T. M. S., McIntosh, F. C., and Mason, S. G. (1966). *Can. J. Physiol. Pharmacol.* **44**, 115.

Charles, M., Coughlin, R. W., Allen, R. B., Paruchuri, E. K., and Hasselberger, F. X. (1974). *In* "Immobilized Biochemicals and Affinity Chromatography" (B. R. Dunlap, ed.), p. 213. Plenum, New York.

Chibata, T., Tosa, T., Sato, T., Mori, T., and Matsuo, Y. (1972). *In* "Fermentation Technology Today" (G. Teru, ed.), p. 383. Soc. Ferment. Technol., Tokyo.

Chibata, I., Tosa, T., and Sato, T. (1974a). *Appl. Microbiol.* **27**, 878.

Chibata, I., Tosa, T., and Matsuo, Y. (1974b). *In* "Enzyme Engineering" (E. K. Pye and L. B. Wingard, eds.), Vol. 2, p. 229. Plenum, New York.

Clark, L. C. (1972). *In* "Enzyme Engineering" (L. B. Wingard, ed.), Vol. 1, p. 377. Wiley (Interscience), New York.

Connett, R. J., and Blum, J. J. (1971). *Biochemistry* **10**, 3299.

Connett, R. J., and Blum, J. J. (1972). *J. Biol. Chem.* **25**, 5199.

Connett, R. J., Wittels, B., and Blum, J. J. (1972). *J. Biol. Chem.* **247**, 2657.

Cooney, C. L., Weaver, J. C., Tannenbaum, S. R., Faller, D. V., Shields, A., and Jahnke, M. (1974). *In* "Enzyme Engineering" (E. K. Pye and L. B. Wingard, eds.), Vol. 2, p. 411. Plenum, New York.

Cuatrecasas, P. (1972). *Adv. Enzymol. Relat. Areas Mol. Biol.* **36**, 29.

Cuatrecasas, P., and Anfinsen, C. B. (1971). *Annu. Rev. Biochem.* **40**, 259.

Cuatrecasas, P., Wilchek, M., and Anfinsen, C. B. (1968). *Proc. Natl. Acad. Sci. U.S.A.* **61**, 636.

Dahlquist, A., Mattiasson, B., and Mosbach, K. (1973). *Biotechnol. Bioeng.* **15**, 395.

Darnall, D. W., and Klotz, I. M. (1975). *Arch. Biochem. Biophys.* **166**, 651.

Dinelli, D. (1972). *Process Biochem.* **7**, 9.

Dunlap, R. B., ed. (1974). *Adv. Exp. Med. Biol.* **42**.

Engasser, J., and Horvath, C. (1973). *J. Theor. Biol.* **42**, 137.

Engasser, J. M., and Horvath, C. (1974a). *Biochemistry* **13**, 3845, 3849, 3855.

Engasser, J., and Horvath, C. (1974b). *Biochim. Biophys. Acta* **358**, 178.

Estermann, E. F., Conn, E. E., and McLaren, A. D. (1959). *Arch. Biochem. Biophys.* **85**, 103.

Fritz, H., Schult, H., Neudecker, M., and Werle, E. (1966). *Angew. Chem., Int. Ed. Engl.* **5**, 735.

Fritz, H., Schult, H., Hutzel, M., Wiederman, M., and Werle, E. (1967). *Hoppe-Seyler's Z. Physiol. Chem.* **348**, 308.

Fritz, H., Hochstrasser, K., Werle, E., Brey, E., and Gebhardt, B. M. (1968). *Z. Anal. Chem.* **243**, 452.

Fritz, H., Brey, B., Schmal, A., and Werle, E. (1969). *Hoppe-Seyler's Z. Physiol. Chem.* **350**, 617.

Fukui, S., Ikeda, S., Fujimura, M., Yamada, H., and Kumagai, H. (1975). *Eur. J. Biochem.* **51**, 155.

Gardner, C. R., Colton, C. K., Langer, R. S., Hamilton, B. K., Archer, M. C., and Whitesides, G. M. (1974). *In* "Enzyme Engineering" (E. K. Pye and L. B. Wingard, eds.), Vol. 2, p. 209. Plenum, New York.

Gestrelius, S., Mattiasson, B., and Mosbach, K. (1972). *Biochim. Biophys. Acta* **276**, 339.

Gestrelius, S., Mattiasson, B., and Mosbach, K. (1973). *Eur. J. Biochem.* **36**, 89.

Gladishev, P. P., Goryaev, M. I., Allambergenova, S. T., and Shapolov, Y. A. (1973). *Izv. Akad. Nauk Kaz. SSR, Ser. Khim.* **23**, 42.

Goldman, R. (1973). *Biochimie* **55**, 953.

Goldman, R., and Katchalski, E. (1971). *J. Theor. Biol.* **32**, 243.

Goldman, R., Silman, H. I., Caplan, S. R., Kedem, O., and Katchalski, E. (1965). *Science* **150**, 758.

Goldman, R., Kedem, O., Silman, H. I., Caplan, S. R., and Katchalski, E. (1968a). *Biochemistry* **7**, 486.

Goldman, R., Kedem, O., and Katchalski, E. (1968b). *Biochemistry* **7**, 4518.

Goldman, R., Kedem, O., and Katchalski, E. (1971a). *Biochemistry* **10**, 165.

Goldman, R., Goldstein, L., and Katchalski, E. (1971b). *In* "Biochemical Aspects of Reactions on Solid Supports" (G. R. Stark, ed.), p. 1. Academic Press, New York.

Goldstein, L. (1969). *In* "Fermentation Advances" (D. Perlman, ed.), p. 391. Academic Press, New York.

Goldstein, L. (1970). *In* "Proteolytic Enzymes" (G. Perlmann and L. Lorand, eds.), Methods in Enzymology, Vol. 19, p. 935. Academic Press, New York.

Goldstein, L. (1972). *Biochemistry* **11**, 4072.
Goldstein, L. (1973). *Isr. J. Chem.* **11**, 379.
Goldstein, L., and Katchalski, E. (1968). *Z. Anal. Chem.* **243**, 375.
Goldstein, L., Levin, Y., and Katchalski, E. (1964). *Biochemistry* **3**, 1913.
Goldstein, L., Freeman, A., and Sokolovsky, M. (1974). *Biochem. J.* **143**, 497.
Green, D. E., and Silman, I. (1967). *Annu. Rev. Plant Physiol.* **18**, 147.
Gregor, H. P., and Rauf, P. W. (1975). *Biotechnol. Bioeng.* **17**, 445.
Gregoriadis, G. (1974). In "Insolubilized Enzymes" (H. Salmona, C. Saronio, and S. Garattini, eds.), p. 165. Raven, New York.
Gregoriadis, G., Leathwood, F. D., and Ryman, B. E. (1971). *FEBS Lett.* **14**, 95.
Grubhofer, N., and Schleith, L. (1953). *Naturwissenschaften* **40**, 508.
Grubhofer, N., and Schleith, L. (1954). *Hoppe-Seyler's Z. Physiol. Chem.* **297**, 108.
Guilbault, G. C. (1970). *Crit. Rev. Anal. Chem.* **1**, 377.
Guilbault, G. C. (1974). In "Enzyme Engineering" (E. K. Pye and L. B. Wingard, eds.), p. 377. Plenum, New York.
Hamilton, B. K., Gardner, C. R., and Colton, C. K. (1974a). In "Immobilized Enzymes in Food and Microbiol Processes" (A. C. Olson and C. L. Cooney, eds.), p. 205. Plenum, New York.
Hamilton, B. K., Gardner, C. R., and Colton, C. K. (1974b). *AIChE J.* **20**(3), 503.
Hamilton, B. K., Colton, C. K., and Cooney, C. L. (1974c). In "Immobilized Enzymes in Food and Microbial Processes" (A. C. Olson and C. L. Cooney, eds.), p. 85. Plenum, New York.
Hamilton, B. K., Montgomery, J. P., and Wang, D. I. C. (1974d). In "Enzyme Engineering" (E. K. Pye and L. B. Wingard, eds.), Vol. 2, p. 153. Plenum, New York.
Hasselberger, F. X., Allen, B., Paruchuri, E. K., Charles, M., and Coughlin, R. W. (1974). *Biochem. Biophys. Res. Commun.* **57**, 1054.
Hédén, C. G. (1973). *Biotechnol. Bioeng. Symp.* **4**, 1003.
Hofstee, B. H. J., and Otillio, N. F. (1973). *Biochem. Biophys. Res. Commun.* **53**, 1137.
Hornby, W. E., and Filippusson, H. (1970). *Biochim. Biophys. Acta* **220**, 343.
Hornby, W. E., Lilly, M. D., and Crook, E. M. (1968). *Biochem. J.* **107**, 669.
Hervagault, J. F., Joly, G., and Thomas, D. (1975). *Eur. J. Biochem.* **51**, 19.
Hueper, F., Rauenbusch, E., Schmidt-Kastner, G., Boemer, B., and Bartl, H. (1973a). Ger. Patent 2,215,687.
Hueper, F., Rauenbusch, E., and Schmidt-Kastner, G. (1973b). Ger. Patent 2,215,539.
Husted, G. O., Richardson, T., and Olson, N. F. (1973). *J. Dairy Sci.* **56**, 118.
Inman, D. J., and Hornby, W. H. (1972). *Biochem. J.* **129**, 255.
Inman, J. K., and Dintzis, H. M. (1969). *Biochemistry* **8**, 4074.
Isliker, H. C. (1953). *Ann. N.Y. Acad. Sci.* **57**, 225.
Isliker, H. C. (1957). *Adv. Protein Chem.* **12**, 387.
Jakoby, W. B., and Wilchek, M., eds. (1974). "Affinity Techniques: Enzyme Purification," Methods in Enzymology, Vol. 34, Part B. Academic Press, New York.
Johansson, A. C., and Mosbach, K. (1974a). *Biochim. Biophys. Acta* **370**, 339.
Johansson, A. C., and Mosbach, K. (1974b). *Biochim. Biophys. Acta* **370**, 348.
Kamogashira, T., Kawaguchi, T., Miyazaki, W., and Doi, T. (1972a). Jpn. Patent 7,228,190.
Kamogashira, T., Mihara, S., Tamaoka, H., and Doi, T. (1972b). Jpn. Patent 7,228,187.
Kaplan, N. O., Everse, J., and Dixon, J. E. (1974). *Proc. Natl. Acad. Sci. U.S.A.* **71**, 3450.
Kasche, V., and Bergwall, M. (1974). In "Insolubilized Enzymes" (M. Salmona, C. Saronio, and S. Garattini, eds.), p. 77. Raven, New York.

Katchalski, E. (1962). *In* "Polyamino Acids, Polypeptides, Proteins" (M. A. Stahman, ed.), p. 283. Univ. of Wisconsin Press, Madison.

Katchalski, E., Silman, I., and Goldman, R. (1971). *Adv. Enzymol. Relat. Areas Mol. Biol.* **34**, 445.

Kay, G., and Lilly, M. D. (1970). *Biochim. Biophys. Acta* **198**, 276.

King T. E., and Klingenberg, M. (1971). "Electron and Coupled Energy Transfer in Biological Systems," Vols. 1 and 2. Dekker, New York.

Kobayashi, T., and Laidler, K. J. (1973). *Biochim. Biophys. Acta* **302**, 1.

Laidler, K. J., and Sundaram, P. V. (1971). *In* "Chemistry of the Cell Interface" (H. D. Brown, ed.), Part A, p. 255. Academic Press, New York.

Landsteiner, K., and Van der Scheer, J. (1936). *J. Exp. Med.* **63**, 325.

Langmuir, I., and Schaefer, V. J. (1938). *J. Am. Chem. Soc.* **60**, 1351.

Langmuir, I., and Schaefer, V. J. (1939). *Chem. Rev.* **24**, 181.

Lecoq, D., Hervagault, J. F., Brown, G., Joly, G., Kernevez, J. P., and Thomas, D. (1975). *J. Biol. Chem.* **250**, 5496.

Lerman, L. S. (1953a). *Nature (London)* **172**, 635.

Lerman, L. S. (1953b). *Proc. Natl. Acad. Sci. U.S.A.* **39**, 232.

Levin, Y., Pecht, M., Goldstein, L., and Katchalski, E. (1964). *Biochemistry* **3**, 1905.

Levitzki, A., Heller, J., and Schramm, M. (1964). *Biochim. Biophys. Acta* **81**, 101.

Liang, T., Raugi, G. J., and Blum, J. J. (1973). *J. Biol. Chem.* **248**, 8073.

Lilly, M. D., and Dunnill, P. (1971). *Process Biochem.* **6**(8), 29.

Lilly, M. D., and Dunnill, P. (1972). *In* "Enzyme Engineering" (L. B. Wingard, ed.), Vol. 1, p. 221. Wiley (Interscience), New York.

Lilly, M. D., and Sharp, A. K. (1968). *Chem. Eng. (London)*, CE 12.

Lilly, M. D., Money, C., Hornby, W. E., and Crook, E. M. (1965). *Biochem. J.* **95**, 45P.

Lilly, M. D., Hornby, W. E., and Crook, E. M. (1966). *Biochem. J.* **100**, 718.

Lilly, M. D., Kay, G., Wilson, R. J. H., and Sharp, A. K. (1970). Brit. Patent 1,183,260.

Lilly, M. D., Balasingham, K., Warburton, D., and Dunnill, P. (1972). *In* "Fermentation Technology Today" (G. Teru, ed.), p. 379. Soc. Ferment. Technol., Tokyo.

Lilly, M. D., Regan, D. L., and Dunnill, P. (1974). *In* "Enzyme Engineering" (E. K. Pye and L. B. Wingard, eds.), Vol. 2, p. 245. Plenum, New York.

Lowey, S., Goldstein, L., and Luck, S. (1966). *Biochem. Z.* **345**, 248.

Lowey, S., Goldstein, L., Cohen, C., and Luck, S. M. (1967). *J. Mol. Biol.* **23**, 287.

Lowey, S., Slayter, H. S., Weeds, G., and Baker, H. (1968). *J. Mol. Biol.* **42**, 1.

McLaren, A. D. (1960). *Enzymologia* **21**, 356.

McLaren, A. D., and Babcock, K. L. (1959). *In* "Subcellular Particles" (T. Hayashi, ed.), p. 23. Ronald Press, New York.

McLaren, A. D., and Estermann, E. F. (1957). *Arch. Biochem. Biophys.* **68**, 157.

McLaren, A. D., and Packer, L. (1970). *Adv. Enzymol. Relat. Areas Mol. Biol.* **33**, 245.

Manecke, G. (1962). *Pure Appl. Chem.* **4**, 507.

Manecke, G. (1974). *Chimia* **28**(9), 467.

Manecke, G. (1975). *Proc. Int. Symp. Macromol., Rio de Janeiro, 1974* (E. B. Mano, ed.), p. 397. Elsevier, Amsterdam.

Manecke, G., and Gillert, K. E. (1955). *Naturwissenschaften* **42**, 212.

Manecke, G., and Singer, S. (1960). *Makromol. Chem.* **37**, 119.

Manecke, G., Singer, S., and Gillert, K. E. (1958). *Naturwissenschaften* **45**, 440.

Manecke, G., Günzel, G., and Förster, H. J. (1970). *J. Poly. Sci., Part C* **30**, 607.

Marconi, W., Gulinelli, S., and Morisi, F. (1974). *In* "Insolubilized Enzymes" (M. Salmona, C. Saronio, and S. Garattini, eds.), p. 51. Raven, New York.

Marshall, D. L. (1974a). *In* "Immobilized Biochemicals and Affinity Chromatography" (R. B. Dunlap, ed.), p. 345. Plenum, New York.
Marshall, D. L. (1974b). *In* "Enzyme Engineering" (E. K. Pye and L. B. Wingard, eds.), Vol. 2, p. 223. Plenum, New York.
Mattiasson, B., and Mossbach, K. (1971). *Biochim. Biophys. Acta* **235**, 253.
Melrose, G. J. H. (1971). *Rev. Pure Appl. Chem.* **21**, 83.
Messing, R. A., ed. (1975). "Immobilized Enzymes for Industrial Reactors." Academic Press, New York.
Micheel, F., and Evers, J. (1949). *Makromol. Chem.* **3**, 200.
Mitz, M. A. (1971). *In* "Chemical Evolution and the Origin of Life" (R. Buvet and C. Ponnamperuma, eds.), p. 355. North-Holland Publ., Amsterdam.
Mitz, M. A., and Summaria, L. J. (1961). *Nature (London)* **189**, 576.
Moo-Young, M., and Kobayashi, T. (1972). *Can. J. Chem. Eng.* **50**, 162.
Morisi, F., Pastore, M., and Viglia, A. (1973). *J. Dairy Sci.* **56**, 1123.
Morris, D. L., Campbell, J., and Hornby, W. E. (1975). *Biochem. J.* **147**, 593.
Mosbach, K. (1970). *Acta Chem. Scand.* **24**, 2093.
Mosbach, K. (1974). *In* "Affinity Techniques: Enzyme Purification" (W. B. Jacoby and M. Wilchek, eds.), Methods in Enzymology, Vol. 34, Part B, p. 229. Academic Press, New York.
Mosbach, K., and Gestrelius, S. (1974). *FEBS Lett.* **42**, 200.
Mosbach, K., and Mattiasson, B. (1970). *Acta Chem. Scand.* **24**, 2093.
Mosbach, K., and Mosbach, R. (1966). *Acta Chem. Scand.* **20**, 2807.
Mosbach, K., Guilford, H., Larsson, P. O., Ohlson, R., and Scott, M. (1971). *Biochem. J.* **125**, 20.
Mosbach, K., Guilford, H., Ohlson, R., and Scott, M. (1972). *Biochem. J.* **127**, 625.
Mosbach, K., Mattiasson, B., Gestrelius, S., and Srere, P. A. (1974a). *In* "Enzyme Engineering" (E. K. Pye and L. B. Wingard, eds.), Vol. 2, p. 143. Plenum, New York.
Mosbach, K., Mattiasson, B., Gestrelius, S., and Srere, P. A. (1974b). *In* "Insolubilized Enzymes" (M. Salmona, C. Saronio, and S. Garattini, eds.), p. 123. Raven, New York.
Naparstek, A., Thomas, D., and Caplan, S. R. (1973). *Biochim. Biophys. Acta* **323**, 643.
Naparstek, A., Romette, J. L., Kernevez, J. P., and Thomas, D. (1974). *Nature (London)* **249**, 490.
Nelson, J. M., and Hitchcocks, D. I. (1921). *J. Am. Chem. Soc.* **43**, 1956.
Nernst, W. (1904). *Z. Phys. Chem., Stoechiom. Verwandschaftslehre* **47**, 52.
Olson, A. C., and Cooney, C. L., eds. (1974). "Immobilized Enzymes in Food and Microbial Processes." Plenum, New York.
O'Neill, S. P. (1972). *Rev. Pure Appl. Chem.* **22**, 133.
Orth, M. D., and Brümmer, W. (1972). *Angew. Chem., Int. Ed. Engl.* **11**, 249.
Pecht, M., and Levin, Y. (1972). *Biochem. Biophys. Res. Commun.* **46**, 2054.
Perlman, D., ed. (1969). "Fermentation Advances." Academic Press, New York.
Porath, J., Axén, R., and Ernback, S. (1967). *Nature (London)* **215**, 1491.
Pye, E. K. (1974). *In* "Immobilized Enzymes in Food and Microbial Processes" (A. C. Olson and C. L. Cooney, eds.), p. 1. Plenum, New York.
Pye, E. K., and Wingard, L. B., eds. (1974). "Enzyme Engineering," Vol. 2. Plenum, New York.
Raugi, G. J., Liang, T., and Blum, J. J. (1973a). *J. Biol. Chem.* **248**, 8064.
Raugi, G. J., Liang, T., and Blum, J. J. (1973b). *J. Biol. Chem.* **248**, 8079.
Raugi, G. J., Liang, T., and Blum, J. J. (1975). *J. Biol. Chem.* **250**, 445.
Rechnitz, G. A. (1975). *Chem. Eng. News* Jan. 27, pp. 29–35.

Roughton, F. J. W. (1959). *Prog. Biophys. Mol. Biol.* **9**, 55.

Royer, G. P., Andrews, J. P., and Uy, R. (1973). *Enzyme Technol. Dig.* **1**, 99.

Salmona, M., Saronio, C., and Garattini, S., eds. (1974). "Insolubilized Enzymes." Raven, New York.

Schramm, M., and Loyter, A. (1962). *Biochim. Biophys. Acta* **65**, 200.

Schramm, M., and Loyter, A. (1966). *In* "Complex Carbohydrates" (E. F. Neufeld and V. Ginsburg, eds.), Methods in Enzymology, Vol. 8, p. 533. Academic Press, New York.

Selegny, E., Avrameas, S., Broun, G., and Thomas, D. (1968). *C. R. Acad. Sci., Ser. C* **266**, 1431.

Selegny, E., Broun, G., and Thomas, D. (1971). *Physiol. Veg.* 9(1), 25.

Shuler, M. L., Aris, R., and Tsuchiya, H. M. (1972). *J. Theor. Biol.* **35**, 67.

Silman, I. H., and Katchalski, E. (1966). *Annu. Rev. Biochem.* **35**, 873.

Slayter, H. S., and Lowey, S. (1967). *Proc. Natl. Acad. Sci. U.S.A.* **58**, 1611.

Smiley, K. L., and Strandberg, G. W. (1972). *Adv. Appl. Microbiol.* **15**, 13.

Solomon, B., and Levin, Y. (1974). *Biotechnol. Bioeng.* **16**, 1161.

Spencer, B., ed. (1974). *Ind. Aspects Biochem., Proc. Fed. Eur. Biochem. Soc., Spec. Meet., Dublin, 1974* Vols. 1 and 2. North-Holland Publ., Amsterdam.

Srere, P. A., and Mosbach, K. (1974). *Annu. Rev. Microbiol.* **28**, 61.

Srere, P. A., Mattiasson, B., and Mosbach, K. (1973). *Proc. Natl. Acad. Sci. U.S.A.* **70**, 2534.

Stanley, W. L., and Palter, R. (1973). *Biotechnol. Bioeng.* **15**, 596.

Stark, G. R., ed. (1971). "Biochemical Aspects of Reactions on Solid Supports." Academic Press, New York.

Sundaram, P. V., Tweedale, A., and Laidler, K. J. (1970). *Can. J. Chem.* **48**, 1498.

Teru, G., ed. (1972). "Fermentation Technology Today." Soc. Ferment. Technol., Tokyo.

Thomas, D., and Broun, G. (1973). *Biochimie* **55**, 975.

Thomas, D., and Caplan, S. R. (1976). *In* "Membrane Separation Process" (P. Meares, ed.), Elsevier, Amsterdam.

Thomas, D., Broun, A., Gellf, G., and Domurado, D. (1972). *In* "Enzyme Engineering" (L. B. Wingard, ed.), Vol. 1, p. 299. Wiley (Interscience), New York.

Thomas, D., Bourdillon, C., Broun, G., and Kernevez, J. P. (1974). *Biochemistry* **13**, 2995.

Tosa, T., Mori, T., Fuse, N., and Chibata, I. (1966). *Enzymologia* **31**, 214.

Tosa, T., Mori, T., Fuse, N., and Chibata, I. (1967). *Biotechnol. Bioeng.* **9**, 603.

Tosa, T., Sato, T., Mori, T., and Chibata, I. (1974). *Appl. Microbiol.* **27**, 886.

Updike, S. J., and Hicks, G. P. (1967). *Nature (London)* **214**, 986.

Vandegaar, J. E., ed. (1974). "Microencapsulation, Processes and Application." Plenum, New York.

Vieth, W. R., and Venkatasubramanian, K. (1973). *Chem. Technol.* **3**, 677.

Vieth, W. R., and Venkatasubramanian, K. (1974). *Chem. Technol.* **4**, 309, 434.

Warburton, D., Dunnill, P., and Lilly, M. D. (1973). *Biotechnol. Bioeng.* **15**, 13.

Weetall, H. H. (1969). *Science* **166**, 615.

Weetall, H. H. (1970). *Biochim. Biophys. Acta* **212**, 1.

Weetall, H. H. (1973). *Food Prod. Dev.* **1**, 46.

Weetall, H. H., ed. (1975). "Immobilized Enzymes, Antigens, Antibodies and Peptides, Preparation and Characterization." Dekker, New York.

Weetall, H. H., and Suzuki, S., eds. (1975). "Immobilized Enzyme Technology." Plenum, New York.

Weetall, H. H., Havewala, N. B., Pitcher, W. H., Detar, C. C., Vann, W. P., and Yaverbaum, S. (1974). *Biotechnol. Bioeng.* **16**, 295.

Weibel, M. K. (1974). *In* "Enzyme Engineering" (E. K. Pye and L. B. Wingard, eds.), Vol. 2, p. 385. Plenum, New York.

Weibel, M. K., Fuller, C. W., Stadel, J. M., Buckmann, A. F. E. P., Doyle, T., and Brigh, H. J. (1974). *In* "Enzyme Engineering" (E. K. Pye and L. B. Wingard, eds.), Vol. 2, p. 203. Plenum, New York.

Wharton, C. W., Crook, E. M., and Brocklehurst, K. (1968). *Eur. J. Biochem.* **6**, 572.

Whitesides, G. M., Chmurny, A., Garrett, P., Lamotte, A., and Colton, C. K. (1974). *In* "Enzyme Engineering" (E. K. Pye and L. B. Wingard, eds.), Vol. 2, p. 217. Plenum, New York.

Wilchek, M. (1974). *In* "Immobilized Biochemicals and Affinity Chromatography" (R. Dunlap, ed.), p. 15. Plenum, New York.

Wilcheck, M., and Hexter, C. S. (1976). *Methods Biochem. Anal.* **23**, 347.

Wingard, L. B. (1972a). *Adv. Biochem. Eng.* **2**, 1.

Wingard, L. B., ed. (1972b). "Enzyme Engineering," Vol. 1. Wiley (Interscience), New York.

Wolodko, T. W., and Kay, C. M. (1975). *Can. J. Biochem.* **53**, 175.

Woychik, J. H., and Wondolowski, M. V. (1973). *J. Milk Food Technol.* **36**, 31.

Zaborsky, O. R. (1973). "Immobilized Enzymes." CRC Press, Ceveland, Ohio.

Zaborsky, O. R. (1974). *In* "Enzyme Engineering" (E. K. Pye and L. B. Wingard, eds.), Vol. 2, p. 115. Plenum, New York.

The Chemistry of Enzyme Immobilization

Leon Goldstein

Department of Biochemistry, The George S. Wise Center for Life Sciences,
Tel Aviv University, Tel Aviv, Israel

and

Georg Manecke

Institut für Organische Chemie der Freien Universität Berlin and
Fritz-Haber-Institut der Max-Planck-Gesellschaft, Berlin, Germany

I. INTRODUCTION

Interest in immobilized enzyme derivatives stems primarily from our growing awareness of their potential as industrial catalysts and as a new type of model system for the investigation of isolated aspects of complex biological phenomena.

Enzymes immobilized on or within a solid matrix by conjugation with synthetic water-insoluble polymeric supports can serve in the laboratory as reusable and removable highly specific reagents, which often possess improved storage and operational stability. Continuous large-scale processes can be carried out in immobilized-enzyme reactors. Immobilized enzymes in conjunction with a detector have led to the development of

23

highly specific electrode systems and similar analytical and monitoring devices. Immobilized enzymes are also being explored for clinical application in the form of extracorporeal shunts or microcapsules. Moreover, the clarification of some of the principles underlying the kinetic behavior of immobilized enzyme systems, i.e., effects of the microenvironment imposed by the chemical nature of the support material and the effects of diffusional restrictions on the translocation of substrate and product, make possible in principle the modulation of the properties of a bound enzyme by its conjugation to a support of predetermined chemical and physical characteristics.

The physicochemical, engineering, and industrial aspects of immobilized enzyme systems are dealt with in other chapters in this volume. In this chapter the authors will concentrate on the chemical aspects of enzyme immobilization, viz., on the description and evaluation of the methods most commonly used for the fixation of enzymes onto solid supports.

The majority of the methods available for the immobilization of enzymes and other biologically active proteins can be grouped in four main classes:

1. Adsorption on inert supports or ion-exchange resins
2. Entrapment, by occlusion within cross-linked gels or by encapsulation within microcapsules, hollow fibers, liposomes, and fibers
3. Cross-linking by bi- or multifunctional reagents, often following adsorption or entrapment within a structure of defined geometry
4. Covalent binding to polymeric supports, via functional groups nonessential for the biological activity of the protein

The covalent binding approach has been by far the most widely investigated. A large number of techniques for covalent coupling of enzymes to water-insoluble as well as to water-soluble polymeric carriers of a variety of chemical and physical characteristics have been described in the recent literature; moreover, comparative studies on enzymes covalently bound to a number of support materials by several methods have supplied valuable information on the effects of chemical modification on the activity and stability of immobilized enzymes as well as some hints on the nature of protein–matrix interactions. Hence, this chapter will be rather strongly oriented toward methods based on the covalent fixation of enzymes. Other approaches will be only briefly surveyed, the reader being referred to Tables I–IV and to several comprehensive reviews (Manecke, 1962, 1964, 1975; Weliky and Weetall, 1965; Silman and Katchalski, 1966; Campbell and Weliky, 1967; Goldstein, 1968, 1970; Goldstein and Katchalski, 1968; Lindsey, 1969; Crook, 1970; Crook *et al.*, 1970; Brown and Hasselberger, 1971;

Goldman *et al.*, 1971a,b; Katchalski *et al.*, 1971; Melrose, 1971; Mosbach, 1971; Weetall, 1971, 1975; Boguslaski *et al.*, 1972; Orth and Brümmer, 1972; Smiley and Strandberg, 1972; Royer *et al.*, 1973; Epton, 1973; Zaborsky, 1973; Gutcho, 1974; Stanley and Olson, 1974; Vieth and Venkatasubramanian, 1974; Kennedy, 1974b).

II. IMMOBILIZATION OF ENZYMES BY ADSORPTION

Historically the earliest method of protein immobilization—adsorption—is also the easiest way of preparing solid-supported enzyme conjugates (for review of early work, see Zittle, 1953; Silman and Katchalski, 1966; Goldman *et al.*, 1971b; James and Augenstein, 1966; McLaren and Packer, 1970; Zaborsky, 1973). Adsorption of an enzyme can be achieved by simply bringing an enzyme solution in contact with the absorbent surface. The binding forces between protein and support are in most cases relatively weak. Moreover, nonspecific adsorption of enzymes has been shown in a number of cases to lead to partial or total inactivation. A suitable adsorbent should thus possess high affinity for the enzyme and yet cause minimal denaturation.

Enzymes have been adsorbed nonspecifically on minerals, ion-exchange resins, and neutral polymeric supports. High concentrations of salt or substrate have been shown to enhance the rate of desorption of the protein. Adsorption techniques are thus of limited reliability when absolute immobilization of an enzyme is desired (see Zaborsky, 1973). Fixation of the enzyme could be achieved, however, by cross linking the adsorbed proteins with bifunctional reagents, such as bisdiazobenzidine-2,2'-disulfonic acid or glutaraldehyde (Goldman *et al.*, 1965, 1968, 1971a; Haynes and Walsh, 1969; Walsh *et al.*, 1970; Broun *et al.* 1969, 1973; Van Leemputten and Horisberger, 1974b; Wheeler *et al.*, 1969). This approach will be more fully discussed in the following sections.

Various minerals and other inorganic supports have been made use of as adsorbents for enzymes. These include kaolinite for the adsorption of chymotrypsin (McLaren and Estermann, 1956, 1957; McLaren 1960; McLaren and Packer, 1970), bentonite for catalase and β-amylase (Velikanov *et al.*, 1971), calcium phosphate gel for leucine aminopeptidase and amylase (Usami and Taketomi, 1965; Schwabe, 1969; Koelsch *et al.*, 1970), porous glass for trypsin, chymotrypsin, and RNase (Messing, 1969, 1970a,b,c), alumina for glucose oxidase, catalase, and amylase (Miyamoto *et al.*, 1971; Usami and Taketomi, 1965), carbon or silica coated with phospholipids, such as lecithin or cephalin, for acid phosphatase, phosphoglucomutase, and catalase

(Vorobeva and Poltorak, 1966; Goldfeld *et al.*, 1966; Nikolaev *et al.*, 1973a; Khoricova *et al.*, 1973; Chukhrai and Poltorak, 1973), and silica gel for trypsin (Haynes and Walsh, 1969; Walsh *et al.*, 1970). In several cases inactivation of the adsorbed enzyme has been reported (Vorobeva and Poltorak, 1966; for review, see Zaborsky, 1973). The nature of the interactions leading to the adsorption of enzymes on such carriers, and their effect on the activity of the bound enzyme, has received only scant attention. The protein-binding capacity of mineral and similar inorganic adsorbents is usually rather low [of the order of less than 1 mg per gram of adsorbent (see Messing, 1969, 1970a,b,c)].

Recently stainless steel particles (100–200 μm in diameter) activated by coating with titanium oxide have been shown to be powerful adsorbents for β-galactosidase [about 17 mg of protein per gram of support (Hasselberger *et al.*, 1974; Charles *et al.*, 1975)].

Cellulose powders have also been used as adsorbents (Fletcher and Okada, 1955, 1959; Wheeler *et al.*, 1969; Barker and Fleetwood, 1957; Barker *et al.*, 1970a; Epton, 1973). It is believed that adsorption of biological macromolecules occurs at sites on the surface where partial disruption of the microcrystalline structure of the matrix has taken place (Epton, 1973). Stronger adsorption on cellulose has been observed with the glucoside hydrolases (Barker and Fleetwood, 1957; Barker *et al.*, 1970a). This has been attributed to the structural similarity between the polysaccharide carrier and the substrates of such enzymes. In this context, it is of interest to note that enzymes adsorbed on specific "affinity supports" have been reported to retain their catalytic activity (Steers *et al.*, 1971).

A more systematic investigation of adsorbed enzymes has been carried out with inert hydrophilic supports based on cellulose, swollen cellophane and collodion membranes (Goldman *et al.*, 1965, 1968, 1971a; Goldman and Lenhoff, 1971; Broun *et al.*, 1969, 1973). Studies on the mechanism of adsorption of papain, alkaline phosphatase, and glucose-6-phosphate dehydrogenase on collodion membranes have indicated that these proteins form a monomolecular layer on the surface of the pores of the collodion matrix. [The adsorption capacity of the collodion membranes was about 70 mg of protein per square centimeter of membrane (Goldman *et al.*, 1965, 1968, 1971a).] In the case of glucose-6-phosphate dehydrogenase, desorption from the collodion membranes was shown to be highly dependent on the presence of specific substrates, i.e., NADP and glucose 6-phosphate (Goldman and Lenhoff, 1971).

Ion-exchange resins have been most commonly used as adsorbents. Several catalytically active conjugates have been prepared by adsorb-ing enzymes on carboxymethyl (CM) cellulose, DEAE-cellulose, and DEAE-Sephadex as well as on synthetic anion and cation exchangers in a medium of low ionic strength (Mitz and Schlueter, 1959; Miyamoto *et al.*, 1971, 1973; Tosa *et al.*, 1966a,b, 1967a,b, 1969a,b; Bachler *et al.*, 1970; Smiley, 1971; Barnett and Bull, 1959; Usami and Taketomi, 1965) (see Table I). The protein-binding capacity of the more extensively investigated polysaccharide ion-exchange resins is relatively high [50–150 mg of protein per gram of support (Tosa *et al.*, 1966a,b, 1967a; Bachler *et al.*, 1970)].

Aminoacylase adsorbed on DEAE-Sephadex and DEAE-cellulose was the first immobilized enzyme to be used commercially for the continuous resolution of racemic mixtures of *N*-acetyl-DL-amino acids (Tosa *et al.*, 1966a,b, 1967b, 1969a,b; see also Chibata and Tosa, this volume). The commercial potential of α-amylase and invertase ionically bound to DEAE-cellulose has been recently demonstrated (Smiley, 1971; Maeda *et al.*, 1973c).

A major handicap of immobilization of proteins based solely on electrostatic attraction to charged supports lies in the tendency of such conjugates to dissociate upon increasing the ionic strength or varying the pH or the temperature of the medium. Some of these disadvantages can be overcome by increasing the charge on the protein by chemical modification, as demonstrated recently by Bessmertnaya and Antonov (1973), Yarovaya *et al.* (1975), and Solomon and Levin (1974a). These authors prepared polyanionic derivatives of enzymes by covalently coupling α-chymotrypsin and trypsin to a water-soluble copolymer of acrylic acid and maleic anhydride or by covalent coupling of amyloglucosidase to a water-soluble copolymer of ethylene and maleic acid. The polyanionic enzyme derivatives adsorbed strongly and practically irreversibly on cationic resins such as DEAE-cellulose and DEAE-Sephadex to give stable complexes, which in the case of amyloglucosidase (Solomon and Levin, 1974a) could be used for the continuous conversion of starch to glucose without significant loss in activity for up to 3 weeks. It is worth mentioning that native amyloglucosidase when added directly to DEAE-Sephadex adsorbed rather poorly.

The adsorption of enzymes on collagen has attracted considerable attention (Vieth *et al.*, 1972,a,b; Saini *et al.*, 1972; Constantinides *et al.*, 1973; Wang and Vieth, 1973; Suzuki *et al.*, 1972, 1974, 1976; Karube and Suzuki, 1972a,b) Collagen conjugates of β-D-fructofuranosidase, lysozyme, urease, glucose oxidase, and penicillin amidase were prepared either by direct impregnation of preswollen collagen membranes or by electrodeposition from a collagen dispersion containing dissolved

TABLE I

SUPPORTS FOR IMMOBILIZATION OF ENZYMES BY ADSORPTION

Adsorbent	Capacity (mg protein/ gm adsorbent)	References
Alumina	—	Miyamoto et al. (1971); Usami and Taketomi (1965)
Bentonite	—	Velikanov et al. (1971)
Calcium carbonate	—	Velikanov et al. (1971)
Calcium phosphate gel	—	Usami and Taketomi (1965); Koelsch et al. (1970); Schwabe (1969)
Carbon	—	Miyamoto et al. (1971)
Carbon coated with cephalin or lecithin	—	Usami and Taketomi (1965); Goldfeld et al. (1966); Vorobeva and Poltorak (1966)
Cellulose	—	Fletcher and Okada (1955, 1959)
Cellulose (Millipore or Sartorius filters)	—	Wheeler et al. (1969)
Cellulose, carboxymethyl ether (CM-cellulose)	—	Miyamoto et al. (1971); Nikolaev and Mardashev (1961)
Cellulose, diethylaminoethyl ether (DEAE-cellulose)	50–150	Tosa et al. (1966a,b, 1967a,b); Miyamoto et al. (1971); Becker and Pfeil (1966); Bachler et al. (1970); Maeda et al. (1973c); Smiley (1971); Suzuki et al. (1966); Mitz and Schlueter (1959); Nikolaev (1962); Nikolaev et al. (1973b); Solomon and Levin (1974a); Usami et al. (1971); Barth and Maskova (1971)
Cellulose acetate membrane	—	Broun et al. (1969, 1973)
Clay	—	Miyamoto et al. (1971); Usami et al. (1967); Usami and Shirasaki (1970)
Collagen membrane	6–50	Vieth et al. (1972a,b); Saini et al. (1972); Constantinides et al. (1973); Wang and Vieth (1973); Suzuki et al. (1972, 1974, 1976); Karube and Suzuki (1972a,b)
Collodion membrane	~200 $\mu g/cm^2$	Goldman et al. (1965, 1968, 1971a); Goldman and Lenhoff (1971)
Glass, porous (pore diameter 900 Å)	0.1–0.33	Messing (1969, 1970a,b,c, 1975)
Hydroxyapatite	—	Traub et al. (1969)

(Continued)

TABLE I (*Continued*)

Adsorbent	Capacity (mg protein/ gm adsorbent)	References
Ion-exchange resins	—	Barnett and Bull (1959); Usami and Taketomi (1965); Miyamoto *et al.* (1971, 1973); Bachler *et al.* (1970)
Kaolinite	—	Velikanov *et al.* (1971); McLaren and Estermann (1956, 1957); McLaren (1960)
Sephadex, diethylaminoethyl ether (DEAE-Sephadex)	100	Tosa *et al.* (1966a, 1967b, 1969a,b, 1971); Solomon and Levin (1974a)
Sephadex, carboxymethyl ether (CM-Sephadex)	—	Miyamoto *et al.* (1971)
Silica gel	—	Tveritinova *et al.* (1969); Zhirkov *et al.* (1971); Usami and Taketomi (1965); Nikolaev *et al.* (1973a,b)
Silica gel coated with lecithin or cephalin	—	Tveritinova *et al.* (1969); Goldfeld *et al.* (1966); Vorobeva and Poltorak (1966); Nikolaev *et al.* (1973a)
Stainless steel particles (100–200 μm) activated by coating with titanium oxide (TiO$_2$)	17	Hasselberger *et al.* (1974); Charles *et al.* (1975)
Specific adsorption		
N-Alkyl agarose derivatives	—	Hofstee and Otillio (1973); Visser and Strating (1975)
Concanavalin A–agarose	—	Sulkowski and Laskowski (1974)

enzyme (5–50 mg of enzyme per gram of collagen could be immobilized by these methods). The forces stabilizing the enzyme–collagen conjugates have been ascribed to multiple salt linkages, hydrogen bonds, and van der Waals interactions (Vieth *et al.*, 1972a,b; Saini *et al.*, 1972).

The principles of "hydrophobic chromatography" (Er-El *et al.*, 1972; Shaltiel and Er-El, 1973; Hofstee, 1973; Shaltiel, 1974a) have been recently applied to effect the virtually irreversible adsorption of several enzymes onto N-alkyl derivatives of Sepharose (Hofstee and Otillio, 1973; Visser and Strating, 1975). The bound enzymes, all of which have isoelectric points in the acid region (xanthine oxidase, lactate dehydrogenase, alkaline phosphatase, and urease), were not easily desorbed even by 1 M NaCl. The strong binding of acidic proteins to N-alkyl-Sepharose, known to carry residual positive charges, could be attributed to a combination of electrostatic and hydrophobic interactions (in this context, see Section V,A,3 on cyanogen bromide activation of polysaccharides; see also Svensson, 1973; Jost *et al.*, 1974; Wilchek *et al.*, 1975).

A method based on biospecific adsorption has been recently described by Sulkowski and Laskowski (1974), who made use of a concanavalin A–Sepharose conjugate to immobilize two glycoprotein enzymes, venom exonuclease (phosphodiesterase) and 5'-nucleotidase. Concanavalin A, which is capable of agglutinating red blood cells, is known to bind specifically to mono- and oligosaccharide substituents present on cell surfaces; moreover, it specifically precipitates polysaccharides and glycoproteins from solution (see Sharon and Lis, 1972).

III. IMMOBILIZATION OF ENZYMES BY ENTRAPMENT

In principle, all entrapment methods are based on the occlusion of an enzyme within a constraining structure tight enough to prevent the protein from diffusing into the surrounding medium, while still allowing penetration of substrate. The obvious advantage of such methods is in their generality, since the enzyme molecule itself does not participate directly in the formation of the water-insoluble constraining structure. Except where an enzyme is adversely affected in the course of the entrapment reaction or where protein-supporting polymer interactions lead to denaturation (Zaborsky, 1973; Epton, 1973; Degani and Miron, 1970; Ohmiya *et al.*, 1975; Maeda *et al.*, 1975), the common occlusion techniques have been successful with most enzyme systems tested (see, e.g., Goldman *et al.*, 1971b; Melrose, 1971; Epton, 1973; Zaborsky, 1973). The generality of the occlusion techniques is limited by the fact that they are suitable mainly for enzymes that utilize substrates of molecular weights low enough to diffuse through the matrix; moreover, diffusional resistances to the penetration of substrate usually lead to perturbed kinetics, which in extreme cases may be erroneously interpreted as intimating rather low specific activities for the immobilized enzyme (Sundaram, 1973; Kobayashi and Laidler, 1973; Fink, 1973; Poznansky and Chang, 1974; Knights and Light, 1974; Marconi *et al.*, 1974a,b; Korus and O'Driscoll, 1974; Kasche and Bergwall, 1974; see also Engasser and Horvath, this volume).

Occlusion within cross-linked polyacrylamide gels has been the most widely used entrapment technique. The method is based on the polymerization of acrylamide in the presence of varying amounts of *N,N'*-methylene bis(acrylamide) as cross-linker, in an aqueous medium containing the dissolved enzyme. The resulting polymeric gel can be mechanically dispersed into particles of defined size and stored in suspension or in the form of lyophilized powders (Bernfeld and Wan, 1963; Bernfeld *et al.*, 1968, 1969; Hicks and Updike, 1966; Wieland *et al.*, 1966; Mosbach and Mosbach, 1966; Mosbach, 1970; Mosbach and Mattiasson, 1970). Acrylamide gels can be charged with considerable amounts of protein (10–100 mg per gram of monomer— see Table II).

The acrylamide-gel entrapment technique suffers from one intrinsic drawback, leakage of enzyme, particularly pronounced with proteins of relatively low molecular weight. This leakage has been attributed to local variations in permeability, ensuing from the broad distribution of pore sizes encountered in cross-linked gels of the polyacrylamide type (see Fawcett and Morris, 1966). This serious practical disadvantage can be only partially overcome by optimizing the composition of the gel, i.e., concentration and degree of crosslinking. Fawcett and Morris (1966), who calculated the dependence of pore size on the degree of cross-linking of acrylamide gels, showed that at constant monomer concentration the effective pore radius of a gel is minimal at 5% cross-linking. At higher degrees of cross-linking there is an increase in gel permeability. This rather unexpected phenomenon has been ascribed to the stacking of fibers into strands, resulting in widening of interstrand distances. These conclusions were confirmed by Degani and Miron (1970) in their study on the entrapment of cholinesterase in acrylamide gels of varying composition. These authors also showed that the amount of entrapped protein increased with increasing monomer concentration, in agreement with the expected decrease in gel porosity (Fawcett and Morris, 1966). The enzymic activity of entrapped cholinesterase decreased sharply, however, in gels prepared at monomer concentrations higher than 15%, implying that acrylamide acts as a denaturing agent, similarly to urea (see, e.g., Martinek *et al.*, 1975). These findings thus seem to define the limits within which the permeability properties of gels could be manipulated. A way of circumventing the problem of protein retention within open gel structures has been suggested by Mosbach, who entrapped glucose-6-phosphate dehydrogenase within a cross-linked acrylamide-acrylic acid gel and then fixed the entrapped enzyme covalently by carbodiimide activation of the carboxyl groups on the support (Mosbach, 1970; Mosbach and Mattiasson, 1970; see also Jaworek, 1974). Polyacrylamide gels have been recently employed for the entrapment of whole microbial cells (see Chibata and Tosa, this volume).

A method of entrapment by a bead-polymerization procedure simi-

TABLE II

SUPPORTS FOR THE IMMOBILIZATION OF ENZYME BY ENTRAPMENT

Entrapment matrix	Capacity (mg/gm conjugate)	References
Polyacrylamide; cross-linked gel	6–100	Walton and Eastman (1973); Mori et al. (1972); Mosbach and Mosbach (1966); Mosbach (1970); Mosbach and Mattiasson (1970); Nilsson et al. (1972); Bernfeld and Wan (1963); Bernfeld et al. (1968, 1969); Bernfeld and Bieber (1969); Hicks and Updike (1966); Degani and Miron (1970); Strandberg and Smiley (1971); Wieland et al. (1966); Nadler and Updike (1974); Guilbault and Das (1970); Dobo (1970); Guilbault and Montalvo (1970); Brown et al. (1968a,b); Maeda et al. (1973a); Yamamoto et al. (1974a,b); Tosa et al. (1974)
Polyacrylamide, cross-linked beads	2–5	Nilsson et al. (1972)
2-Hydroxyethyl methacrylate; cross-linked gel	—	Miyamura and Suzuki (1972); O'Driscoll et al. (1972)
Poly(2-hydroxyethylacrylate), radiation cross-linked gel	—	Maeda et al. (1975)
Polyvinyl alcohol; radiation cross-linked gel	5–10	Maeda et al. (1973b)
Polysiloxane (silicon rubber, Silastic)	—	Pennington et al. (1968a,b); Brown et al. (1968b)
Starch gel	—	Guilbault and Das (1970); Bauman et al. (1965, 1967); Guilbault and Kramer (1965)
Polyvinylpyrrolidone; radiation cross-linked gel	—	Denti (1974); Maeda et al. (1974); Maeda (1975)
Silica gel	—	Johnson and Whateley (1971)
Nylon microcapsules	—	Chang (1964, 1966, 1967, 1971a,b, 1972a,b,c, 1974); Chang et al. (1966); Kitajima et al. (1969); Chang and Poznansky (1968)
Fibers	—	Dinelli (1972); Marconi et al. (1974a,b); Giovenco et al. (1973)
Liposomes	—	Gregoriadis (1974); Gregoriadis and Ryman (1972a,b); Gregoriadis et al. (1971); Gregoriadis and Buckland (1973)
Liquid membranes	—	May and Li (1972); Mohan and Li (1974)

lar to that described for the preparation of cross-linked polyacrylamide used in gel chromatography (Hjerten, 1962c; Hjerten and Mosbach, 1962) has been published recently (Nilsson *et al.*, 1972; Johansson and Mosbach, 1974a,b). In this procedure an aqueous solution containing enzyme and acrylic monomers is dispersed in a hydrophobic phase and polymerized, resulting in well defined spherical beads. The beads contain entrapped active enzyme and show good mechanical stability and high flow rates in column processes. Another possibility for entrapping enzymes was the γ-ray irradiation of polyvinylpyrrolidone in the presence of an enzyme (Maeda *et al.*, 1974; Maeda, 1975; Denti, 1974).

Entrapment by other methods has been less successful. Practical application of cholinesterase entrapped in starch gel was possible only when the soft conjugates were dispersed in polyurethane foam pads to improve their mechanical stability (Guilbault and Kramer, 1965; Bauman *et al.*, 1967; Guilbault and Das, 1970). The use of a silicon polymer, Silastic, as a cross-linked network for entrapment of cholinesterase has been reported (Pennington *et al.*, 1968a,b; Brown *et al.*, 1968b). The Silastic-entrapped enzyme exhibited improved thermal stability. The permeability of the Silastic–enzyme conjugate to substrates and inhibitors, however, seems to be low on the basis of the published data.

The inclusion of whole droplets of enzyme solution within semipermeable nylon microcapsules has attracted considerable attention in view of the potential medical applications of such preparations (Chang, 1964, 1966, 1967, 1969, 1971a,b, 1972a,b,c, 1974; Chang and Poznansky, 1968; Chang *et al.*, 1966; Kitajima *et al.*, 1969; Poznansky and Chang, 1974). Microencapsulation is usually achieved by dispersing an aqueous enzyme solution containing 1,6-diaminohexane (hexamethylenediamine) into a solution of hexanedioic acid dichloride (adipoyl chloride) in an organic solvent immiscible with water (e.g., chloroform, carbon tetrachloride, toluene). The diamine and acid dichloride polymerize upon contact at the water–organic solvent interface, forming a thin polyamide (nylon-6,6) membrane around aqueous droplets of enzyme solution. The stability of microencapsulated enzymes has been found to be similar in most cases to that of the corresponding free enzymes in solution. Enhancement of stability could be obtained in several cases by encapsulating enzymes in the presence of an inert protein, followed by cross-linking with glutaraldehyde. Details of the microencapsulation method can be found in several recent publications and review articles (see Chang, 1969, 1972c, 1974; see also Table II). By means of a similar approach, enzymes have also been encapsulated within liposomes (Gregoriadis, 1974; Gregoriadis and Ryman 1972a,b; Gregoriadis and Buckland, 1973; Gregoriadis *et al.*, 1971).

A process for the immobilization of enzymes by entrapment within synthetic fibers has been recently described (Dinelli, 1972; Giovenco *et al.*, 1973; Marconi *et al.*, 1974a,b). In this process, well suited for large-scale application, an emulsion is formed with an aqueous enzyme solution and a solution of a synthetic polymer in an organic solvent (e.g., cellulose acetate or polyvinyl chloride in methylene chloride). The emulsion is extruded through a spinneret into a precipitant, droplets of enzyme solution being trapped in the fiber. By varying the condition of precipitation, the pore size of the fiber can be controlled. The fibers can be braided or woven into cloth, depending on the reactor configuration envisaged. Using very fine fibers, reasonably high area : weight ratios can be obtained; the loading capacities of the fibers are quite high; e.g., 1500 mg of invertase per gram of polymer with 20% retention of activity has been reported. High retention of activity (up to 60%) and good operational stabilities have been found with several enzymes. The applicability of the process is limited by the necessity of using water-immiscible liquids as polymer solvents and precipitants; such liquids may in some cases cause inactivation of the enzyme (Dinelli *et al.*, 1975).

Both microencapsulated and fiber-entrapped enzymes exhibit diffusion-limited kinetics and hence are best suited for enzyme systems that work on low-molecular-weight substrates.

IV. IMMOBILIZATION OF ENZYMES BY COVALENT CROSS-LINKING WITH BI- OR MULTIFUNCTIONAL REAGENTS

Insolubilization of enzymes, solely by their intermolecular cross-linking into large aggregates, has found rather limited application because of the difficulties encountered in controlling such reactions to give well characterized products in terms of aggregate size and mechanical properties (Silman and Katchalski, 1966; Goldman *et al.*, 1971b; Zaborsky, 1973). Carrying out the cross-linking reaction on precipitates obtained by the addition of sodium or ammonium sulfate or an organic solvent such as acetone has permitted in a few cases a certain degree of control on the properties of the final product (Habeeb, 1967; Ogata *et al.*, 1968; Jansen and Olson, 1969; Schejter and Bar-Eli, 1970; Jansen *et al.*, 1971; Ottesen and Svensson, 1971; Glassmeyer and Ogle, 1971).

Intermolecular cross-linking by bi- or multifunctional reagents has been used routinely for the fixation of enzyme crystals and similar macromolecular aggregates of well defined structure and in histochemistry for the preservation of cellular ultrastructure, e.g., in the preparation of specimens for electron microscopy (Sabatini *et al.*, 1963; Quiocho and Richards, 1964, 1966; Marfrey and King, 1965; Bishop *et al.*, 1966; Sluyterman and deGraaf, 1969; Hopwood, 1972; Josephs *et al.*, 1973). Reagents containing two functional groups of different reactivities (e.g., toluene 2-isocyanate-4-isothiocyanate and toluene 2,4-diisocyanate) have been considered for the two-step covalent linking of dissimilar proteins, e.g., enzyme–antibody and ferritin–antibody conjugates (Schick and Singer, 1961; Avrameas, 1969; Wold, 1972; Weir, 1973). Currently, low-molecular-weight bifunctional reagents are used mainly for the covalent fixation of proteins adsorbed on solid supports or entrapped within gels or microcapsules (Haynes and Walsh, 1969; Goldman *et al.*, 1965, 1968, 1971a,b; Walsh *et al.*, 1970; Chang, 1971a,b; Broun *et al.*, 1973; Broun, 1976; see also preceding section). Another approach that has found wide application is the chemical modification of performed polymeric supports by means of bi- or multifunctional low-molecular-weight reagents. In this method, one of the molecule's functional groups forms a covalent link with the support; the other functional group or groups can then be used to bind a protein (Kay and Crook, 1967; Surinov and Manoilov, 1966; Habeeb, 1967; Glassmeyer and Ogle, 1971; Kay and Lilly, 1970; Stanley *et al.*, 1975; Ternynck and Avrameas, 1972; Johansson and Mosbach, 1974a,b; Goldstein, 1973b; Hornby *et al.*, 1972; Inman and Hornby, 1974). The most common cross-linking reagents together with their major uses are listed in Table III. General reviews on cross-linking procedures are available (Fasold *et al.*, 1971; Wold, 1972).

Of the considerable number of cross-linking agents described in the literature, only two have found widespread use for enzyme immobilization: bisdiazobenzidine-2,2′-disulfonic acid and glutaraldehyde (see Table III). Bisdiazobenzidine-2,2′-disulfonic acid was used for immobilization of enzymes within collodion membranes (Goldman *et al.*, 1965, 1968, 1971a). Glutaraldehyde, which is by far the most commonly used cross-linking reagent, has been employed for the fixation of crystals and histochemical preparations (Sabatini *et al.*, 1963; Quiocho and Richards 1964, 1966; Bishop *et al.*, 1966; Hopwood, 1972; Josephs *et al.*, 1973), for cross-linking enzymes adsorbed on solid supports (Haynes and Walsh, 1969; Walsh *et al.*, 1970; Broun *et al.*, 1969, 1973; Liu *et al.*, 1975) for co-cross-linking enzymes with an

TABLE III

MULTIFUNCTIONAL REAGENTS FOR THE CROSS-LINKING AND IMMOBILIZATION OF ENZYMES

Multifunctional reagent	Method of immobilization	Capacity (mg protein/gm)	References
Glutaraldehyde	Impregnation of cellophane membranes with enzyme followed by cross-linking	0.1 mg protein/cm² membrane	Broun et al. (1969, 1973)
Glutaraldehyde	Enzyme cross-linked in solution, then included in agarose-polyacrylamide gel	7	Broun et al. (1973)
Glutaraldehyde	Enzyme co-cross-linked in solution with inert protein, e.g., albumin, then spread on glass plate to obtain membrane	7–8	Broun et al. (1973)
Glutaraldehyde	Enzyme co-cross-linked in solution with inert protein, e.g., albumin, then frozen at −30°C and warmed slowly to obtain spongelike conjugate	70–80	Broun et al. (1973)
Glutaraldehyde	Enzyme co-cross-linked with inert protein, e.g., gelatin, in the presence of fillers (bentonite, alumina, silica gel or Celite)	50–500	Solomon and Levin (1974b)
Glutaraldehyde	Enzyme co-cross-linked with chitin	30	Stanley et al. (1975)
Glutaraldehyde	Enzyme cross-linked with inert protein, bovine serum albumin		Avrameas and Ternynck (1969); Avrameas and Guilbert (1971)
Glutaraldehyde	Enzyme adsorbed on magnetite (Fe_2O_3) followed by cross-linking	4–36	Van Leemputten and Horisberger (1974b)
Glutaraldehyde	Enzyme adsorbed on carbon followed by cross-linking		Liu et al. (1975)

Glutaraldehyde	Adsorption on collodion membranes followed by cross-linking	—	Goldman et al. (1965, 1968, 1971a)
Glutaraldehyde	Adsorption on colloidal silica particles (210–230 m²/gm) followed by cross-linking	300	Haynes and Walsh (1969); Walsh et al. (1970)
Glutaraldehyde	Microencapsulation within collodion or nylon microcapsules followed by cross-linking	—	Chang (1971a,b)
Glutaraldehyde	Cross-linking of whole crystals	—	Quiocho and Richards (1964, 1966); Bishop et al. (1966); Sluyterman and deGraaf (1969)
Glutaraldehyde	Cross-linking of glutamate dehydrogenase aggregates	—	Josephs et al. (1973)
Glutaraldehyde	Enzyme cross-linked in the presence of sodium sulfate, ammonium sulfate, or acetone	—	Habeeb (1967); Ogata et al. (1968); Jansen and Olsen (1969); Jansen et al. (1971); Ottesen and Svensson (1971); Scheiter and Bar-Eli (1970)
Glutaraldehyde	Adsorption on nylon floc, nylon membranes or pellicular nylon followed by cross-linking	—	Inman and Hornby (1972); Reynolds (1974); Horvath (1974)
Glutaraldehyde	Activation of aminoethyl cellulose followed by coupling of protein	—	Habeeb (1967); Glassmeyer and Ogle (1971)
Glutaraldehyde	Activation of 1-amino-6-hexamido derivatives of cross-linked ethylene–maleic acid copolymers, followed by coupling of protein	100	Goldstein (1973b)
Glutaraldehyde	Activation of partially hydrolyzed nylon, followed by coupling of protein	—	Sundaram and Hornby (1970); Filippusson et al. (1972); Hornby et al. (1972); Inman and Hornby (1972, 1974)

(Continued)

TABLE III (Continued)

Multifunctional reagent	Method of immobilization	Capacity (mg protein/gm)	References
Glutaraldehyde	Activation of alkylamino derivatives of silanized porous-glass beads followed by coupling of protein	12–16	Robinson et al. (1971, 1973); Dixon et al. (1973); Weetall and Filbert (1974)
Glutaraldehyde	Activation of macroreticular poly(p-aminostyrene), followed by coupling of protein	—	Baum (1975)
Bisdiazobenzidine-2,2'-disulfonic acid	Adsorption of enzyme on collodion membranes followed by cross-linking	—	Goldman et al. (1965, 1968); Wheeler et al. (1969)
Ethyl chloroformate	Cross-linking	—	Avrameas and Ternynck (1967); Rao et al. (1970)
Bifunctional isocyanates and isothiocyanates	Inter- and intramolecular cross-linking		
Xylylene diisocyanate		—	Schick and Singer (1961); Wold (1972)
Toluene 2,4-diisocyanate		—	Schick and Singer (1961); Wold (1972)
Toluene 2-isocyanate 4-isothiocyanate		—	Schick and Singer (1961); Wold (1972)
3-Methoxydiphenylmethane 4,4'-diisocyanate		—	Wold (1972)
2,2'-Dicarboxy 4,4'-azophenyldiisocyanate or diisothiocyanate		—	Fasold (1964, 1965); Fasold et al. (1971); Ozawa (1967a)
Hexamethylene diisocyanate		—	
Diphenyl-4,4'-diisothiocyanate-2,2'-disulfonic acid		—	Manecke and Günzel (1967b)

38

Reagent	Cross-linking	References
Bifunctional imidoesters		
Diethyl malonimidate	Inter- and intramolecular cross-linking	Dutton et al. (1966)
Dimethyl adipimidate	—	Hartman and Wold (1966, 1967); Wold (1972); Zaborsky (1974b)
Dimethyl suberimidate	—	Davies and Stark (1970); Carpenter and Harriston (1972); Handschumacher and Gaumond (1972)
Dithiobispropionimidate		Rüaho et al. (1975)
Bifunctional iodoacetamides		
1,6-Bisiodoacetamidohexane	Intramolecular cross-linking	Ozawa (1967b)
1,4-Bisiodoacetamidobutane	—	Ozawa (1967b); Reiner et al. (1975)
4,4'-Bisiodoacetamidobenzene	—	Ozawa (1967b); Reiner et al. (1975)
1,4-Bisiodoacetamido-2,2'-dicarboxyazobenzene		Fasold et al. (1973)
Bifunctional maleimide derivatives		
N,N'-(1,3-phenylene)bismaleimide	Inter- and intramolecular cross-linking	Moore and Ward (1956)
Azophenyl-p-N,N'-dimaleimide	—	Fasold et al. (1963)
N,N'-(Hexamethylene)bismaleimide	—	Zahn and Lumper (1968)
Bis(N-maleimidomethyl) ether	—	Tawney et al. (1961); Simon and Konigsberg (1966); Freedberg and Hardman (1971)
Bifunctional aryl halides		
4,4'-Difluoro-3,3'-dinitrodiphenylsulfone	Inter- and intramolecular cross-linking	Zahn (1955); Wold (1972)
1,5-Difluoro-2,4-dinitrobenzene		Zahn (1955); Wold (1972); Marfrey and King (1965)

inert protein in the presence or in the absence of a solid support or filler (Broun *et al.*, 1969, 1973; Solomon and Levin, 1974b; Stanley *et al.*, 1975) and for the cross-linking of enzymes enclosed in microcapsules (Chang, 1971a,b) as well as for the chemical modification of materials such as aminoethyl cellulose (Habeeb, 1967; Glassmeyer and Ogle, 1971), partially hydrolyzed nylon (Sundaram and Hornby, 1970; Allison *et al.*, 1972; Filippusson *et al.*, 1972; Bunting and Laidler, 1974; Hornby *et al.*, 1972; Inman and Hornby, 1974; Morris *et al.*, 1975; Campbell *et al.*, 1975), and other polymeric supports containing primary amino groups (Robinson *et al.*, 1971, 1973; Goldstein, 1973b; Dixon *et al.*, 1973; Weetall and Filbert, 1974; Baum, 1975). The linkages formed between a protein and glutaraldehyde are irreversible and survive extremes of pH and temperature. The straightforward participation of the reagent's aldehyde groups to form aldimine (Schiff's base) bonds with protein amino groups has been questioned. Since aqueous glutaraldehyde solutions are reported to contain appreciable amounts of α,β-unsaturated oligomeric condensation products (Richards and Knowles, 1968; Hardy *et al.*, 1969), it has been suggested that the glutaraldehyde reaction most probably involves conjugate-addition of protein amino groups to ethylenic double bonds of α,β-unsaturated oligomers (Richards and Knowles, 1968; Quiocho, 1974) (see Fig. 13). The chemistry of glutaraldehyde is discussed in detail in Section V,A,7 (see also Richards and Knowles, 1968; Hardy *et al.*, 1969; Whipple and Ruta, 1974; Monsan *et al.*, 1975).

The possibility of enhancing the conformational and operational stability of enzymes by incorporation of additional cross-links is being explored. Such studies, aimed at finding the appropriate combination of functional groups and chain length that will result in preparations of increased stability, have led to the synthesis of a wide variety of new bifunctional reagents and to several novel approaches to selective intramolecular cross-linking of proteins, to covalent fixation of subunits, and to intermolecular cross-linking of protein complexes (Fasold, 1964, 1965; Darlington and Keay, 1965; Hartman and Wold, 1966, 1967; Dutton *et al.*, 1966; Ozawa, 1967a,b; Fasold *et al.*, 1971, 1973; Handschumacher and Gaumond, 1972; Carpenter and Harriston, 1972; Wold, 1972; Wetz *et al.*, 1974; Zaborsky, 1974b; Reiner *et al.*, 1975; Ruaho *et al.*, 1975; see also Table III).

After the work of Hartman and Wold (1966, 1967), who employed a bifunctional imidoester, dimethyl adipimidate, to map distances between lysyl residues of ribonuclease, bifunctional imidoesters of varying chain length were used for the intramolecular cross-linking of enzymes. In a few cases increased stability as well as increased activity toward low-molecular-weight substrates have been reported (Wold, 1972; Zaborsky, 1974b; Hartman and Wold, 1967; Davies and Stark, 1970; Dutton *et al.*, 1966; Handschumacher and Gaumond, 1972; Ruaho *et al.*, 1975). Cross-linking with bifunctional isocyanates, isothiocyanates, and azides and with bisiodoacetamides of different length has been investigated (Fasold, 1964, 1965; Fasold *et al.*, 1971, 1973; Wetz *et al.*, 1974; Ozawa, 1967a,b). The use of such reagents is limited by the fact that increasing the chain length by increasing the number of hydrocarbon residues of the backbone decreases their solubility; moreover, the increased hydrophobicity of such reagents may often have deleterious effects on the activity of the enzyme owing to local denaturation effects. Recently, hydrophilic bifunctional reagents based on an oligoproline backbone and varying from 30 Å to more than 100 Å in length have been used to cross-link hemoglobin tetramers (Wetz *et al.*, 1974).

To identify the amino acid residues participating in the formation of a cross-link, bifunctional reagents that contain a labile link, e.g., azo or disulfide, have been developed in several laboratories (Fasold *et al.*, 1971, 1973; Wold, 1972; Wetz *et al.*, 1974; Ruaho *et al.*, 1975). The cross-linked enzyme can be digested with a protease and the bridged peptides isolated. After mild reduction to break the labile azo or disulfide bonds, the component peptides are then separated and characterized.

In most cases investigated, intramolecular cross-linking involves only a limited number of sites on the protein, unless high concentrations of reagent and relatively harsh conditions are used (Fasold *et al.*, 1971; Wold, 1972). A new approach leading to fairly extensive intramolecular cross-linking has been recently described (Reiner *et al.*, 1975). By this method, ribonuclease is initially polythiolated with *N*-acetylhomocysteine thiolactone and then cross-linked via the newly formed sulfhydryl groups either by oxidation with ferricyanide or by reaction with bifunctional iodoacetamides.

V. IMMOBILIZATION OF ENZYMES BY COVALENT COUPLING TO POLYMERIC SUPPORTS

Covalent binding to polymeric supports has been the most thoroughly investigated approach to enzyme immobilization. Nevertheless, the compositional and structural complexity of proteins has not allowed, except in a very limited number of cases, the application of general rules by means of which the method best suited for a specific task could be predicted. Accumulated experience in the field has stressed the importance of two main factors that have to be considered when choosing a method for the covalent immobilization of an enzyme: (a) the type of functional groups on the protein through which the covalent bonds with the support material are formed and hence the chemical reaction to be employed; (b) the physical and chemical characteristic of the support material onto which chemically reactive groups are to be grafted.

The methods for covalent binding that have gained acceptance, classified according to support materials, are listed in Table IV.

A. Coupling Reactions

The type of functional groups on the protein through which the covalent bond with the support is to be formed should naturally be nonessential for the catalytic activity of the enzyme; moreover, binding reactions that can be carried out under relatively mild conditions and in essentially aqueous media should be preferred. Such reactions should exhibit, under ideal conditions, relatively high specificity toward one type of functional group on the protein and minimal side reactions with other functional groups or with the aqueous medium; in practice such a situation is seldom if ever realized (see, e.g., Hirs, 1967; Cohen, 1968; Vallee and Riordan, 1969; Shaw *et al.*, 1969; Phillips *et al.*, 1970; Stark, 1970; Desnuelle *et al.*, 1970; Boyer, 1970; Means and Feeney, 1971; Hirs and Timasheff, 1972). In selecting the appropriate coupling reaction and functionalized polymer to be employed for the immobilization of a given enzyme, all the available information on the amino acid composition, the amino acids involved in the active site, the effects of specific chemical modifications on activity, the protection of the active site region by specific chemical agents or inhibitors, as well as the three-dimensional structure of the enzyme, when known, should be considered.

The protein functional groups that can be utilized in principle for the covalent binding of enzymes to polymeric supports include: (1) *amino groups*, the ε-amino groups of lysine and the α-NH₂ groups of the N-termini of the polypeptide chains; (2) *carboxyl groups*, the β- and γ-carboxyl groups of aspartic and glutamic acid, respectively, and terminal α-carboxyls; (3) *phenol rings* of tyrosine; (4) *sulfhydryl groups* of cysteine; (5) *hydroxyl groups* of serine, threonine, and tyrosine; (6) *imidazole* groups of histidine; (7) *indole* groups of tryptophan. In practice, most of the common covalent coupling reactions involve amino groups, carboxyls, or the aromatic rings of tyrosine and histidine.

The major classes of reactions used for the immobilization of enzymes will be discussed in this section, with emphasis on the type of functional groups on the proteins participating in the formation of co-valent bonds with the support. Methods of mainly historical interest have been listed in Table IV, but will not be discussed in detail. The older literature has been summarized in several reviews (Silman and Katchalski, 1966; Manecke, 1962, 1964, 1975; Weliky and Weetall, 1965; Brown and Hasselberger, 1971; Goldman *et al.*, 1971b; Zaborsky, 1973; Goldstein and Katchalski-Katzir, this volume).

The support materials most commonly used as parent polymers, their properties, and methods available for their preparation and chemical modification will be discussed in the next section.

1. Acylation Reactions

Coupling via amino groups on the protein has been commonly car-ried out with polymers containing pendant acylating groups, such as acyl azide or acid anhydride (Mitz and Summaria, 1961; Epstein and Anfinsen, 1962; Hornby *et al.*, 1966; Brown *et al.*, 1966; Wharton *et al.*, 1968a,b; Crook *et al.*, 1970; Brümmer *et al.*, 1972; Barker *et al.*, 1970a,b; Levin *et al.*, 1964; Goldstein *et al.*, 1964; Goldstein, 1970, 1972a,b; Fritz *et al.*, 1968a,b; Zingaro and Uziel, 1970; Conte and Lehmann, 1971), by polymers containing carboxychloride or sulfonyl chloride groups (see Silman and Katchalski, 1966; Zaborsky, 1973), by active esters of carboxylate polymers, such as hydroxysuccinimide es-ters (Cuatrecasas and Parikh, 1972), and by polymers containing car-boxyl groups that have been activated with N,N'-disubstituted car-bodiimides (Weliky and Weetall, 1965; Weetall and Weliky, 1966; Hoare and Koshland, 1966, 1967; Weliky *et al.*, 1969; Martensson and Mosbach, 1972; Johansson and Mosbach, 1974a) and other condensing reagents, e.g., N-alkyl-5-phenylisoxazolium salts (e.g., Woodward's Reagent K, N-ethyl-5-phenylisoxazolium 3'-sulfonate) (Woodward and Olofson, 1961; Woodward *et al.*, 1961, 1966; Patel and Price, 1967; Patel *et al.*, 1967), N-ethoxycarbonyl-2-ethoxy-1,2-dihydroquinoline (Sundaram, 1974; Bartling *et al.*, 1974b), and carbonyldiimidazole (Bartling *et al.*, 1974a).

The common mechanistic feature in acylation reactions is the attack of a nucleophile (in the case of proteins —NH₂, —OH, or —SH groups) at an activated carbonyl group (e.g., acyl azide or acid anhydride). Nucleophiles are most effective in their unprotonated forms (R—NH₂, PhO⁻, RS⁻), i.e., at pH values above their pK_a values; high pH might cause irreversible denaturation of the protein, however, as well as fast hydrolysis of the reagent. Acylation reactions are hence commonly carried out at intermediate pH values (7.5–8.5), where at least a frac-

TABLE IV
POLYMERIC SUPPORTS FOR THE IMMOBILIZATION OF ENZYMES BY COVALENT BONDS

Parent polymer	Modification of polymer	Method of coupling	Capacity (mg protein/gm conjugate)	References
POLYSACCHARIDES				
Cellulose	—	Cyanogen bromide activation	60–300	Axén and Ernback (1971); Patel et al. (1972); Maeda and Suzuki (1972b); Bartling et al. (1972); Coughlan and Johnson (1973)
Cellulose	—	Activation with 2,4,6-trichloro-s-triazine (cyanuric chloride); arylation of amino groups on protein	200–300	Kay and Crook (1967); Surinov and Manoilov (1966); Kay et al. (1968); Self et al. (1969)
Cellulose	—	Activation with 2-amino-4,6-dichloro-s-triazine; arylation of amino groups on protein	4	Kay and Lilly (1970)
Cellulose	Coated with diazotized m-diaminobenzene (Bismarck Brown)	Azo bond formation	50–60	Gray et al. (1974)
Cellulose	—	Activated with dichloro-s-triazinyl dyestuffs (Procion dyes); arylation of amino groups on protein	—	Wilson et al. (1968a,b); Stasiw et al. (1972)
Cellulose	Carboxymethyl ether (CM-cellulose)	Activation with 2-amino-4,6-dichloro-s-triazine; arylation of amino groups on protein	113	Kay and Lilly (1970); Wykes et al. (1971)
Cellulose	Carboxymethyl ether (CM-cellulose)	Activation of support carboxyl groups with N-ethyl-5-phenylisoxazolium	300–500	Patel et al. (1967, 1969)

Support	Derivative	Method	Capacity	References
Cellulose	Carboxymethyl ether (CM-cellulose)	3'-sulfonate (Woodwards Reagent K); peptide bond formation with protein amino groups. Activation of support carboxyls with N,N'-disubstituted carbodiimides; peptide bond formation with protein amino groups	40	Weliky and Weetall (1965); Weliky et al. (1969)
Cellulose	Carboxymethyl ether (CM-cellulose)	Activation of support carboxyls with N-ethoxycarbonyl-2-ethoxy-1,2-dihydroquinoline	—	Sundaram (1974)
Cellulose	Carboxymethyl ether hydrazide (CM-cellulose hydrazide)	Activation of support hydrazide groups by conversion to azide; peptide bond formation with protein amino groups	50–400	Micheel and Ewers (1949); Mitz and Summaria (1961); Epstein and Anfinsen (1962); Hornby et al. (1966); Wharton et al. (1968a,b); Brümmer et al. (1972); Crook et al. (1970)
Cellulose	Aminoethyl ether	Activation of the support amino groups with glutaraldehyde	—	Habeeb (1967); Glassmeyer and Ogle (1971)
Cellulose	Diethylaminoethyl ether (DEAE-cellulose)	Activation with 2-amino-4,6-dichloro-s-triazine; arylation of amino groups on protein	130–220	Kay and Lilly (1970)
Cellulose	Diethylaminoethyl ether (DEAE-cellulose)	Activation with dichloro-s-triazinyl dyestuffs (Procion dyes); arylation of amino groups of protein	~15	Stasiw et al. (1970, 1972)
Cellulose	4-Aminobenzyl ether (PAB-cellulose)	Activation by diazotization; azo bond formation mainly with tyrosyl residues on protein	100	Mitz and Summaria (1961); Lilly et al. (1965); Goldstein et al. (1970); Datta et al. (1973)
Cellulose	(4-Aminophenyl-sulfonyl) ethyl ether	Activation by diazotization; azo bond formation, mainly with tyrosyl residues and protein	—	Li et al. (1973)

(Continued)

45

TABLE IV (Continued)

Parent polymer	Modification of polymer	Method of coupling	Capacity (mg protein/ gm conjugate)	References
Cellulose (microcrystalline)	3-(4-Aminophenoxy)-2-hydroxypropyl ether	Activation by diazotization; azo bond formation mainly with tyrosyl residues on protein	10–40	Barker et al. (1968, 1969)
Cellulose	(3-Aminobenzyloxy) methyl ether	Activation by diazotization, mainly with tyrosyl residue on protein	180–200	Surinov and Manoilov (1966)
Cellulose (microcrystalline)	3-(4-Isothiocyanatophenoxy)-2-hydroxypropyl ether	Thiocarbamylation of amino groups on protein	10–18	Barker et al. (1968, 1969)
Cellulose	Haloacetyl esters	Alkylation, mainly of amino groups of protein	—	Jagendorf et al. (1963); Robbins et al. (1967); Shaltiel et al. (1970); Sato et al. (1971); Maeda and Suzuki (1972a)
Cellulose	trans-2,3-Cyclic carbonate	Formation of urethan bonds with amino groups of protein	10–15	Barker et al. (1971a,b); Kennedy and Zamir (1973); Kennedy et al. (1972, 1973); Kennedy and Rosevear (1974a); Kennedy (1974a)
Cellulose	Periodate oxidized (dialdehyde cellulose)	Schiff's base formation with amino groups of protein	—	Flemming et al. (1973a); Van Leemputten and Horisberger (1974a)
Cellulose	—	Activation with transition metal salts (e.g., $TiCl_4$)	—	Barker et al. (1971b)
Starch	Periodate oxidized (dialdehyde-starch)	Schiff's base formation with amino groups of protein	—	Weakley and Mehltretter (1973)
Starch	Aminoaryl derivative (prepared from oxi-	Activation by diazotization; azo bond formation, mainly with	100	Goldstein et al. (1970)

	dized starch and p,p'-diaminodiphenyl-methane)	tyrosyl residues of protein		
Agarose beads (Sepharose)	—	Cyanogen bromide activation	70–330	Porath et al. (1967, 1973); Axén et al. (1969, 1971a); Axén and Ernback (1971); Cuatrecasas (1970, 1972a,b); Gabel et al. (1970, 1971); March et al. (1974); Kristiansen et al. (1969); Sundberg and Kristiansen (1972); Jost et al. (1974); Wilchek (1974); Porath (1974); Wilchek et al. (1975)
Agarose beads (Sepharose)	—	Activation with 2-amino-4,6-dichloro-s-triazine; arylation of amino groups of protein	390	Kay and Lilly (1970)
Agarose beads (Sepharose)	Polylysine conjugate converted to the bromoacetyl derivative	Coupling via alkylation of amino groups of protein	—	Wilchek (1974); Wilchek and Miron (1974b)
Agarose beads (Sepharose)	Polylysine conjugate converted to the p-aminobenzamido derivative	Activation by diazotization; azo bonds mainly with tyrosyl residues of protein	—	Wilchek (1974); Wilchek and Miron (1974b)
Agarose beads (Sepharose)	Bis-oxirane [1,4-bis(2,3-epoxy propoxy) butane] derivative	Alkylation of amino groups of protein	100–120	Sundberg and Porath (1974); Vretblad (1974)
Agarose beads (Sepharose)	Cross-linked with epichlorohydrine	Cyanogen bromide activation	100–120	Axén et al. (1971); Porath et al. (1971)

(Continued)

47

TABLE IV (*Continued*)

Parent polymer	Modification of polymer	Method of coupling	Capacity (mg protein/ gm conjugate)	References
Agarose beads (Sepharose)	Arylamino derivative	Coupling to protein carboxyl groups by four-component condensation in the presence of acetaldehyde and dimethyl-aminopropyl isocyanide	—	Vretblad and Axén (1971)
Agarose beads (Sepharose)	Glutathione-2-pyridyl disulfide derivative	Coupling to cysteine residues of protein by a thiol–disulfide interchange reaction	100–180	Carlsson et al. (1974)
Agarose beads (Sepharose)	Dithiobis-5,5'-(2-nitrobenzoic acid) (DTNB) derivative	Coupling to cysteine residues of protein by a thiol–disulfide interchange reaction	—	Lin and Foster (1975)
Agarose beads (Sepharose)	—	Activation with p-benzoquinone; coupling by arylation of amino groups of protein	70–80	Brandt et al. (1975)
Dextran; cross-linked beads (Sephadex)	—	Cyanogen bromide activation	100–280	Axén et al. (1967, 1970); Gabel et al. (1970, 1971); Axén and Ernback (1971)
Dextran; cross-linked beads (Sephadex)	3-(4-Isothiocyanato-phenoxy)-2-hydroxypropyl ether	Thiocarbamylation of amino groups of protein	—	Axén and Porath (1966)
Dextran; cross-linked beads (Sephadex)	Carboxymethyl ether	Activation of support carboxyls with N-ethoxycarbonyl-2-ethoxy-1,2-dihydroquinoline; peptide bond formation with protein amino groups	—	Sundaram (1974)
Dextran; cross-linked beads (Sephadex)	Periodate oxidized (dialdehyde epidex)	Schiff's base formation with amino groups of protein	—	Flemming et al. (1973b,c)

Support	Derivative	Method	Capacity	Reference
Dextran; cross-linked beads (Sephadex)	—	Activation with 2-amino-4,6-dichloro-s-triazine; arylation of amino groups of protein	212	Kay and Lilly (1970)
Dextran, linear (water-soluble)	—	Activation with 2-amino-4,6-dichloro-s-triazine; arylation of amino groups on protein	60–80	Wykes et al. (1971)
Dextran, linear (water-soluble)	Diethylaminoethyl ether (DEAE-dextran)	Activation with 2-amino-4,6-dichloro-s-triazine; arylation of amino groups on protein	90	Wykes et al. (1971)
Polygalacturonic acid	—	Activation of support carboxyls with N-ethyl-5-phenylisoxazolium 3'-sulfonate (Woodward's Reagent K); peptide bond formation with amino groups of protein	600–800	Patel et al. (1969)
Alginic acid	—	Activation with transition metal salts ($TiCl_4$)	—	Kennedy and Doyle (1973)
Chitin	—	Activation with transition metal salts ($TiCl_4$)	—	Kennedy and Doyle (1973)
Chitin	—	Activation with glutaraldehyde	30	Stanley et al. (1975)

VINYL POLYMERS

Support	Derivative	Method	Capacity	Reference
Polyacrylamide; cross-linked beads	—	Activation with glutaraldehyde	30–100	Weston and Avrameas (1971); Ternynck and Avrameas (1972); Johansson and Mosbach (1974a,b)
Polyacrylamide; cross-linked beads (Bio-Gel)	Acyl hydrazide derivative	Activation of support hydrazide groups by conversion to azide; peptide bond formation with amino groups of protein	160	Inman and Dintzis (1969); Inman (1974)
Polyacrylamide; cross-linked beads (Bio-Gel)	p-Aminobenzamido-ethyl derivative	Activation by diazotization. Azo bond formation, mainly with tyrosine residues of protein	300	Inman and Dintzis (1969); Zabriskie et al. (1973); Datta et al. (1973); Inman (1974)

(Continued)

TABLE IV (*Continued*)

Parent polymer	Modification of polymer	Method of coupling	Capacity (mg protein/ gm conjugate)	References
Copolymer of acrylamide and acrylic acid; cross-linked	—	Activation of support carboxyls with water-soluble carbodiimide	114	Martensson and Mosbach (1972); Mosbach (1970); Mosbach and Mattiasson (1970)
Copolymer of acrylamide and 2-hydroxyethyl-methacrylate; cross-linked	—	Cyanogen bromide activation	30	Mosbach (1970); Turkova et al. (1973); Johansson and Mosbach (1974a); Turkova (1974)
Copolymer of acrylamide and methacrylic acid anhydride; cross-linked	—	Peptide bond formation with amino groups of protein	200–500	Krämer et al. (1974)
Copolymer of acrylamide and p-amino-acrylanilide; cross-linked (Enzacryl AA)	—	Activation by diazotization; azo bond formation, mainly, with tyrosine residues of protein	10–30	Barker et al. (1970a,b)
Copolymer of acrylamide and p-amino-acrylanilide; cross-linked (Enzacryl AA)	Conversion to the isothiocyanato derivative by treatment with thiophosgene	Thiocarbamylation of amino groups of protein	20–30	Barker et al. (1970a,b)
Copolymer of acrylamide and acryloyl hydra-zide; cross-linked (Enzacryl AH)	—	Activation of support hydrazide groups by conversion to azide; peptide bond formation with amino groups of protein	2	Barker et al. (1970a,b)
Copolymer of acrylamide and N-acryloylcys-	—	Formation of —S—S— bonds by oxidative coupling in the	—	Barker and Epton (1970)

Support material	Derivative	Method		Reference
teine, cross-linked (Enzacryl polythiol)		presence of potassium ferricyanide		
Poly(N-acryloylamino-acetaldehyde dimethyl acetal cross-linked (Enzacryl polyacetal)	—	Coupling to amino groups of protein to give aminol and possibly azomethine (Schiff's base) linkages	—	Epton et al. (1972)
Poly(N-acryloyl-4-amino salicylic acid); cross-linked	—	Activation with transition metal salts ($TiCl_4$)	—	Kennedy and Epton (1973)
Polyvinyl alcohol (cross-linked with terephthalaldehyde)	—	Activation with 2,4,6-trichloro-s-triazine (cyanuric chloride); arylation of amino groups on protein	320	Manecke (1975); Manecke and Vogt (1976)
Polyvinyl alcohol (cross-linked with terephthalaldehyde)	p-Aminobenzyl ether	Activation by diazotization; azo bond formation, mainly with tyrosine residues of protein	480	Manecke (1975); Manecke and Vogt (1976)
Polyvinyl alcohol (cross-linked with terephthalaldehyde)	Arylamino derivative obtained by treatment with 2-(m-aminophenyl)-1,3-dioxolane	Activation by diazotization; azo bond formation mainly with tyrosine residues of protein	300–650	Manecke (1975); Manecke and Vogt (1976)
Polyacrylic acid; cross-linked	Acid hydrazide derivative	Activation of support hydrazide groups by conversion to azide; peptide bond formation with amino groups of protein	—	Erlanger et al. (1970)
Polyacrylic acid	—	Activation of support carboxyls with N-ethyl-5-phenylisoxazolium 3'-sulfonate (Woodward's Reagent K); peptide bond formation with amino groups of protein	450	Patel et al. (1967)

51

(Continued)

TABLE IV (Continued)

Parent polymer	Modification of polymer	Method of coupling	Capacity (mg protein/ gm conjugate)	References
Poly(4-methacryloxy-benzoic acid; cross-linked	—	Activation of support carboxyls with N-ethoxycarbonyl-2-ethoxy-1,2-dihydroquinoline	—	Bartling et al. (1974b)
Copolymer of acrylic acid and 3- or 4-isothio-cyanatostyrene	—	Thiocarbamylation of amino groups of protein		Manecke et al. (1970); Manecke and Günzel (1967a); Manecke (1975)
Cross-linked gel			200–450	
Macroreticular structure			1500–1700	
Polymethacrylic acid anhydride	—	Peptide bond formation with amino groups of protein	—	Conte and Lehmann (1971)
Polymethacrylic acid	Esters of ω-iodo-n-alcohols (polyiodals)	Coupling by alkylation of cysteine —SH groups of protein	10–20	Brown et al. (1970, 1971); Brown and Racois (1971a,b)
Poly(hydroxyalkyl methacrylate); cross-linked gels (Spheron)	—	Activation with cyanogen bromide	—	Coupek et al. (1973)
Poly(glycidyl meth-acrylate); cross-linked macroreticular structure	—	—	—	Svec et al. (1975)
Copolymer of meth-acrylic acid and 3-	—	Coupling by arylation of amino groups of protein	—	Manecke and Förster (1966); Manecke et al. (1970)

Carrier	Coupling method		Capacity	References
fluoro-4,6-dinitro-styrene (2:1) Cross-linked gel Macroreticular structure			500–1100 3000	Manecke (1962, 1964; Manecke and Singer (1960b); Manecke and Günzel (1962); Manecke et al. (1960, 1970)
Copolymer of methacrylic acid and methacrylic acid-3-fluoro-4,6-dinitro anilide (2:1) Cross-linked gel Macroreticular structure	Coupling by arylation of amino groups of protein	—	400–1100 3500	
Copolymer of methacrylic acid and 3-isothiocyanatostyrene Macroreticular structure	Coupling by thiocarbamylation of amino groups of protein	—	330–1600	Manecke et al. (1970); Manecke and Günzel (1967a)
Poly(allyl carbonate)	Formation of urethane bonds with amino groups of protein	—	2–5	Kennedy et al. (1972)
Copolymer of allyl alcohol and vanillin methacrylate (vanacryl)	Coupling to amino groups of protein to give aldimine (Schiff's base) linkages	—	20–140	Brown and Racois (1972, 1974a,b)
Copolymer of maleic anhydride and ethylene (EMA)	Peptide bond formation with amino groups of protein	—	100–800	Levin et al. (1964); Goldstein et al. (1964); Goldstein (1970, 1972a,b); Solomon and Levin (1974a); Ong et al. (1966); Westman (1969); Fritz et al. (1968a,b, 1969a,b); Zingaro and Uziel (1970); Weetall (1970)

(Continued)

TABLE IV (*Continued*)

Parent polymer	Modification of polymer	Method of coupling	Capacity (mg protein/ gm conjugate)	References
Copolymer of maleic anhydride and acrylic acid	—	Peptide bond formation with amino groups of protein	—	Bessmertnaya and Antonov (1973); Yarovaya *et al.* (1975)
Copolymer of maleic anhydride and butandiol divinyl ether	—	Peptide bond formation with amino groups of protein	200	Brümmer *et al.* (1972)
Copolymer of maleic anhydride and methyl vinyl ether	—	Peptide bond formation with amino groups of protein	—	Zingaro and Uziel (1970)
Copolymer of maleic anhydride and styrene	—	Peptide bond formation with amino groups of protein	—	Goldstein *et al.* (1971); Solomon and Levin (1974a)
Copolymer of maleic anhydride and acrylamide; cross-linked	—	Peptide bond formation with amino groups of protein	0.5–25	Jaworek (1974)
Copolymer of maleic acid and ethylene; cross-linked	1-Amino-6-hexamido derivative	Activation with glutaraldehyde	100	Goldstein (1973b)
Copolymer of maleic acid and ethylene; cross-linked	1-Amino-6-hexamido derivative	Coupling to carboxyl groups of protein by carbodiimide activation	100	Goldstein (1973b)
Copolymer of maleic acid and ethylene; cross-linked	4-Amino-4′-amido-diphenylmethane derivative	Activation by diazotization; azo bond formation, mainly with tyrosine residues of protein	100–300	Goldstein (1973a)
Copolymer of maleic acid and ethylene; cross-linked	Acid hydrazide derivative	Activation of support hydrazide groups by conversion to azide; peptide bond formation with amino groups of protein	100–400	Goldstein (1973a)

Polystyrene; macroreticular		Glutaraldehyde activation	—	Baum (1975)
Poly(p-aminostyrene)	Nitration and reduction	Activation by diazotization; azo bonds mainly with tyrosine residues of protein	2–14	Grubhofer and Schleith (1953, 1954); Manecke and Gillert (1955); Manecke (1962, 1964); Filippusson and Hornby (1970); Ledingham and Hornby (1969); Ledingham and Ferreira (1973); Manecke and Gillert (1955)
Poly(isocyanatostyrene)	—	Coupling by carbamylation of amino groups of protein	—	Manecke (1962, 1964); Brandenberger (1955); Manecke and Singer (1960a); Manecke et al. (1958)
Poly(4-isothiocyanatostyrene)	—	Coupling by thiocarbamylation of amino groups of protein	12–15	Manecke (1975); Manecke and Günzel (1967a); Manecke et al. (1970)
Copolymer of styrene and 4-vinylbenzoic acid	—	Activation of carrier carboxyls with N,N'-carbonyldiimidazole; coupling of enzyme in organic solvent (dimethylformamide); formation of peptide bonds with amino groups of protein	50	Bartling et al. (1974a)
Polypropylene	Radiation grafted poly-(p-nitrostyrene), reduced and converted to the isothiocyanato derivative	Coupling by thiocarbamylation of amino groups of protein	—	Garnett et al. (1974); Liddy et al. (1975)

55

(Continued)

TABLE IV (*Continued*)

Parent polymer	Modification of polymer	Method of coupling	Capacity (mg protein/ gm conjugate)	References
POLYAMINO ACIDS AND PROTEINS				
Polyglutamic acid	—	Activation of support carboxyls with N-ethyl-5-phenylisoxazolium 3'-sulfonate (Woodward's Reagent K); peptide bond formation with amino groups of protein	600–800	Patel *et al.* (1967)
Copolymer of L-glutamic acid and L-alanine	—	Activation of support carboxyls with N-ethyl-5-phenylisoxazolium 3'-sulfonate (Woodward's Reagent K); peptide bond formation with amino groups of protein	330	Patel *et al.* (1969)
Copolymer of *p*-amino-DL-phenylalanine and L-leucine	—	Activation by diazotization; azo bond formation, mainly with tyrosine residues of protein	100–300	Bar-Eli and Katchalski (1960, 1963); Alexander *et al.* (1965); Cebra *et al.* (1961); Silman *et al.* (1966)
Collagen	Acid hydrazide derivative	Activation of support hydrazide groups by conversion to azide; peptide bond formation with amino groups of protein	—	Julliard *et al.* (1971); Coulet *et al.* (1974)
POLYAMIDES				
Polyamides (nylon-6, nylon-6,6)	Acid hydrazide derivative of partially hydrolyzed nylon	Activation of support hydrazide groups by conversion to azide; peptide bond formation with amino groups of protein	—	Hornby and Filippusson (1970)

56

Support material	Method of activation/coupling		Reference
Polyamides (nylon-6, nylon-6,6)	4-Amino-4'-amido diphenyl derivative of partially hydrolyzed nylon	Activation by diazotization; azo bond formation, mainly with tyrosine residues of protein	—	Hornby and Filippusson (1970)
Polyamide (nylon-6, nylon-6,6)	Isocyanate derivative of partially hydrolyzed nylon	Coupling by carbamylation of amino groups of protein	—	Horvath and Solomon (1972)
Polyamides (nylon-6, nylon-6,6)	Partial hydrolysis	Activation with 2,4,6-trichloros-triazine (cyanuric chloride); arylation of amino groups of protein	—	Horvath and Solomon (1972)
Polyamides (nylon-6, nylon-6,6)	Partial hydrolysis	Activation with glutaraldehyde	—	Sundaram and Hornby (1970); Allison et al. (1972); Filippusson et al. (1972); Bunting and Laidler (1974)
Polyamides (nylon-6, nylon-6,6)	Partial depolymerization by transamidation with N,N-dimethyl-1,3-propanediamine	Activation with glutaraldehyde	—	Hornby et al. (1972); Inman and Hornby (1974)
Polyamides (nylon-6, nylon-6,6)	O-Alkylation of peptide bonds with dimethyl sulfate to form polymeric imidate salts	Coupling via formation of amidines with amino groups of protein	—	Campbell et al. (1975); Hornby et al. (1974)
Polyamides (nylon-6, nylon-6,6)	O-Alkylation of peptide bonds with triethyl oxonium tetrafluoroborate to form imidate salts	Coupling via formation of amidines with amino groups of proteins	—	Morris et al. (1975)

57

(Continued)

TABLE IV (Continued)

Parent polymer	Modification of polymer	Method of coupling	Capacity (mg protein/ gm conjugate)	References
Polyamides (nylon-6, nylon-6,6)	Introduction of iso-cyanide side chains by N-alkylation of peptide bonds	Coupling by four-component condensation reaction in the presence of acetaldehyde and acetate (coupling to amino groups of protein) or Tris (coupling to carboxyl groups of protein)	40–90	Goldstein et al. (1974a,b)
Polyamides (nylon-6, nylon-6,6)	Introduction of amino-aryl side chains by N-alkylation of pep-tide bonds	Activation by diazotization; azo bond formation, mainly with tyrosine residues of protein	90	Goldstein et al. (1974a)
INORGANIC SUPPORTS Porous glass	Arylamino derivative	Activation by diazotization; azo bond formation mainly with tyrosine residues of protein	10–20	Weetall (1969a,b, 1970); Weetall and Hersh (1969); Weetall and Baum (1970); Royer and Green (1971); Grove et al. (1971); Mason and Weetall (1972); Weetall and Messing (1972); Dixon et al. (1973); Weetall and Mason (1973); Royer and Uy (1973); Royer and Andrews (1973); Weibel et al. (1973); Weetall and Filbert (1974)
Porous glass	Alkylamino derivative	Activation with glutaraldehyde	12–16	Robinson et al. (1971); Dixon et al. (1973); Weetall and Filbert (1974)

Carrier	Derivative	Method of coupling		References
Porous glass	Alkylamino derivative	Coupling to carboxyl groups of protein by carbodiimide activation	10	Line et al. (1971); Cho and Swaisgood (1972); Baum et al. (1971)
Porous glass	N-Hydroxysuccinimide ester derivative	Coupling by formation of peptide bonds with amino groups of protein	25	Cuatrecasas and Parikh (1972)
Porous glass	Isothiocyanato derivative	Coupling by thiocarbamylation of amino groups of protein	—	Weetall (1970); Weibel et al. (1973)
Nickel/nickel oxide screens	Isothiocyanato derivative	Coupling by thiocarbamylation of amino groups of protein	0.4–0.5	Weetall and Hersh (1970)
Silica-alumina impregnated with nickel oxide	Isothiocyanato derivative	Coupling by thiocarbamylation of amino groups of protein	—	Herring et al. (1972); Traher and Kittrell (1974)
Bentonite	—	Activation of support with 2,4,6-trichloro-s-triazine (cyanuric chloride); coupling by arylation of amino groups of protein	3–15	Monsan and Durand (1971)
Iron oxide powder (magnetite, Fe_2O_3)	Silanization with γ-aminopropyltriethoxysilane and conversion to isocyanate	Coupling by carbamylation of amino groups of protein	7	Van Leemputten and Horisberger (1974b)
Iron oxide powder (magnetite, Fe_2O_3)	Particles coated with cellulose	Cyanogen bromide activation	4	Robinson et al. (1973)
Iron oxide powder (magnetite, Fe_2O_3)	Silanization with γ-aminopropyltriethoxysilane	Activation with glutaraldehyde	4	Robinson et al. (1973); Van Leemputten and Horisberger (1974b)
Iron oxide powder (magnetite, Fe_2O_3)	Particles coated with polyacrylamide and then converted to the acid hydrazide derivative	Activation of support hydrazide groups by conversion to azide; peptide bond formation with amino groups of protein	—	Dunnill and Lilly (1974)

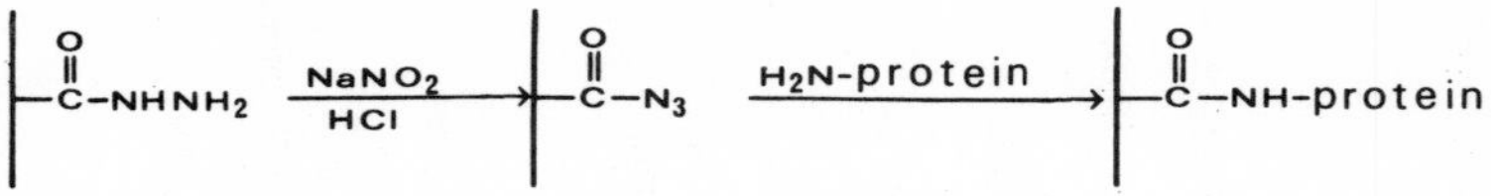

$$\overset{O}{\underset{\|}{-C-NHNH_2}} \xrightarrow[\text{HCl}]{\text{NaNO}_2} \overset{O}{\underset{\|}{-C-N_3}} \xrightarrow{\text{H}_2\text{N-protein}} \overset{O}{\underset{\|}{-C-NH-protein}}$$

Fig. 1. Coupling of proteins to polymeric acyl azides.

tion of the nucleophile will be unprotonated, and at low temperatures (4°C) to slow down all reactions including hydrolysis of reagent. Acid anhydrides and acyl azides react with protein amino groups as shown in Figs. 1 and 2. Substitution of amino groups by these polymeric reagents is strongly preferred in the pH range 7.5 to 8.5; side reactions with aliphatic or aromatic hydroxyl groups or with sulfhydryl groups occur, however, under these conditions (Micheel and Ewers, 1949; Riordan *et al.*, 1965; Gounaris and Perlmann, 1967; Meighen and Schachman, 1970; Cohen, 1968; Vallee and Riordan, 1969; Stark, 1970). Acylation can be limited to amino groups by subsequent treatment with hydroxylamine at neutral pH to hydrolyze ester and thioester bonds. In the case of *N*-hydroxysuccinimide esters of carboxylic polymers, it has been shown that highly preferential reactions with α-amino groups can be obtained by using relatively low pH ($\sim$6.0) and by decreasing the time of the reaction (Cuatrecasas and Parikh, 1972; Parikh *et al.*, 1974). This suggests that it may in principle be possible to devise experimental conditions under which proteins could be preferentially coupled via their N-terminal amino acid, rather than randomly through lysyl residues.

As indicated in Fig. 2, opening of the acid anhydride ring by a nucleophile, the —NH$_2$ groups on the protein, leads to the formation of free carboxyl groups; moreover, unreacted acid anhydride or acyl azide groups will be hydrolyzed spontaneously by OH⁻ ions in the aqueous medium, generating free carboxyl groups. The immobilized protein is therefore located in an environment of excess negative charge; this fact is reflected in the kinetic behavior and stability properties of the bound enzyme (Goldstein *et al.*, 1964; Levin *et al.*, 1964; Wharton *et al.*, 1968a,b; Goldman *et al.*, 1971a; Goldstein, 1972b, 1976; Engasser and Horvath, 1975; see also Engasser and Horvath, this volume).

The most general methods for the activation of carboxyl groups in-

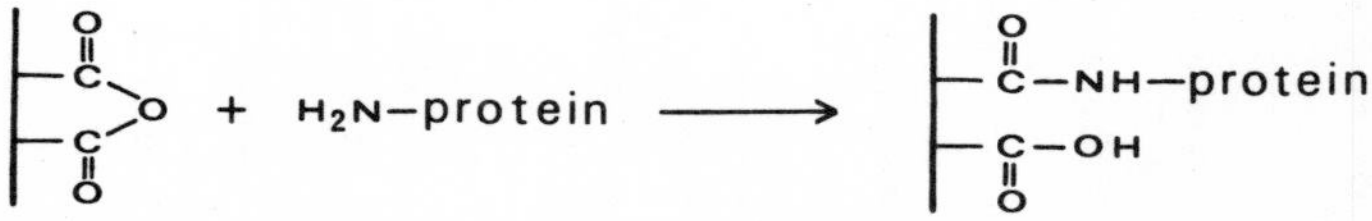

$$\begin{matrix} -\overset{O}{\underset{\|}{C}} \\ \diagdown O \\ -\underset{\|}{C} \\ O \end{matrix} + \text{ H}_2\text{N-protein} \longrightarrow \begin{matrix} -\overset{O}{\underset{\|}{C}}-\text{NH-protein} \\ \\ -\underset{\|}{C}-\text{OH} \\ O \end{matrix}$$

Fig. 2. Coupling of proteins to polymeric acid anhydrides.

volve the use of water-soluble carbodiimides and similar reagents. Carbodiimides react with carboxyl groups at slightly acidic pH values (pH 4.75–5) to give O-acyl isourea derivatives. These highly reactive intermediates can rearrange to an acyl urea or condense with amines to yield the corresponding amides (Fig. 3). The activated carboxyl groups can also react with other nucleophiles, e.g., −OH, −SH, to give different carboxylate derivatives. The relative rates of these reactions, however, are much lower. The chemistry of carbodiimides has been reviewed by Khorana (1953) and by Kurzer and Douraghi-Zadeh (1967). Several water-soluble carbodiimides have been used for the activation of carboxyl groups on polymers: 1-cyclohexyl-3[2-(4-*N*-methylmorpholinium)ethyl] carbodiimide tosylate [(Fig. 3), structure (I)] and 1-ethyl-3-(3-dimethylaminopropyl) carbodiimide [Fig. 3, structure (II)], are now commercially available and are the most widely used. The synthesis of 1-benzyl-3-(3-dimethylaminopropyl) carbodiimide has been described (Hoare and Koshland, 1966, 1967). All three reagents react in the same fashion, although the smaller molecules might be expected to have greater access to sterically restricted regions on polymer or protein.

N-alkyl-5-phenylisoxazolium salts (Woodward *et al.*, 1961, 1966; Patel *et al.*, 1967; Patel and Price, 1967) and *N*-ethoxycarbonyl-2-ethoxy-1,2-dihydroquinoline (Sundaram, 1974; Bartling *et al.*, 1974b)

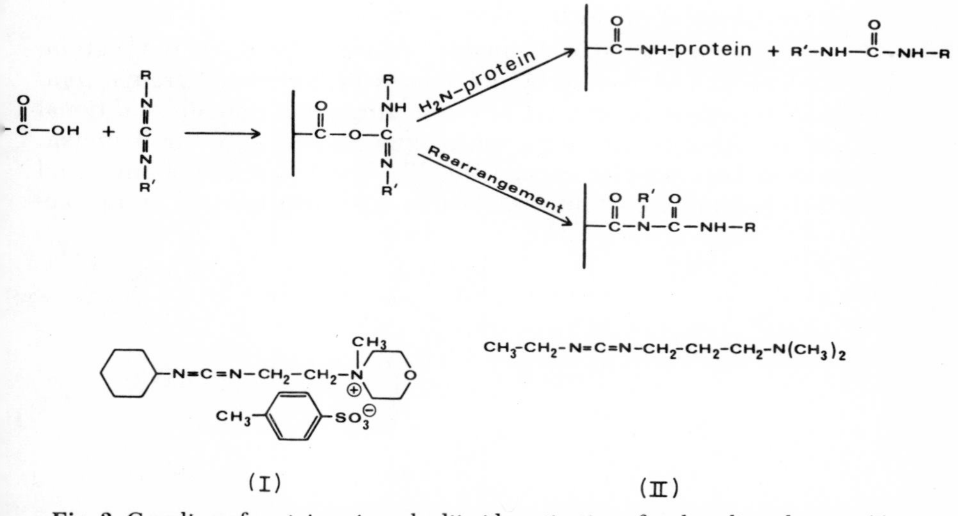

(I)   (II)

Fig. 3. Coupling of proteins via carbodiimide activation of carboxylic polymers. (I) 1-cyclohexyl-3 [2-(4-*N*-methylmorpholinium) ethyl] carbodiimide tosylate; (II) 1-ethyl-3(3-dimethylaminopropyl)carbodiimide.

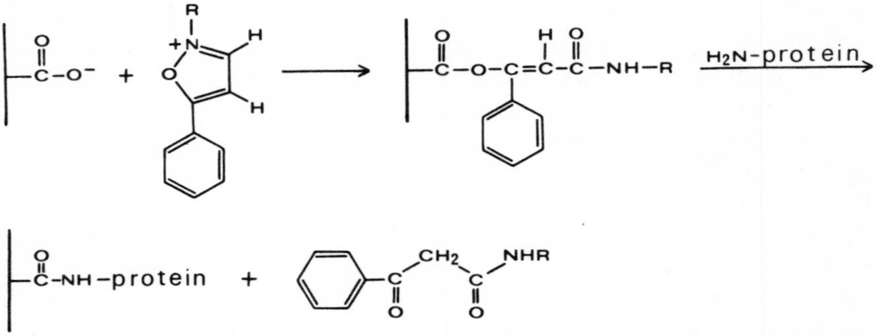

Fig. 4. Couplings of proteins by activation of carboxylic polymers with an *N*-alkyl-5-phenylisoxazolium salt (Patel *et al.*, 1967).

can be used in a similar manner. These compounds have been shown to react rapidly, with carboxyl groups only, at pH values below 4.75. The active intermediates formed in the initial step, enol esters in the case of *N*-alkyl-5-phenylisoxazolium salts (Fig. 4) and mixed anhydrides in the case of *N*-ethoxycarbonyl-2-ethoxy-1,2-dihydroquinoline (Fig. 5), are sufficiently stable to allow the isolation of the activated carboxylate polymer for subsequent reaction with a protein at neutral or slightly alkaline pH (Bodlaender *et al.*, 1969; Feinstein *et al.*, 1969; Sundaram, 1974).

2. Arylation and Alkylation Reactions

Coupling via arylation, presumably of amino groups on the protein, has been primarily effected by polymers with functional groups comprised of halogen-substituted aromatic rings that contain additional activating substituents, e.g., nitro groups (Fig. 6) or halogen-substituted heterocyclic rings (Fig. 7). The most commonly used arylating polymers contain the 3-fluoro-4,6-dinitrophenyl group (co-

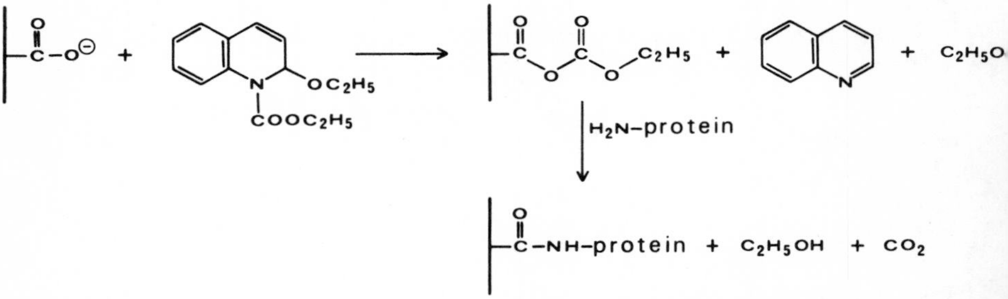

Fig. 5. Coupling of proteins by activation of carboxylic polymers with *N*-ethoxycarbonyl-α-ethoxy-1,2-dihydroquinoline (Sundaram, 1974).

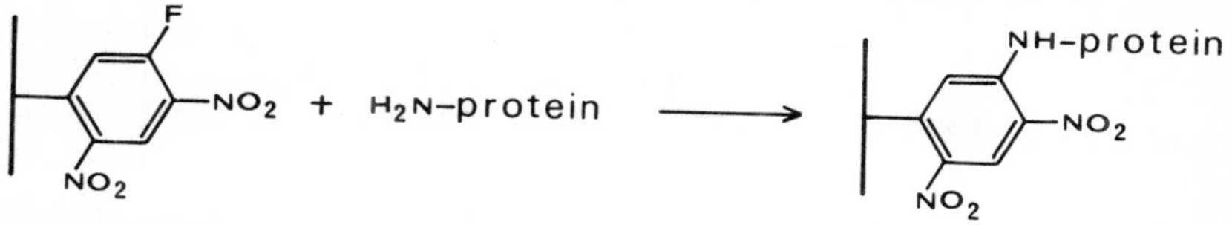

Fig. 6. Coupling of proteins to polymers containing 3-fluoro-4,6-dinitrophenyl functional groups (Manecke and Singer, 1960b; Manecke *et al.*, 1970).

polymers of 3-fluoro-4,6-dinitrostyrene or methacrylic acid-3-fluoro-4,6-dinitroanilide; see Manecke, 1962, 1964, 1975; Manecke and Günzel, 1962; Manecke and Förster, 1966; Manecke *et al.*, 1960, 1970) or the monochloro- or dichloro-*s*-triazinyl group (Surinov and Manoilov, 1966; Kay and Crook, 1967; Kay *et al.*, 1968; Wilson *et al.*, 1968a,b; Self *et al.*, 1969; Sharp *et al.*, 1969; Kay and Lilly, 1970; Stasiw *et al.*, 1970, 1972).

The latter method has been used mainly in conjunction with polysaccharide supports, e.g., cellulose, agarose, and cross-linked dextran. The support is treated with 2,4,6-trichloro-*s*-triazine (cyanuric chloride) or with a 4,6-dichloro-*s*-triazine derivative, e.g., 2-amino-4,6-dichloro-*s*-triazine, or the so-called Procion M dyes, in which one of the chlorine atoms of cyanuric chloride is replaced by a chromophore containing anionic solubilizing groups, and then allowed to react with the protein (Fig. 7).

In a limited number of cases, polymers having monohaloacetyl functional groups have been used in a similar manner for the coupling of enzymes by alkylation of amino groups on protein (Jagendorf *et al.*, 1963; Sato *et al.*, 1971; Maeda and Suzuki, 1972a). Coupling via alkylation of sulfhydryl groups with polymers containing ω-iodoalkyl side chains has also been described (Brown *et al.*, 1970, 1971; Brown and Racois, 1971a,b).

The specificity of polymeric arylating and alkylating reagents toward the various amino acid residues in a protein is still to be investigated. Some insight into the nature of these reactions, however, can be obtained from protein modification studies carried out with low-

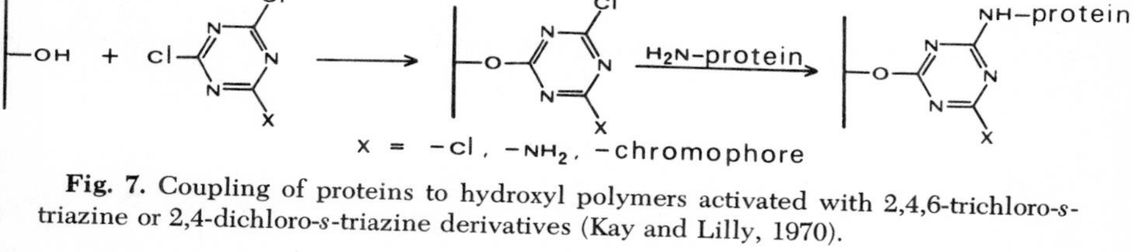

x = −cl , −NH₂ , −chromophore

Fig. 7. Coupling of proteins to hydroxyl polymers activated with 2,4,6-trichloro-*s*-triazine or 2,4-dichloro-*s*-triazine derivatives (Kay and Lilly, 1970).

molecular-weight mono- and bifunctional analogs, such as fluorodinit-robenzene (FDNB), 4,4'-difluoro-3,3'-dinitrophenyl sulfone, or cyanuric halides (see, e.g., Cohen, 1968; Vallee and Riordan, 1969; Means and Feeney, 1971; Wold, 1961, 1972). Amino acid residues that have undergone arylation or alkylation may in most cases be identified by the conventional methods of amino acid analysis by virtue of the stability of arylated or alkylated groups to acid hydrolysis.

Arylation and alkylation reactions are slower than acylation; they require an unprotonated nucleophile and are hence strongly pH dependent. The relative rates of arylation of different functional groups can therefore be controlled to a certain extent by performing the reaction at a pH at which some of the functional groups are essentially fully protonated. Amino groups and to a smaller extent phenolic OH are arylated at higher relative rates at alkaline pH values (pH 8.5–9). Side reactions involving sulfhydryl groups cannot, however, be eliminated. Dinitrophenylation can in principle be directed mainly toward —SH groups by carrying out the reaction at pH values below neutrality (Cohen, 1968; Vallee and Riordan, 1969; Means and Feaney, 1971). In the case of FDNB-treated proteins it has been shown that the sulfhydryl groups can be regenerated by exposure to excess β-mercaptoethanol (Shaltiel and Soria, 1969; Shaltiel, 1974b). Displacement of dinitrophenyl-substituted imidazole and phenolic groups, but not of amino groups, has been accomplished under similar conditions (Shaltiel, 1967, 1974b). The hydrophobic nature of the fluorodinitrophenyl group may lead in some cases to its direction toward lipophilic sites on the protein; it may thus react with sites that are not necessarily highly exposed or highly reactive (Cohen, 1968).

No detailed information on the specificity of chloro-s-triazinyl polymers is available. Protein modification studies with the highly reactive 2,4,6-trifluoro-s-triazine have shown that cyanuric fluoride reacts with all protein nucleophiles, including tryptophan with little or no selectivity (Cohen, 1968; Stark, 1970; Gorbunov, 1970, 1971; Sluyterman and Wijdenes, 1972). The extension of these findings to chloro-s-triazinyl polymers is not self-evident, however, as the reactivity of the chloro-s-triazines is dependent on the substituents on the triazine nucleus; as the ring becomes more highly substituted, the remaining chloride atoms become less reactive, presumably with concomitant increase in selectivity.

3. The Cyanogen Bromide Method

An approach that has gained wide popularity and is probably among the methods most commonly used for the laboratory preparation of immobilized enzyme derivatives as well as for insoluble adsorbents for affinity chromatography, is based on the activation of water-insoluble polysaccharides, cellulose, cross-linked dextran, and agarose, with cyanogen bromide (Axén *et al.*, 1967, 1969, 1970; Porath *et al.*, 1967; Axén and Ernback, 1971; Cuatrecasas, 1970; Gabel *et al.*, 1970, 1971; March *et al.*, 1974; Jost *et al.*, 1974; Porath, 1974; Porath and Kristiansen, 1975; Wilchek *et al.*, 1975).

The polysaccharide is activated with cyanogen bromide at high pH (10–11.5). Compounds containing free amino groups, including peptides and proteins, can then be attached covalently to the activated polymer at mildly alkaline pH values (Axén *et al.*, 1967, 1969, 1970; Porath *et al.*, 1967; Cuatrecasas, 1970; March *et al.*, 1974; Parikh *et al.*, 1974; Porath, 1974). Despite its wide acceptance, the method suffers from two main drawbacks: the nature of the bond formed between the amino group and the carbohydrate is not fully understood; moreover, although in most cases investigated no leakage of enzymic activity has been reported, some contradictory evidence suggesting leaching of covalently bound molecules, particularly low-molecular-weight ligands bound to the support by a single bond, can be found in the literature (see, e.g., Tesser *et al.*, 1972; Oka and Topper, 1974; Kristiansen *et al.*, 1969; Wilchek *et al.*, 1975).

On the basis of work with model peptides, Axén and Porath have suggested that the coupling of amines to cyanogen bromide-activated carbohydrates, proceeding presumably through cyanate as an unstable intermediate, results in the formation of three different types of structures: N-substituted carbamates, N-substituted imidocarbonates, and N-substituted isoureas (Fig. 8). The scheme of Fig. 8 was based mainly on infrared data showing the existence of —CO and —C=N— structures and on the finding that, by allowing ethylimidocarbonate to react with amino acids and amino acid derivatives, the main products were indeed N-substituted imidocarbonates, N-substituted isoureas, and N-substituted carbamic acid esters (Axén *et al.*, 1967; Porath *et al.*, 1967; Porath, 1967, 1968, 1974; Kågedal and Åkerström, 1970; Axén and Ernback, 1971; Axén and Vretblad, 1971a,b; Porath and Kristiansen, 1975). Additional evidence accumulated in the last few years suggests, however, that substituted isourea structures are most probably the major reaction products of cyanogen bromide-activated carbohydrates with amines. Examination of structures (I)–(III) in Fig. 8 shows that N-substituted isoureas [structure (III)] may be expected to carry a protonated nitrogen at physiological pH. Structures (I) and (II), on the other hand, would be electrically neutral under these conditions (Svensson, 1973; Jost *et al.*, 1974). Formation of isourea structures would

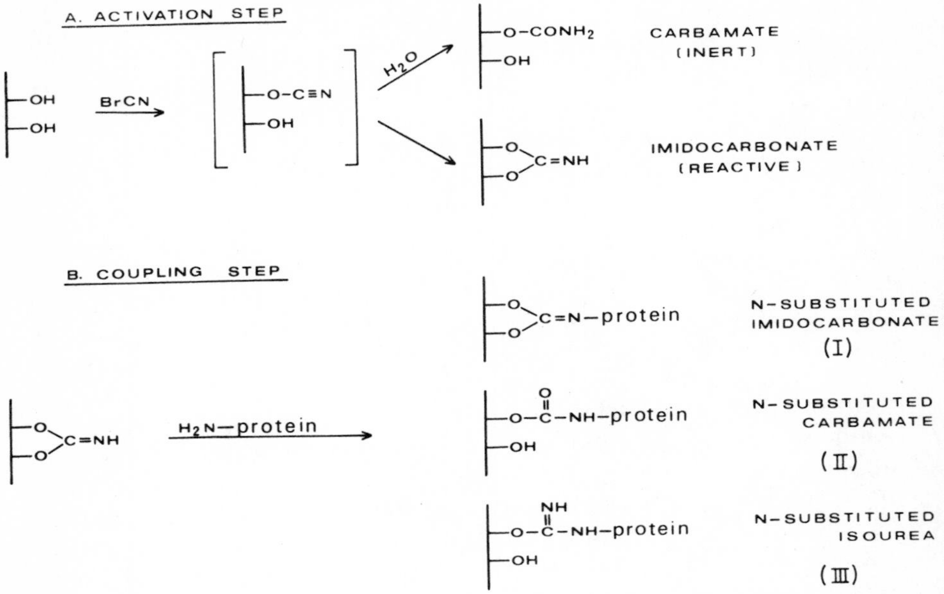

Fig. 8. Cyanogen bromide activation of polysaccharides, according to Axén and Ernback (1971).

preserve the charge of the amine ligand, since the dissociation constants of amidines and of amino groups are of the same order of magnitude (see Svensson, 1973, and references therein). The different types of derivatives that might result from the cyanogen bromide coupling procedure could thus be distinguished by comparing the ionization patterns of the ligand before and after attachment of cyanogen bromide-activated carbohydrate. Evidence along these lines, supporting the relative importance of isourea structures, has been presented by several authors. Svensson (1973) showed on the basis of electrofocusing studies that the net charge of subtilisin Novo was not altered significantly following coupling to cyanogen bromide-activated amylodextrin. Jost *et al.* (1974) demonstrated that alkyl amines coupled to cyanogen bromide-activated agarose retain their charge, forming a strong anion exchanger with an apparent dissociation constant ($pK_a \approx 10$) in agreement with the value expected for a basic amidine nitrogen.

The presence of N-substituted isourea structures was further indicated by the finding that a soluble superactive form of insulin is released from insulin–Sepharose when the latter is treated with bovine serum albumin, while no such material could be detected in the absence of added protein (Oka and Topper, 1974; Topper *et al.*, 1975). The formation of N-substituted isoureas in the coupling of amines to cyanogen bromide-activated Sepharose was directly demonstrated by Wilchek *et al.* (1975), who showed that attack by an amine on an isourea structure results in the release of N^1,N^2-disubstituted guanidines (Fig. 9). These authors also showed by means of ^{125}I-labeled insulin that insulin–Sepharose is also a substituted isourea and that the soluble superactive form of insulin that is released when insulin is treated with bovine serum albumin (BSA) is in effect a covalent BSA–insulin conjugate having an N^1,N^2-disubstituted guanidine structure [compound (IV) in Fig. 9; R = insulin, R' = BSA]. The model summarized in Fig. 9 helps us to understand the contradictory evidence found in the literature regarding the bond stability of conjugates obtained by cyanogen bromide activation of carbohydrates; in studies in which leakage was observed, the buffer usually contained a nucleophile, amine or protein (Tesser *et al.*, 1972; Oka and Topper, 1974; Topper *et al.*, 1975). Buffers containing no compounds that could react with isourea bonds were used in the cases where no leakage occurred (Kristiansen *et al.*, 1969). Multipoint attachment of a protein to cyanogen bromide-activated polysaccharide supports would be expected to mitigate to a large extent the effects inherent in the lability of isourea bonds (Wilchek, 1974; Wilchek and Miron, 1974a).

4. Carbamylation and Thiocarbamylation Reactions

Many polymers containing isocyanate or isothiocyanate functional groups have been used as carbamylating or thiocarbamylating reagents, respectively, to effect coupling of proteins (Manecke, 1962, 1964, 1975; Manecke and Günzel, 1967a; Manecke and Singer, 1960a; Manecke *et al.*, 1958, 1970; Axén and Porath, 1966; Barker *et al.*, 1968, 1969, 1970a,b; Weetall, 1970; Weetall and Hersh, 1970; Horvath and Solomon, 1972;

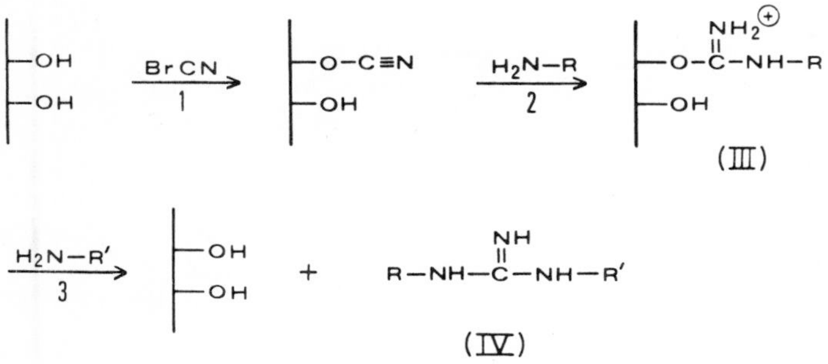

Fig. 9. A probable mechanism of cyanogen bromide activation (Wilchek *et al.*, 1975).

Herring *et al.,* 1972; Traher and Kittrell, 1974; Van Leemputten and Horisberger 1974b; Garnett *et al.,* 1974). Isocyanates and isothiocyanates react with most protein nucleophiles; only the reaction with amino groups, however, results in the formation of stable products, substituted ureas, or thioureas (Fig. 10). The carbamyl and thiocarbamyl derivatives with sulfhydryl, imidazole, aromatic hydroxyl and carboxyl groups are relatively unstable and decompose at mildly alkaline pH values or in the presence of nucleophiles, such as hydrazine (Cohen, 1968; Means and Feeney, 1971; Stark, 1970). The reaction with the isocyanate or isothiocyanate group involves unprotonated amines; hence as a result of their lower pK_a, α-amino groups would react at neutral pH faster than ϵ-amino groups, allowing in principle selective coupling of proteins through the formation of stable bonds mainly with α-amino groups (see Stark, 1965a,b). Polymers containing the isothiocyanate group have gained larger acceptance than those containing isocyanate owing to the higher stability of the former and the relative ease of its preparation (treatment of amino groups with thiophosgene). When isocyanates react in aqueous medium with proteins, the isocyanato groups are mainly hydrolyzed.

5. *Amidination Reactions*

Polymers containing imidoester functional groups have been recently employed for the coupling of proteins (Zaborsky, 1974a). Such reagents could be prepared by treating polymeric nitriles with alcohols and hydrogen chloride (Fig. 11), and recently by the 0-alkylation of backbone peptide bonds in nylon (Campbell *et al.,* 1975; Morris *et al.,* 1975). Imidoesters are readily attacked by nucleophiles and react selectively with α- and ϵ-amino groups of proteins at pH 8.5 and 9.5 to form amidines which are stronger bases than the parent amines [pK_a acetamidine = 12.5; pK_a benzamidine = 11.6 (Hunter and Ludwig, 1962)]. The amidine structures, like the protein amino groups they replace, are thus protonated at physiological pH (Hunter and Ludwig, 1962, 1972; Hand and Jencks, 1962; Cohen, 1968). Amidines are stable in neutral or acidic solutions, but hydrolyze slowly at high pH (Means and Feeney, 1971). The original amino groups can be regenerated by treatment with hydrazine (0.6–1.2 M, pH 9; Ludwig and Byrne, 1962). It

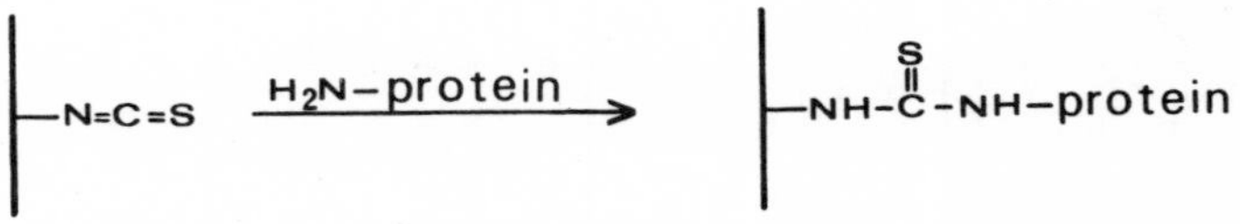

Fig. 10. Coupling of proteins to polymers containing isothiocyanate functional groups.

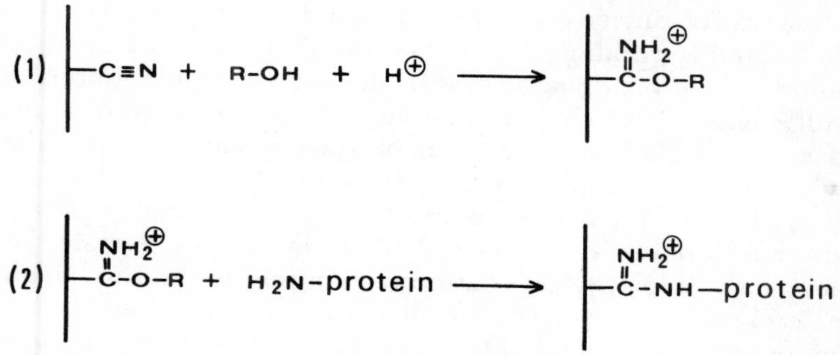

$$(1) \quad -\!C\!\equiv\!N \;+\; R\text{-}OH \;+\; H^{\oplus} \;\longrightarrow\; \underset{\overset{\|}{C}\text{-}O\text{-}R}{\overset{NH_2^{\oplus}}{}}$$

$$(2) \quad \underset{\overset{\|}{C}\text{-}O\text{-}R}{\overset{NH_2^{\oplus}}{}} \;+\; H_2N\text{-}protein \;\longrightarrow\; \underset{\overset{\|}{C}\text{-}NH\text{-}protein}{\overset{NH_2^{\oplus}}{}}$$

Fig. 11. Coupling of proteins to polymers containing imidoester functional groups (Zaborsky, 1974a).

should be recalled that bifunctional imidoesters, e.g., dimethyl suberimidate and dimethyl adipimidate (Table III), have been explored as reagents for intramolecular cross-linking of proteins (Hartmand and Wold, 1966, 1967; Dutton *et al.*, 1966; Davies and Stark, 1970; Handschumacher and Gaumond, 1972; Zaborsky, 1974a,b). Such bifunctional imido esters can in principle be used to introduce the imidoester function on polymers containing side chains with primary amino groups.

6. Reactions with Polymeric Aldehydes

Several types of polymers containing aldehyde functional groups have been described. Polyfunctional aldehydes derived from synthetic polymers [e.g., polyacryloylamino acetaldehyde (Epton *et al.*, 1972) or copolymers of allyl alcohol and vanillin methacrylate (Brown and Racois, 1972, 1974a,b)] as well as polymeric dialdehydes prepared by periodate or dimethyl sulfoxide oxidation of polysaccharides, such as cellulose, starch, and dextran (Flemming *et al.*, 1973a,b,c; Weakley and Mehltretter, 1973; Axén, *et al.*, 1971b; Vretblad and Axén 1973a; Van Leemputten and Horisberger, 1974a), have been explored. Aldehydes react with amino groups on the protein to form aminol or Schiff's base (azomethine, aldimine) linkages (Fig. 12). Sulfhydryl and imidazole groups may undergo similar reactions (see Means and Feeney, 1971). Such side reactions could have deleterious effects on the activity of the bound enzyme. Although the reaction of polymeric aldehydes with

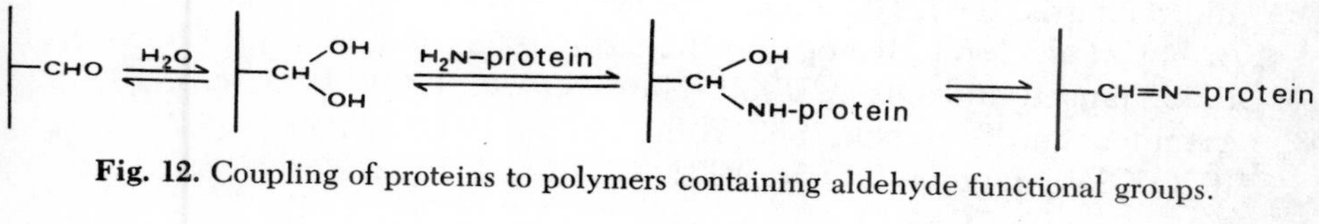

$$-\!CHO \underset{\longleftarrow}{\overset{H_2O}{\longrightarrow}} -\!CH\!\!\begin{array}{c}{}^{OH}\\{}_{OH}\end{array} \underset{\longleftarrow}{\overset{H_2N\text{-}protein}{\longrightarrow}} -\!CH\!\!\begin{array}{c}{}^{OH}\\{}_{NH\text{-}protein}\end{array} \rightleftharpoons -\!CH\!\!=\!\!N\text{-}protein$$

Fig. 12. Coupling of proteins to polymers containing aldehyde functional groups.

proteins can be carried out under mild conditions, the method has found only limited application, mainly because of the uncertainty about the stability of the bonds formed with the protein amino groups. Since Schiff's base formation is reversible and the equilibrium is usually unfavorable for aldimine formation in aqueous solution, particularly at low pH values (Jencks, 1964, 1969), such structures would not be expected to survive a wide range of pH and temperature. Moreover, many nucleophiles can reverse the equilibrium to generate the free amine; for example, Schiff's bases involving aniline and arylaldehydes are attacked by semicarbazide to yield a semicarbazone and free amine (Cordes and Jencks, 1962). Fixation of the —CH=N— bonds by reduction to stable alkylamino groups, might be undesirable, since it could lead to cleavage of disulfide bridges and hence to loss of enzymic activity (see, however, Royer *et al.*, 1975).

7. *Reactions with Glutaraldehyde*

Glutaraldehyde, initially introduced as an intermolecular cross-linking agent, to produce stable and insoluble three-dimensional networks of proteins (Sabatini *et al.*, 1963; Quiocho and Richards, 1964; Habeeb, 1967; Avrameas, 1969; Avrameas and Ternynck, 1969; Hopwood, 1972; see also Table III) has recently found extensive use for immobilizing enzymes onto a wide variety of polymeric supports. This bifunctional aldehyde can be reacted with polymers containing primary amino groups to yield matrices containing the aldehyde function. Proteins are bound irreversibly to the glutaraldehyde-treated polymer by a reaction presumably analogous to that occurring during cross-linking with the low-molecular-weight bifunctional reagent (see Green and Crutchfield, 1969; Avrameas, 1969; Avrameas and Ternynck, 1969; Sundaram and Hornby, 1970; Glassmeyer and Ogle, 1971; Robinson *et al.*, 1971; Weston and Avrameas, 1971; Ternynck and Avrameas, 1972; Allison *et al.*, 1972; Filippusson *et al.*, 1972; Hornby *et al.*, 1972; Goldstein, 1973b; Broun *et al.*, 1973; Dixon *et al.*, 1973; Bunting and Laidler, 1974; Inman and Hornby, 1972, 1974; Johansson and Mosbach, 1974a,b; Stanley *et al.*, 1975).

The nature of the reaction of glutaraldehyde with proteins and synthetic polymers is not fully understood. The attempts to characterize the reaction have brought forth several facts (Sabatini *et al.*, 1963; Quiocho and Richards, 1964, 1966; Habeeb, 1967; Habeeb and Hiramoto, 1968; Bishop *et al.*, 1966; Bishop and Richards, 1968; Richards and Knowles, 1968; Jansen and Olson, 1969; Hardy *et al.*, 1969; Avrameas, 1969; Avrameas and Ternynck, 1969; Wang and Tu, 1969; Schejter and Bar-Eli, 1970; Glassmeyer and Ogle, 1971; Hopwood, 1972; Korn *et al.*, 1972;

Quiocho, 1974; Whipple and Ruta, 1974; Monsan *et al.*, 1975): (a) The reaction can be carried out in aqueous solution within a rather wide range of pH values (pH 5–9); the rate of the reaction increases with increasing pH. (b) The modification usually carried out at pH values above 7 is apparently irreversible and survives treatment with semicarbazide and fairly concentrated mineral acids. (c) The reaction seems to involve. mainly amino groups on the protein. Amino acid analysis of acid hydroly- zates of modified protein shows loss of lysine residues only. The effi- ciency of the reaction as judged by the number of missing lysines increases with increasing pH. The majority of the above findings have been confirmed in model experiments with ^{14}C-labeled amino acids (Monsan *et al.*, 1975).

The data in their entirety are not consistent with the unstable aldimine (Schiff's base) structure that would be normally assigned to the adduct of aldehyde with an amino group (see preceding section and Fig. 12); this has led to attempts to relate the anomalous properties of the system to pH-induced changes in the structure of a reactive species presumably present in glutaraldehyde solutions. Richards and Knowles (1968) have suggested on the basis of nuclear magnetic resonance (NMR) mea- surements the existence of α,β-unsaturated oligomers of glutaraldehyde in commercial solutions of the reagent; such oligomers would be formed by aldol condensation processes via elimination of water. According to these authors, coupling of protein amino groups to "glutaraldehyde" takes place via a Michael-type condensation reaction, viz., conjugate- addition of an amine (R-NH$_2$) to the ethylenic double bond of an α,β- unsaturated carbonyl compound

$$\text{R—NH}_2 + \text{—CH} = \overset{\overset{\text{R}'}{|}}{\text{C}}-\overset{\overset{\text{O}}{\|}}{\text{C}}- \;\rightarrow\; -\text{CH}-\overset{\overset{\text{R}'}{|}}{\underset{\underset{\text{NH—R}}{|}}{\text{CH}}}-\overset{\overset{\text{O}}{\|}}{\text{C}}-$$

to give an alkylamino derivative. The N-alkyl bond is resistant to acid hydrolysis [for analogous reactions of ϵ-amino groups of proteins with α,β-unsaturated compounds, see, e.g., Cavins and Friedman (1967)]. A simplified version of the mechanism suggested by Richards and Knowles is shown in Fig. 13.

This mechanism has been contested by several groups on the basis of NMR studies of highly purified solutions of glutaraldehyde, the various hydrated forms and the cyclic hemiacetal of monomeric glutaraldehyde being indicated as the predominant species (Hardy *et al.*, 1969; Korn *et al.*, 1972; Whipple and Ruta, 1974). Such structures cannot, however, explain the unusual chemical properties of the reagent; moreover, it has been shown that, on increasing the alkalinity of the medium, glutaral-

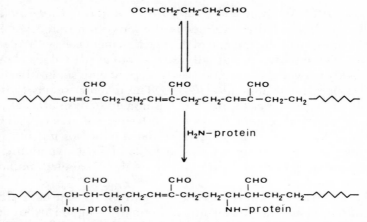

Fig. 13. A probable mechanism for the reaction of glutaraldehyde with proteins.

dehyde rapidly polymerizes, eventually separating out as an insoluble solid, polyglutaraldehyde, the spectroscopic properties of which are consistent with the unsaturated structure of Fig. 13 (Hardy *et al.*, 1969; Monsan *et al.*, 1975). It thus remains still highly probable that reactions with macromolecules (proteins or supports) may involve such unsaturated polymers. The nature of the reaction of glutaraldehyde with amines has been recently reinvestigated using the model compounds crotonaldehyde, CH_3—CH=CH—CHO, which contains the essential structural features suggested by Richards and Knowles for the unsaturated oligomers, i.e., the aldehyde function and the adjacent ethylenic double bond in the α,β-position (Monsan, *et al.*, 1975). Of the two possible pathways for the reaction of crotonaldehyde with an amine, conjugate addition to the double bond or aldimine formation [Fig. 14, structures (I) and (II)], the aldimine, structure (II), was the sole product upon reaction of stoichiometric amounts of aldehyde and amine. In the presence of excess amine, the product resulting from double addition was found [Fig. 14, structure (III)]. The data thus suggested that,

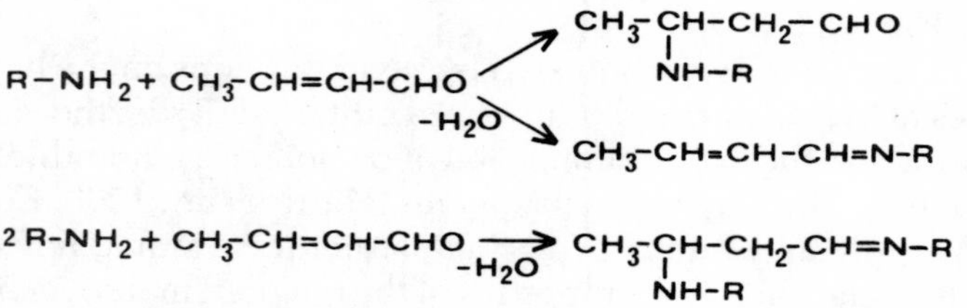

$$R-NH_2 + CH_3\text{-}CH=CH\text{-}CHO \xrightarrow{} \underset{\underset{NH-R}{|}}{CH_3\text{-}CH\text{-}CH_2\text{-}CHO} \quad \text{(I)}$$

$$\xrightarrow{-H_2O} CH_3\text{-}CH=CH\text{-}CH=N\text{-}R \quad \text{(II)}$$

$$2\,R\text{-}NH_2 + CH_3\text{-}CH=CH\text{-}CHO \xrightarrow[-H_2O]{} \underset{\underset{NH-R}{|}}{CH_3\text{-}CH\text{-}CH_2\text{-}CH=N\text{-}R} \quad \text{(III)}$$

Fig. 14. Reactions of crotonaldehyde with amines (Monsan *et al.*, 1975).

contrary to the assumptions of Richards and Knowles, unsaturated oligomers of glutaraldehyde would most probably react with amines, primarily through their aldehyde function, not through the ethylenic double bond. In the case of macromolecules (proteins and supports), it seems probable that, owing to steric and proximity effects, double addition may occur. The stabilization of aldimine structures by resonance with adjacent ethylenic double bonds has been indicated in work reported by Salomaa (1966), who showed that the condensation product of aniline with cinnamaldehyde, ϕ—CH=CH—CH=N—ϕ, resisted acid hydrolysis.

8. Diazotization Reactions

Azo coupling of proteins can be effected by polymers containing aryldiazonium functional groups. Polydiazonium reagents are prepared by treating polymers with pendent arylamino side chains with nitrous acid (Fig. 15). The electrophilic aryldiazonium ion attacks mainly activated aromatic rings, such as phenols (tyrosine) or imidazole (histidine), to form the corresponding azo derivatives (see Vallee and Riordan, 1969; Means and Feeney, 1971). Historically the coupling of proteins to diazotized polymers is among the oldest described in the literature (Campbell *et al.*, 1951; Lerman, 1953; Grubhofer and Schleith, 1953, 1954; Manecke and Gillert, 1955; Mitz and Summaria, 1961; Bar-Eli and Katchalski, 1960, 1963; Cebra *et al.*, 1961; for reviews, see Manecke, 1962, 1964, 1975; Silman and Katch-

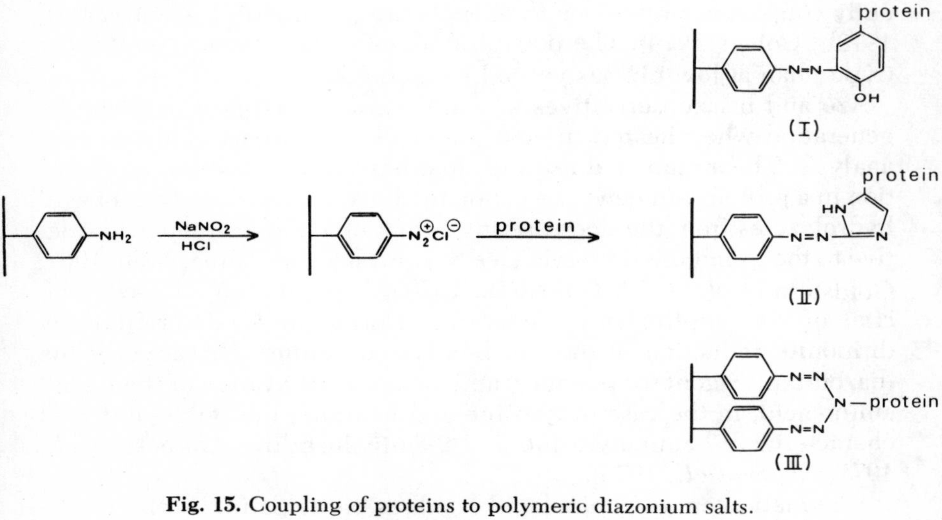

Fig. 15. Coupling of proteins to polymeric diazonium salts.

alski, 1966; Brown and Hasselberger, 1971; Zaborsky, 1973; Goldstein and Katchalski-Katzir, this volume). A large number of diazotizable polymers have been described (Lilly *et al.*, 1965; Surinov and Manoilov, 1966; Silman *et al.*, 1966; Barker *et al.*, 1968, 1969, 1970a,b; Inman and Dintzis, 1969; Weetall, 1969a,b, 1970; Goldstein *et al.*, 1970, 1974a; Ledingham and Hornby, 1969; Hornby and Filippusson, 1970; Zabriskie *et al.*, 1973; Datta *et al.*, 1973; Goldstein, 1973a; Li *et al.*, 1973; Gray *et al.*, 1974; Weetall and Filbert, 1974; Wilchek, 1974; Wilcheck and Miron, 1974a,b).

The specificity of azo coupling, as determined with low-molecular-weight aryldiazonium compounds, is rather broad; diazotizing reagents have been shown to attack, in addition to tyrosine and histidine, several other amine acids in proteins (Sokolovsky and Vallee, 1966, 1967; see also Means and Feeney, 1971; Cohen, 1968, 1974; Vallee and Riordan, 1969). Tyrosine and histidine react at mildly alkaline pH values (pH 8–9) at comparable rates to form monoazo derivatives. Amino groups (α-amines and the ϵ-amino groups of lysine) react under similar conditions with 2 moles of diazonium salt to give the disubstituted bisazo derivatives known as triazenes (Fig. 15). In the presence of excess diazonium reagent, bisazo derivatives of tyrosine and histidine are formed. The guanido group (arginine) and indole (tryptophan) also undergo coupling; their reactions may be slower than those of phenol, imidazole, and amino groups (Howard and Wild, 1957; Higgins and Harrington, 1959; Sokolovsky and Vallee, 1966, 1967; Spande and Glenner, 1973). At pH 5–6, imidazole apparently couples in preference to phenol (Krug *et al.*, 1971; Cuatrecasas, 1972b; Cohen, 1974). The possibility of effecting specific coupling to tryptophan at low pH has yet to be explored.

Azo and bisazo derivatives of amino acids decompose without regeneration when heated in acid under the conditions of amino acid analysis. The amino acid residues that have undergone azo modification in a protein can hence be estimated by amino acid analysis of acid hydrolyzates from the decrease in content of a given amino acid relative to the unmodified protein (see Sokolovsky and Vallee, 1966, 1967; Goldstein *et al.*, 1970; Goldstein, 1973a; Weetall *et al.*, 1974a). The sites of azo coupling in proteins can in principle be determined by dithionite reduction of the azo bond to an amine derivative of the diazonium reagent (or polymer) and an amino derivative of the ligand amino acid, in the case of tyrosine and histidine the stable and well characterized 3-aminotyrosine and 2-aminohistidine (Gorecki *et al.*, 1971; Nagai *et al.*, 1973).

Information on the amino acid residues participating in the formation of azo bonds with various types of polymeric diazonium reagents is still rather incomplete. The available data, based on the estimation of amino acids missing in acid hydrolyzates of immobilized enzyme derivatives, indicate that tyrosine and lysine are mainly involved, histidine and arginine being affected to a considerably smaller extent (see, e.g., Goldstein *et al.*, 1970; Goldstein, 1973a; Weetall *et al.*, 1974a). Such findings reflect indirectly the topography and relative abundance of the different amino acids on the surface of the specific protein investigated, but cannot be interpreted except in terms of the activity retained by an enzyme bound to different types of diazotized carriers and the particular amino acids modified in the coupling process. The fragmentary information available can be summarized in a few empirical generalizations: aryldiazonium groups, which because of their essentially hydrophobic character have adsorptive properties of their own (Cohen, 1974), may tend to attack preferentially hydrophobic, tyrosine-rich regions in a protein; such site-directed specificity could cause irreversible damage if key regions in the enzyme are affected. This property of the aryldiazonium moiety would be enhanced when the groups are attached on an electrically neutral and essentially hydrophobic backbone. In the case of highly hydrophilic support materials, such as polyelectrolyte diazotized resins, or highly solvated polysaccharide and glass surfaces with diazonium side chains, the site-directed specificity of aryldiazonium groups is mitigated to a large extent. These statements can be substantiated in part: trypsin and chymotrypsin coupled to diazotized electrically neutral resins e.g., *p*-aminobenzyl cellulose, copolymers of leucine and *p*-aminophenylalanine and arylamine resins derived from oxidized starch, gave completely inactive derivatives (Bar-Eli and Katchalski, 1960, 1963; Mitz and Summaria, 1961; Goldstein *et al.*, 1970). When the same enzymes were coupled to polyanionic or polycationic diazotized resins or to diazotized arylamine–glass, insoluble preparations of high enzymic activity were obtained (Goldstein, 1973a; Weetall, 1969a,b, 1970). In the case of the enzymically active polyelectrolyte azo derivatives of trypsin and chymotrypsin, far fewer amino acid residues were found to be modified as compared to the conjugates of these enzymes with uncharged supports where complete inactivation was observed (Bar-Eli and Katchalski, 1963; Goldstein *et al.*, 1970; Goldstein, 1973a). These data could serve as a reasonable indication of the modified chemical selectivity of polyelectrolyte diazonium resins, probably owing to the redirection of the charged polymeric reagent toward the more solvated (hydrophilic) parts of the enzyme molecule.

It should be kept in mind, however, that a large number of enzymes, particularly those of relatively hydrophobic character (Bigelow, 1967) and higher tyrosine content, e.g., papain, urease, glucose oxidase, amino acid oxidase, alkaline phosphatase, and β-glucosidase, give active azo derivatives with a wide variety of diazotizable polymeric supports (Cebra *et al.*, 1961; Silman *et al.*, 1966; Weetall, 1969a,b, 1970; Weetall and Hersh, 1969; Weetall and Baum, 1970; Hornby and Filippusson, 1970; Goldstein *et al.*, 1970; Zingaro and Uziel, 1970; Gray *et al.*, 1974).

9. Thiol–Disulfide Interchange Reactions

Polymeric mixed disulfides that can be used for the specific, reversible coupling to protein sulfhydryl groups via thiol–disulfide interchange reactions have been recently described (Carlsson *et al.*, 1974; Lin and Foster, 1975). These reagents are in effect polymeric analogs of low-molecular-weight disulfides used for the quantitative estimation of sulfhydryl groups in proteins, i.e., 2,2'-dipyridyldisulfide (Brocklehurst *et al.*, 1973) and 5,5'-dithiobis (2-nitrobenzoic acid), Ellman's reagent (Ellman, 1959; Butterworth *et al.*, 1967).

The preparation and use of polymeric disulfides is demonstrated in Fig. 16 for one such support (Carlsson *et al.*, 1974). The polymeric mixed disulfide is prepared by treating a thiolated polymer (glutathione-Sepharose) with 2,2'-dipyridyldisulfide (Fig. 16, Scheme 1). The coupling of an enzyme through its —SH groups is accompanied by the liberation of 2-thiopyridone, this offering a means of monitoring the reaction (Fig 16, Scheme 2). The support —S—S— protein bonds are stable under nonreducing conditions; the coupling can, however, be reversed with low-molecular-weight sulfhydryl reagent (Fig. 16, Scheme 3). The same thiol–disulfide interchange reaction has recently

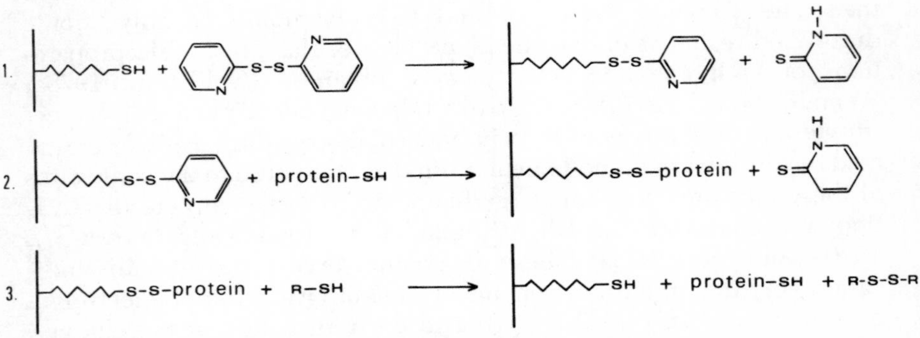

Fig. 16. Coupling of proteins to polymers containing sulfhydryl functional groups by thiol–disulfide interchange reactions (Carlsson *et al.*, 1974).

been used for the "covalent chromatography" of an SH-enzyme (papain), whereby the enzyme, after covalent fixation to the support via —S—S— bonds, is eluted with an SH-reagent (Brocklehurst *et al.*, 1973).

10. Four-Component Condensation Reactions

Four-component condensation (4CC) reactions involving carboxylate, amine, aldehyde, and isocyanide, lead to the formation of an N-substituted amide (Ugi, 1962, 1971, see esp. Ch. 8, p. 145). In this reaction the carboxyl and amine components (R^1 and R^2) combine to form an amide bond, the aldehyde and isocyanide components (R^3 and R^4) appearing as the side chain on the amide nitrogen (Fig. 17). Four-component reactions can be carried out in an aqueous medium at neutral pH and allow for considerable versatility and high selectivity when applied to immobilization of proteins. By proper choice of support material and additives to the reaction mixture, coupling can be directed toward either amino or carboxyl groups on the enzyme, side reactions being avoided. Axén and co-workers used polymers containing amine functional groups to couple proteins through their carboxyl groups in the presence of acetaldehyde and a low-molecular-weight isocyanide (cyclohexyl isocyanide or N-dimethylaminopropyl isocyanide). Conversely, proteins could be coupled through their amino groups, polymers containing carboxyl groups being used, again in a medium containing aldehyde and isocyanide (Axén *et al.*, 1971b; Vretblad and Axén, 1971, 1973a,b). Goldstein and co-workers used a polymer containing isocyanide functional groups; enzymes could be immobilized via the —N≡C group on the polymer by 4CC reactions carried out in the presence of a water-soluble aldehyde (i.e., acetaldehyde). The protein supplied either the amino or carboxyl group, the isocyanide group on the support being steered mainly toward one type of functional group on the enzyme by adding an excess of the missing fourth component to the reaction medium. Thus enzymes could be bound to polyisocyanide supports through their amino groups by a 4CC reaction in the presence of acetaldehyde and acetate; coupling through

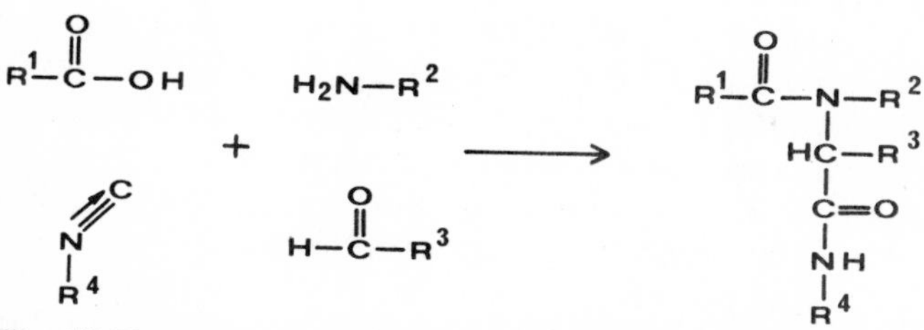

Fig. 17. Four-component condensation, 4CC (Ugi, 1962, 1971).

protein carboxyl groups, on the other hand, could be affected in the presence of acetaldehyde and an amine such as Tris (Goldstein *et al.*, 1974a,b). It should be noted that despite their many attractive points and considerable flexibility, 4CC reactions can be used to advantage only when an enzyme is not sensitive to aldehyde.

B. Polymeric Supports

Numerous natural and synthetic macromolecules have been explored as potential supports for enzyme immobilization; this has led to the development of a wide variety of techniques for (1) grafting of specific functional groups onto preformed polymers, (2) synthesis of tailor-made polymers and copolymers devised to fulfill particular needs, and (3) the use of "parent carrier polymers," which by consecutive chemical manipulations can be transformed into the chemical species best suited for a given task (Goldman *et al.*, 1971b; Zaborsky, 1973; Manecke, 1975; Weetall, 1975; Inman and Dintzis, 1969; Goldstein, 1973a,b; Goldstein *et al.*, 1974a; Filippusson *et al.*, 1972; Campbell *et al.*, 1975; Morris *et al.*, 1975; Hornby, 1976; Wilchek, 1974).

This section surveys the structure and properties of the various classes of materials most commonly used as matrices for the immobilization of enzymes, with strong emphasis on the methods employed to introduce chemically reactive groups and on problems encountered when using a given type of polymeric support; specifics of coupling techniques will be discussed insofar as they relate to the properties of a support or an immobilized enzyme conjugate. The majority of the published covalent-immobilization procedures classified according to "parent polymer" are listed in Table IV. The table also contains information on the type of chemical modification carried out on the parent polymer, the type of bond formed with a protein, and, when available, the protein-binding capacity of the derivatized support. The reader will be referred to Table IV throughout the section. Comprehensive lists of enzymes for which immobilization procedures have been reported are available in several publications (Goldman *et al.*, 1971b; Melrose, 1971; Zaborsky, 1973; Royer *et al.*, 1973; Kennedy, 1974b).

On a trial-and-error basis, it has been established that with most supports rich in hydrophobic groups (e.g., aromatic residues) preparations of low protein content and low enzymic activity are obtained; moreover, such preparations often exhibit low stability presumably due to denaturation effects analogous to those caused by organic solvents. Supports rich in hydrophilic groups bind on the average considerably larger amounts of protein; in such preparations the bound enzyme retains a higher proportion of its activity and usually exhibits higher stability. It has also been demonstrated in a number of cases that the presence of hydrophilic groups on a support can mitigate the deleterious effects of a hydrophobic polymeric environment (Manecke and Singer, 1960b; Manecke and Günzel, 1962, 1967a; Manecke, 1962, 1964, 1975; Silman and Katchalski, 1966; Manecke and Förster, 1966; Crook *et al.*, 1970; Goldstein, 1970, 1972a,b, 1973a,b; Goldstein *et al.*, 1970, 1971; Manecke *et al.*, 1970; Weetall, 1970; Zaborsky, 1973).

Our fragmentary understanding of the factors determining tertiary structure and conformational stability does not allow at this stage more than rough predictions, based on analogy, about the retention of activity by a given enzyme, following immobilization, and its storage and operational stability, even in the case of enzymes that are structurally and functionally similar. This statement can be illustrated with a few examples: The "serine proteases," trypsin and α-chymotrypsin, are remarkably similar in amino acid composition and sequence (57% homology; Hartley, 1970) and in tertiary structure (Matthews *et al.*, 1967; Stroud *et al.*, 1974; Krieger *et al.*, 1974). Studies on the coupling of trypsin and chymotrypsin to cyanogen bromide-activated agarose and Sephadex have indicated significant differences in the behavior of the two enzymes (Axén and Ernback, 1971). In chymotrypsin–agarose the bound enzyme retained 20% of its activity; trypsin bound to the same matrix, was 40–45% active; conversely, with Sephadex as support chymotrypsin retained 30% while trypsin retained only 10% activity. Differences in the amounts of bound protein were also recorded [300 mg per gram of support for chymotrypsin, 100 mg/gm for trypsin (Axén and Ernback, 1971)]. Since the two enzymes have essentially the same molecular weight and number of lysyl residues, and the same method was employed for binding, no rational explanation for the observed differences can be put forth. Chymotrypsin–agarose, chymotrypsin–Sephadex, as well as trypsin–agarose, lost their activity in 8 *M* urea; trypsin–Sephadex on the other hand retained most of its activity under the same conditions. Differences in thermal stability were also observed with the trypsin and chymotrypsin derivatives (Gabel *et al.*, 1970, 1971). Inconclusive observations of the same type have been recorded for CM-cellulose derivatives of ficin and bromelain (Crook *et al.*, 1970; Hornby *et al.*, 1966; Wharton *et al.*, 1968a). In CM-cellulose–ficin (maximal loading, 50 mg of protein per gram of support), the enzyme retained about 12% of its activity; both the protein content and activity of immobilized bromelain prepared under the same conditions were considerably higher (110 mg of protein per gram of support; 55% activity). CM-cellulose–ficin exhibited enhanced thermal stability and retained essentially all its activity upon lyophilization (Hornby *et al.*, 1966), whereas CM-cellulose–bromelain was not stabilized toward heat denaturation; this derivative lost about 50% of its activity after freeze-drying (Wharton *et al.*, 1968a). Collected data on the stability of immobilized enzyme derivatives under different conditions and on the retention of activity by the bound protein can be found in several review articles (Melrose, 1971; Royer *et al.*, 1973; Zaborsky, 1973).

Several of the common support materials, e.g., cross-linked polyacrylamide, cross-linked dextrans, or cross-linked copolymers of maleic, acrylic, and methacrylic acids are xerogels in the broader sense (see Epton, 1973; see also Table IV). They are hydrophilic and thus possess desirable characteristics in terms of the improved stability they might confer on the bound protein. Their mechanical strength, however, is rather poor; removal of solvent by drying or by exposure to high pressures results in the collapse of the swollen three-dimensional networks and to changes in porosity. Rigid dense particles, e.g., quartz, glass, nylon, or microcrystalline cellulose, would not suffer from such drawbacks; their protein-binding capacity, on the other hand, will be low owing to the relatively small surface area. Rigid, preformed macroporous matrices, e.g., porous "silica-rich" alkali–borosilicate glasses (Weetall and Filbert, 1974), do not collapse or change in porosity under normal conditions; their protein-binding capacity and permeability to substrate would be lower as compared to those of highly swollen xerogels, but considerably higher relative to rigid, dense particles. Macroreticular polymers may offer a compromise between the requirement for mechanical rigidity and the need for high loading capacity. Such polymers are prepared by carrying out the polymerization reaction in a solvent in which the polymer chains are sparingly soluble; under such conditions, the spherical submicroscopic particles that separate out fuse together to form eventually within the material a system of pores and crevices. Macroreticular polymers are generally semirigid in texture.

1. Vinyl Polymers

Vinyl polymers, as a class, have been widely explored as supports for enzyme immobilization since by proper choice of monomers and polymerization conditions almost any desired combination of mechanical and chemical properties can be attained in principle in the final product (see, e.g., Sorenson and Campbell, 1968).

Historically, polystyrene was the first synthetic polymer to be used for the immobilization of antibodies, antigens, and enzymes (Grubhofer and Schleith, 1953, 1954; Manecke and Gillert, 1955; Brandenberger, 1955; Gyenes and Sehon, 1960; Manecke, 1962). In these early studies polystyrene was converted to poly(*p*-aminostyrene) by nitration and reduction; the latter was diazotized and coupled to several proteins (Manecke and Gillert, 1955; Manecke and Singer, 1960a; Manecke *et al.*, 1958; Manecke, 1962). Poly(*p*-amino styrene) was also converted to poly(4-isocyanatostyrene) by treatment with phosgene (Brandenberger, 1955; Manecke *et al.*, 1958; and Singer, 1960a). Despite the rather high concentration of reactive groups on these polymers, the amounts of bound protein were rather low, presumably owing to the hydrophobicity of the support (Manecke, 1962).

Recently there has been some renewed interest in polystyrene as support mainly because of its availability and low price: polystyrene tubes have been used, after nitration, reduction, and diazotization of their inner surface, for the immobilization of β-fructofuranosidase and α-amylase; the enzymic activity of these preparations was, however, rather low (Filippusson and Hornby, 1970; Ledingham and Hornby, 1969; Ledingham and Ferreira, 1973); macroreticular poly(*p*-aminostyrene) activated with glutaraldehyde has been employed for the immobilization of papain and amyloglucosidase (Baum, 1975). Polystyrene is also being used for the adsorption of specific antibodies in enzyme-linked immunoadsorbent assay (ELISA) kits (Engvall and Perlmann, 1971; Holmgren and Svennerholm, 1973).

The inherent hydrophobicity of poly(*p*-aminostyrene) and related polymers containing high surface concentrations of aromatic groups could be mitigated to some extent by their "dilution" with a hydrophilic component. Thus Manecke and co-workers reported satisfactory results in terms of bound protein and retention of activity with nitrated copolymers of methacrylic acid and methacrylic acid fluoroanilide in various ratios. These products contained the 3-fluoro-4,6-dinitroanilide group as the chemically reactive component and the carboxylic group as a hydrophilic component (Manecke and Singer, 1960b; Manecke *et al.*, 1960; Manecke and Günzel, 1962) [see Fig. 18, structure (I)]. Parallel investigations were carried out with nitrated copolymers of methacrylic acid and fluorostyrenes (Manecke and Förster, 1966) [Fig. 18, structure (II)]. By a similar approach, hydrophilic copolymers containing isothiocyanate groups [Fig. 18, structure (III)] were prepared by treating copolymers containing *p*-aminostyrene with thiophosgene or by the direct copolymerization of acrylic or methacrylic acid with *m*- and *p*-isothiocyanatostyrenes (divinylbenzene was used in all these preparations as cross-linking agent). Variation of the ratio of monomers gave carriers of different capacities and hydrophilicites (Manecke and Günzel, 1967a; Manecke *et al.*, 1970; Manecke, 1975). All these materials could be prepared in

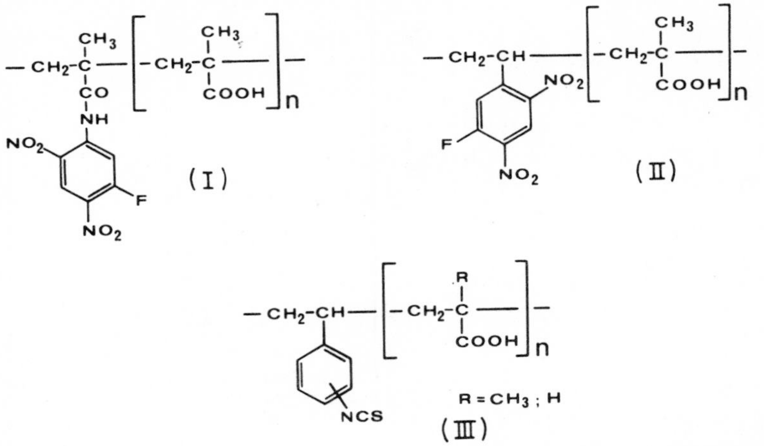

Fig. 18. Some chemically reactive vinyl polymers (Manecke *et al.*, 1960, 1970; Manecke, 1975). (I) Copolymer of methacrylic acid and methacrylic acid-3-fluoro-4,6-dinitroanilide; (II) copolymer of methacrylic acid and 3-fluoro-4,6-dinitrosytrene, (III) copolymer of methacrylic or acrylic acid and 3- or 4-isothiocyanatostyrene.

macroreticular form with concomitant increase in binding capacity (see Table IV). Retention of enzymic activity by the bound protein tested with papain was generally 30–50%; in a few cases papain bound to acrylic acid–*m*-isothiocyanatostyrene copolymers was reported to exhibit higher activities; moreover, increasing the content of hydrophilic groups in the various copolymers, led to an increase in their binding capacity, maximal amounts of bound protein being found with polymers of 3:1 and 4:1 ratios of hydrophilic to hydrophobic component (Manecke *et al.*, 1970). The optimal composition of the carrier depends also on the degree of cross-linking and the macroreticular structure (Manecke, 1975).

Copolymers containing acid anhydride functional groups (Fig. 2) have been used to prepare a variety of polyelectrolyte derivatives of enzymes (Levin *et al.*, 1964; Goldstein *et al.*, 1964; Goldstein, 1970, 1972a,b; Ong *et al.*, 1966; Westman, 1969; Fritz *et al.*, 1968a,b, 1969a,b; Zingaro and Uziel, 1970; Weetall, 1970; Conte and Lehmann, 1971; Brümmer *et al.*, 1972; Bessmertnaya and Antonov, 1973; Solomon and Levin, 1974a; Jaworek, 1974) (see Table IV).

The most commonly used support materials of this type have been the 1:1 copolymers of maleic anhydride and ethylene (EMA) (Levin *et al.*, 1964; Goldstein *et al.*, 1970; Goldstein, 1972a,b) (see Fig. 19). Such linear macromolecules are in effect polyfunctional cross-linking reagents and allow considerable versatility as to the physical properties and total protein content of the insoluble enzyme derivative. With excess enzyme (e.g., 4 : 1 enzyme-to-polymer ratio in the reaction mixture) preparations containing up to 75% protein and of dense flakelike texture were obtained. Excess-polymer combinations led to soft, highly swollen xerogels; with such preparations additional cross-linking with bifunctional low-molecular-weight reagents, e.g., 1,6-diaminohexane, was often necessary to enhance insolubility (Levin *et al.*, 1964; Goldstein, 1970). Other copolymers of maleic anhydride have been used in a similar fashion (see Zingaro and Uziel, 1970; Brümmer *et al.*, 1972; see also Table IV). Maleic anhydride copolymers could be pre-cross-linked in an organic solvent and subsequently allowed to react with a protein in an aqueous buffer (Levin *et al.*, 1964; Jaworek, 1974). The mechanical properties and the porosity of the final product could be regulated by controlling the degree of cross-linking and the chain length of the cross-linking reagent (Levin *et al.*, 1964; Goldstein, 1973a,b; Jaworek, 1974). As shown in Fig. 2, opening of the anhydride ring by protein —NH_2 groups or by spontaneous hydrolysis in the aqueous medium generates free carboxyl groups. The polyelectrolyte microenvironment in which the bound protein is located is reflected in its stability pattern and its kinetic behavior; the polyanionic maleic anhydride copolymer–enzyme conjugates exhibited alkaline shifts in their pH activity and stability profiles and perturbed Michaelis constants with charged substrates (Goldstein *et al.*, 1964; Goldman *et al.*, 1971a). Enzyme derivatives of the acrylic and methacrylic acid copolymers described above behaved in a similar manner (Manecke, 1975). These phenomena have been attributed to changes in the local concentration of charged low-molecular-weight species, hydrogen and hydroxyl ions or charged substrate, stemming from Donnan-type redistribution effects [for a discussion of polyelectrolyte effects in immobilized enzyme systems, see Goldstein *et al.* (1964), Wharton *et al.* (1968a,b), Goldman *et al.* (1971b), Engasser and Horvath, this volume, and Goldstein (1976)].

Maleic anhydride copolymers could serve as starting materials for further chemical modification. The preparation of cross-linked polyanionic and polycationic resins containing arylamino, acid-

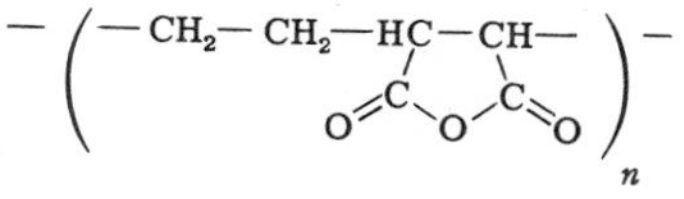

Fig. 19. Copoly(ethylene-maleic anhydride).

hydrazide, or alkylamino functional groups is shown schematically in Fig. 20. The parent polymer for these preparations was ethylene-maleic anhydride copolymer (Goldstein, 1973a,b). As with the charged derivatives described earlier the polyanionic enzyme conjugates shown in Fig. 20 exhibited improved stability at alkaline pH values; conversely the analogous polycationic derivatives showed higher stability at acidic pH values. The pH range in which an immobilized enzyme retains its functional stability could thus be ex-

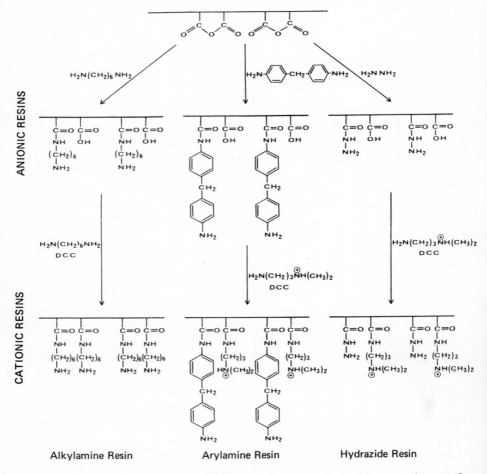

Fig. 20. Synthesis of polyanionic and polycationic resins derived from copolymers of ethylene and maleic anhydride (Goldstein, 1973a,b).

panded by proper selection of the charge characteristics of the support materials.

The binding capacities of maleic anhydride copolymers were found to be strongly dependent on their anhydride content: binding experiments with ethylene–maleic anhydride copolymer samples of known anhydride content showed that optimal recovery of immobilized enzymic activity and bound protein could be attained with polymers containing 50–70% of the maleyl residues in the anhydride form (Goldstein, 1970). The low binding of protein by ethylene–maleic anhydride copolymers of high anhydride content and the concomitant low recovery of enzymic activity in the insoluble material could be attributed to the essentially hydrophobic character of polymers in which all the maleyl residues were in the anhydride form. The properties of polymers of this type could hence be manipulated by controlled hydrolysis, to generate the desired amount of hydrophilic centers before putting the carrier in contact with an enzyme solution (Goldstein, 1972a). The activity retained by enzymes bound to maleic anhydride-based support materials ranged, with a few exceptions, between 30% and 70% (see collected data in Royer *et al.*, 1973).

All vinyl polymers described above had charged groups and would thus normally possess ion-exchange properties leading to nonspecific adsorption of proteins as well as to anomalous kinetics in the case of an immobilized enzyme. The search for hydrophilic electrically neutral vinyl supports that would have the advantage of low nonspecific adsorption of biological macromolecules and offer an inert matrix that would impose a minimum of kinetic perturbations on an immobilized enzyme has centered mainly around materials based on polyacrylamide and polyvinyl alcohol.

The structural network of cross-linked polyacrylamide or polyvinyl alcohol gels consists mainly of —CH_2—CHX— segments with alternate backbone carbon atoms bearing primary amide (X= —$CONH_2$) or alcohol groups (X = —OH). The great abundance of these groups is responsible for the hydrophilic character of the polymers and their low adsorption of macromolecules. The linear polymers are water soluble. Insoluble gel networks are formed by including some bifunctional monomer in the polymerization mixture, to produce cross-links, e.g., *N,N'*-methylenebisacrylamide in the case of polyacrylamide gels.

Chemically modified polyacrylamides have been prepared by (1) derivatization of preformed polyacrylamide gel or beads, e.g., Bio-Gel (Inman and Dintzis, 1969; Zabriskie *et al.*, 1973; Weston and Avrameas, 1971; Ternynck and Avrameas, 1972; Datta *et al.*, 1973; Inman, 1974; Johansson and Mosbach, 1974a,b); (2) copolymerization of acrylamide and cross-linker with a functional group-bearing acrylic or vinyl monomer (Mosbach, 1970; Mosbach and Mattiasson, 1970; Barker *et al.*, 1970a,b; Barker and Epton, 1970; Martensson and Mosbach, 1972; Ohno and Stahmann, 1972a,b; Johansson and Mosbach, 1974a,b; Jaworek, 1974).

The direct chemical modification of commercially available crosslinked polyacrylamide beads was first described by Inman and Dintzis (Inman and Dintzis, 1969; Inman, 1974). By this method Bio-Gel beads were treated with hydrazine or with ethylenediamine to obtain the corresponding acyl hydrazide or aminoethyl derivatives (Fig. 21). The functional group density could be regulated by the concentration of reagent and the time and temperature of the reaction (Inman, 1974).

The acyl hydrazide could be activated with nitrous acid to form the acyl azide; the aminoethyl derivative could be further modified to the diazotizable *p*-aminobenzoylaminodethyl derivative or used directly for coupling via carboxyl groups on a protein in the presence of a carbodiimide or via amino groups on a protein after treatment with glutaraldehyde (Inman, 1974). The protein-binding capacity of the derivatized polyacrylamide beads was 100–300 mg per gram of support depending on the porosity of the beads and the size of the protein. In the few cases documented in the literature the bound enzymes retained 30–40% of their activity and displayed essentially unperturbed kinetic behavior (Inman and Dintzis, 1969; Erlanger *et al.*,. 1970). More recently the direct activation of polyacrylamide with glutaraldehyde has been described (Weston and Avrameas, 1971; Ternynck and Avrameas, 1972; Johansson and Mosbach, 1974a,b). The

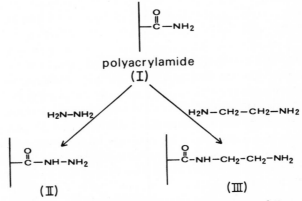

Fig. 21. Chemical modifications of polyacrylamide (Inman and Dintzis, 1969).

nature of the reaction of glutaraldehyde with polyacrylamide is not understood; both the protein-binding capacity and the retention of activity by the bound enzyme, however, are quite satisfactory (see Johansson and Mosbach, 1974a,b; see also Table IV).

A variety of polymers in which a vinyl or acrylic monomer carrying chemically reactive groups is copolymerized with acrylamide in the presence of a crosslinker have been described.

Copolymers of acrylamide and 2-hydroxyethylmethacrylate, activated with cyanogen bromide by a procedure analogous to that employed for polysaccharides (see Figs. 8 and 9) were used for the immobilization of chymotrypsin, trypsin, and β-glucosidase (Mosbach, 1970; Mosbach and Mattiasson, 1970; Turkova *et al.*, 1973; Turkova, 1974; Johansson and Mosbach, 1974a); 20–30% of the activity of the bound enzymes was retained (see also Table IV).

Mosbach used copolymers of acrylamide and acrylic acid for the coupling of enzymes via carbodiimide activation of the support carboxyl groups. The same author copolymerized acrylamide and acrylic acid in the presence of enzyme and then fixed the protein entrapped within the gel, again by carbodiimide activation of the carboxyl groups on the gel matrix (Mosbach, 1970; Mosbach and Mattiasson, 1970; Martensson and Mosbach, 1972; Johansson and Mosbach, 1974a). Copolymers of acrylamide and methacrylic acid anhydride or maleic anhydride of different degrees of cross-linking were similarly used to immobilize several enzymes (Krämer *et al.*, 1974; Jaworek, 1974). A different approach to covalent fixation of enzymes within cross-linked gels, the so-called "protein-copolymerization" method has been recently described (Jaworek, 1974). By this method protein amino groups are alkylated with reagents which in addition to the alkylating moiety contain acrylic double bonds, e.g., acrylic acid- or methacrylic acid-2,3-epoxypropyl ester (Jaworek, 1974; Cremonesi *et al.*, 1975); the alkylated protein when copolymerized with acrylamide becomes an integral part of the cross-linked copolymer network. The protein-copolymerization method is well suited in principle for the coimmobilization of several enzymes, particularly if the latter consist of subunits. In all these methods high retention of enzymic activity and improved stability have been reported. The protein-binding capacities of the various acrylamide copolymers are given in Table IV.

Several workers have reported the use of hydrophilic, electrically neutral copolymers based on acrylamide (Barker *et al.*, 1970a,b; Barker and Epton, 1970; Calam and Thomas, 1972; Epton *et al.*, 1972; Ohno and Stahmann, 1972a,b). One such type of acrylamide copolymers, called enzacryls, utilizes *p*-aminoacrylanilide, acryloyl hydrazide, N-acryloylcysteine, or N-acryloyl-cysteinethiolactone as comonomers (Barker *et al.*, 1970a,b; Epton *et al.*, 1972; Barker and Epton, 1970). The enzacryls offer in principle considerable versatility; copoly(acrylamide-*p*-aminoacrylanilide) could be activated by diazotization or converted into the corresponding isothiocyanate with thiophosgene (see Figs. 15 and 10); copoly(acrylamide-acryloyl hydrazide) could be activated with nitrous acid to the polymeric azide (Fig. 1) while copoly(acrylamide-N-acryloylcysteine) could be employed for the reversible attachment of proteins via their —SH groups (compare with Fig. 16). The somewhat limited published data on the efficiency of these supports suggests, however, that both the protein-binding capacity and the retention of activity by the bound protein are low; with α- and β-amylase, a maximum of 32 mg of enzyme bound per gram of support with a retention of activity of only 16% was reported by Barker and co-workers (Barker *et al.*, 1970a,b). The heat and storage stabilities of enzacryl enzyme conjugates have been claimed to be greatly improved relative to the corresponding free enzymes. Carriers, which consisted of cross-linked polyglycidyl methacrylates (reactive epoxy group), were optimized with respect to their macroreticular structure and their mechanical stability (Švec *et al.*, 1975).

Polyvinyl alcohol (PVA) cross-linked with terephthalaldehyde to give mechanically stable xerogels, is being explored as starting material for the preparation of hydrophilic, nonbiodagradable supports, devoid of charged groups (Manecke, 1975; Manecke and Vogt, 1976). Cross-linked PVA gels could be activated directly with 2,4,6-trichloro-*s*-triazine [Fig. 22, structure (I)] or converted to a diazotizable derivative, the *p*-aminobenzyl ether by treatment with 4-nitrobenzylchloride followed by reduction [Fig. 22, structure (II)]. A novel type of diazotizable support was prepared by transacetalation of PVA with 2-(*m*-aminophenyl)-1,3-dioxolane [Fig. 22, structure (III)]. The latter carrier had superior properties in terms of protein-binding capacity and enzymic activity (300–650 mg of papain bound per gram of support, 15–22% active; Manecke and Vogt, 1976).

2. Polysaccharides

Naturally occurring polysaccharides have been widely used as support materials, foremost among them being cellulose, starch, agarose, and cross-linked dextrans (Fig. 23). Historically cellulose was one of the first materials to be employed for the covalent fixation of enzymes and other biologically active macromolecules (Micheel and Ewers, 1949; Mitz and Summaria, 1961; Jagendorf *et al.*, 1963; Weliky and Weetall, 1965; Surinov and Manoilov, 1966; Hornby *et al.*, 1966;

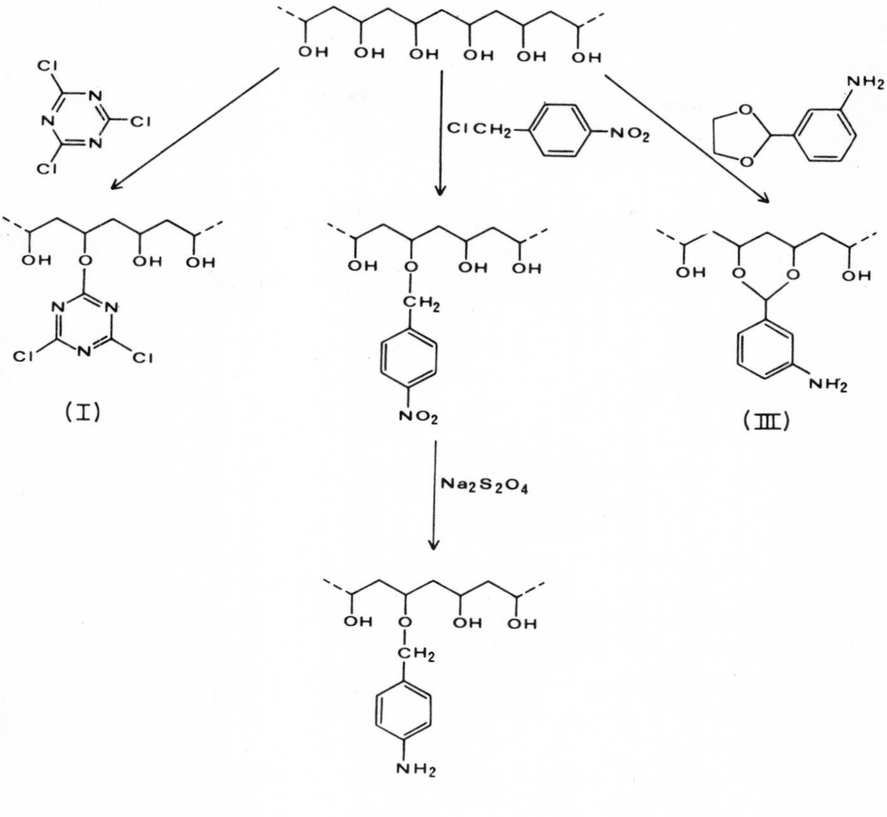

Fig. 22. Chemical modifications of polyvinyl alcohol (Manecke, 1975; Manecke and Vogt, 1976).

Campbell and Weliky, 1967). The popularity of cellulose even today stems as much from reasons of tradition and comparatively low price as from the fact that a large number of relatively simple and well established procedures are available for its chemical modification. Agarose and cross-linked dextran beads, while of more recent origin as supports, offer several advantages not found in cellulose; moreover, in devising methods for the chemical modification of these polysaccharides one could draw on the vast experience accumulated by the cellulose chemists (see, e. g., Weliky and Weetall, 1965; Campbell and Weliky, 1967; Bikales and Segal, 1971; Tesoro and Willard, 1971; Kennedy, 1974b).

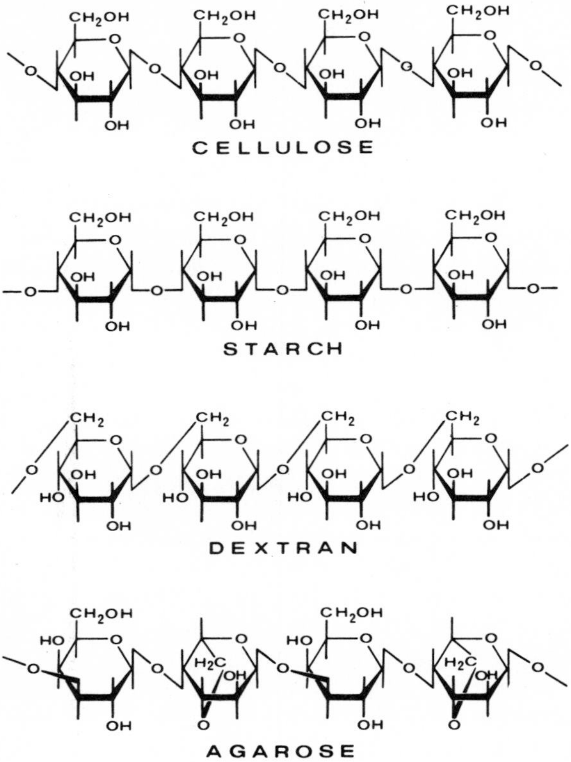

Fig. 23. Some common polysaccharides.

Cellulose consists of linear chains of 1,4-linked β-D-glucose residues (Fig. 23) organized in fibers of a high degree of crystallinity (Jones, 1964; Bikales and Segal, 1971; Jahn, 1971). Local irregularities in the alignment of individual chains are reflected in the structure of the whole fiber which consists of microcrystalline aggregates separated by amorphous regions. Swelling of the cellulose matrix involves presumably a certain degree of constrained dissolution of the glycan chains in the amorphous regions. This would suggest that it is in these regions of low three-dimensional molecular order and hence of higher accessibility that chemical reactions, and particularly the conjugation of macromolecules, are most likely to occur. This view is supported by a large body of indirect evidence indicating that treatments that lead to a decrease in the crystallinity of cellulose result in an increase in the degree of substitution and enhanced binding capacity for proteins. Swelling accompanied by increase in the amorphous character and accessibility of cellulose has been accomplished by treatment with

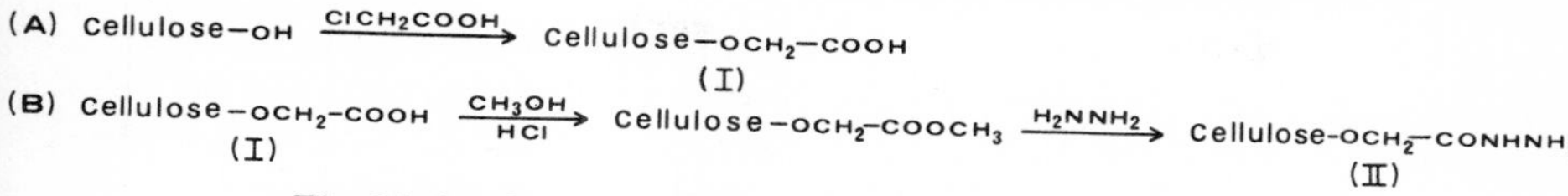

(A) Cellulose—OH $\xrightarrow{ClCH_2COOH}$ Cellulose—OCH$_2$—COOH

(I)

(B) Cellulose—OCH$_2$—COOH $\xrightarrow[HCl]{CH_3OH}$ Cellulose—OCH$_2$—COOCH$_3$ $\xrightarrow{H_2NNH_2}$ Cellulose—OCH$_2$—CONHNH$_2$

(I)   (II)

Fig. 24. Synthesis of *O*-(carboxymethyl) cellulose hydrazide.

30–50% sodium hydroxide, with pyridine or with glacial acetic acid. Swelling of cellulose could also be effected with certain organic solvents, notably ethylamine and dimethyl sulfoxide. Macroporous reconstituted celluloses, of open structures that allow the penetration of macromolecules and hence exhibit higher binding capacities, have been prepared by precipitating cellulose out of ammoniacal copper, zinc, or cadmium solutions (Weliky and Weetall, 1965; Bikales and Segal, 1971; Epton, 1973; Kennedy, 1974a,b; Royer *et al.*, 1973).

Paradoxically, particulate microcrystalline cellulose preparations obtained by controlled acid-digestion of some of the amorphous regions in the native polysaccharide, have been reported to afford satisfactory supports for the covalent immobilization of enzymes, particularly in terms of the flow properties of the final product (Barker *et al.*, 1968, 1969). Here the diminution in number of binding sites is probably offset by some increase in porosity, although the overall binding capacity of these materials is still rather low as compared to reconstituted celluloses and related polysaccharide supports (see Table IV).

Cellulose undergoes all characteristic reactions of polyhydric alcohols: oxidation, esterification, ether formation, halogenation, as well as reactions with unsaturated compounds, such as acrylonitrile, isocyanates, and ketene (Weliky and Weetall, 1965; Bikales and Segal, 1971; Kennedy, 1974b). The graft copolymerization of cellulose with acrylic and vinyl monomers has been extensively investigated (Arthur, 1970; Moore, 1972; Kennedy, 1974b). Several types of chemically modified celluloses are commercially available (see Table IV). The preparation of two such derivatives, the carboxymethyl ether of cellulose (CM-cellulose) and its hydrazide and the *p*-aminobenzyl ether of cellulose (PAB-cellulose) is shown schematically in Figs. 24 and 25. A discussion of the chemistry of cellulose and similar natural polysac-

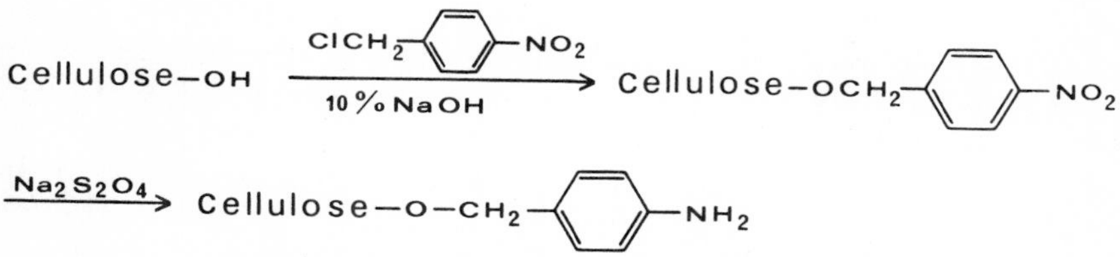

Cellulose—OH $\xrightarrow[10\% \, NaOH]{ClCH_2-\langle\!\!\!\bigcirc\!\!\!\rangle-NO_2}$ Cellulose—OCH$_2$-$\langle\!\!\!\bigcirc\!\!\!\rangle$-NO$_2$

$\xrightarrow{Na_2S_2O_4}$ Cellulose—O—CH$_2$-$\langle\!\!\!\bigcirc\!\!\!\rangle$-NH$_2$

Fig. 25. Synthesis of *O*-(4-aminobenzyl) cellulose.

charides is outside the scope of this article (for comprehensive reviews, the reader is referred to Weliky and Weetall, 1965; Pigman *et al.*, 1970; Bikales and Segal, 1971; Kennedy, 1974b). Specific aspects of the chemistry of polysaccharides will be dealt with below insofar as they bear upon the preparation of functionalized supports.

Several porous polysaccharide matrices possessing molecular sieving properties, have gained wide acceptance as supports for enzyme immobilization. These materials based on dextran and agar were originally developed as supports for gel-filtration chromatography; they are available commercially, in grades characterized by their water-regain and molecular exclusion limits (see, e.g., Reiland, 1971; Munier, 1973).

Dextran, a linear water-soluble polysaccharide composed of 1,6-linked α-D-glucose residues (Fig. 23) is produced by microorganisms of the genus *Leuconostoc*. The commercially available dextran gels (Sephadex) are prepared by cross-linking the linear polysaccharide with epichlorohydrin (Porath and Flodin, 1959; Porath, 1960, 1962; Flodin, 1961; Flodin and Porath, 1961) (see Fig. 26). By control of the chain length of the starting material, linear dextran, and the degree of cross-linking, gels of well defined water regain and molecular sieving properties are obtained (Porath and Flodin, 1959; Gelotte, 1960; Porath, 1960; Flodin, 1961). The work on cross-linked dextran gels has been summarized in several reviews (see, e.g., Porath, 1962; Reiland, 1971; Munier, 1973; Kennedy, 1974b).

Agarose is one of the components of agar, a complex mixture of polysaccharides, extracted from several species of the Rhodophyceae family of red sea-water algae (Araki, 1956; Polson, 1961; Hjerten, 1961, 1962a; Andrews, 1962; Hjerten and Porath, 1962; Duckworth and Yaphe, 1971; Izumi, 1971). Agarose is composed of alternating 1,3-linked β-D-galactose and 1,4-linked 3,6-anhydro-α-L-galactose residues (Araki, 1956; Hjerten, 1961, 1962a,b, 1963, 1964; Hjerten and Porath, 1962; Russell *et al.*, 1964; Araki and Arai, 1967; Duckworth and Yaphe, 1971; Duckworth *et al.*, 1971; Izumi, 1971; see also Fig. 23). Other fractions isolated from agar have been collectively named agaropectin. They have essentially the same backbone structure, but with varying amounts of the residues shown in Fig. 23 replaced by the pyruvic acid ketal, 4,6-O-(1-carboxyethylidene)-D-galactopyranose or by methylated or sulfated sugar units in a manner that maintains the alternating sequence of 3-linked β-residues and 4-linked α-residues. For most agars a clean separation between agarose and agaropectin cannot be achieved by the conventional separation techniques. Hence the polysaccharide is more conveniently described as a continuously

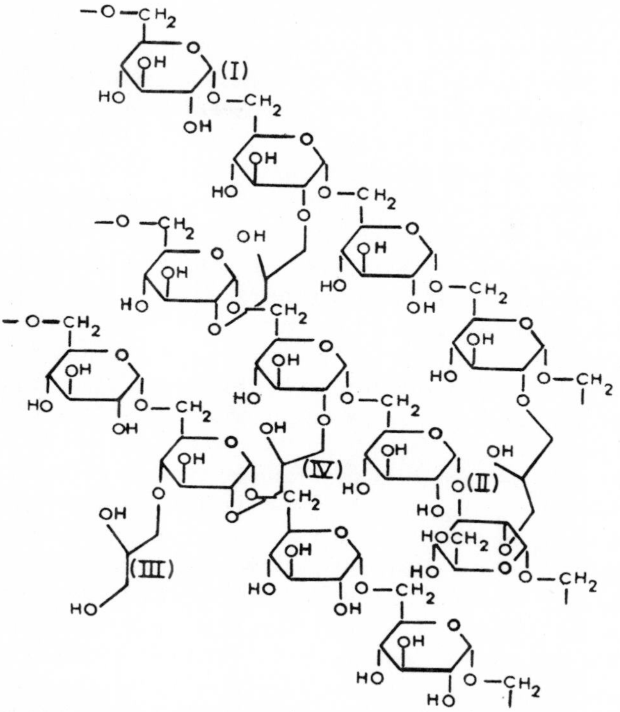

Fig. 26. Epichlorhydrine-cross-linked dextran gel. (After Porath, 1962.) (I) 1,6-α-glucosidic bonds; (II) 1,3-α-glucosidic bonds; (III) glycerol side chains (side reaction); (IV) glucose residues connected through glycerol bridges.

varying covalent structure, agarose being one idealized extreme (Duckworth *et al.*, 1971). The most effective methods for the preparation of truly neutral agarose involve the removal of charged groups by chemical means (Porath *et al.*, 1971). When an aqueous solution of agarose is cooled below 50°C, gelation occurs. This fact has been utilized for the preparation of agarose gels in bead and pellet form (Polson, 1961; Hjerten, 1962a, 1964; Hjerten and Porath, 1962; Bengtsson and Philipson, 1964). Agarose gels are mechanically more stable and have greater pore size than other xerogels of comparable matrix dilution; they can therefore be used for the separation of macromolecules of considerably wider range of molecular weights by gel filtration chromatography (see, e.g., Gelotte and Porath, 1967; Determann, 1968; Munier, 1973).

The unusual properties of agarose gels have been attributed to the stacking of hydrogen-bonded polysaccharide chains into strands or bundles of ordered conformation, the gel structure being maintained through "junction zones" involving noncovalent cooperative binding (Laurent, 1967; Joustra, 1969; Rees, 1969; Dea *et al.*, 1972; Ghetie and Schell, 1971; Arnott *et al.*, 1974; Amsterdam *et al.*, 1974, 1975). Such aggregation of agarose into a "network phase" in a gel that may contain up to 100 parts of water for each part of polysaccharide would leave relatively large voids through which large molecules and particles could diffuse. In contrast, any gel of comparable concentration, but prepared by random cross-linking of single-polymer chains would lead to a meshwork in which the pore size would be distributed about a much smaller mean value (cf. Fawcett and Morris, 1966; Arnott *et al.*, 1974 and references therein). The structure of agarose gel beads and the distribution of bound protein within such beads have been recently visualized by electron microscopy (Amsterdam *et al.*, 1975; Lasch *et al.*, 1975).

Agarose gels, despite their superior properties as porous hydrophilic and nonadsorbing supports, suffer from several disadvantages: they cannot be heat-sterilized, since they melt upon heating; agarose gels disintegrate in strong alkaline solutions and in organic solvents; even at neutral pH values the possibility of at least slight solubilization following coupling of protein, due to disruption of interchain hydrogen bonds, cannot be completely eliminated. Furthermore, agarose gels must be stored in wet form since they shrink irreversibly on drying (see, e.g., Schell and Ghetie, 1968, 1971; Ghetie and Schell, 1971; Porath *et al.*, 1971; Porath and Sundberg, 1972; Wilchek, 1974; Wilchek and Miron, 1974b). Most of these deficiencies could be eliminated by cross-linking of preformed agarose-gel beads with epihalohydrins or bisepoxides (Schell and Ghetie, 1968, 1971; Porath *et al.*, 1971; Axén *et al.*, 1971a; Porath and Sundberg, 1972; Sundberg and Porath, 1974) (compare Fig. 26). It was reported that cross-linked agarose beads not only resisted strong alkali and high temperatures, but could also be freeze-dried and then reswollen in water to essentially their original size while retaining their spherical shape and molecular sieving properties; moreover cross-linked agarose beads could be exposed to organic solvents, such as dimethyl sulfoxide, dioxane, or hexane, without apparent damage to the size and shape of the particles (Porath *et al.*, 1971).

One of the first polysaccharide derivatives to be used as support for enzyme immobilization was the acyl hydrazide of O-(carboxymethyl) cellulose (CM-cellulose hydrazide). Treatment of the methyl ester of CM-cellulose with hydrazine gives the corresponding hydrazide (Fig. 24), which can be converted to the azide with nitrous acid and used for the binding of proteins mainly through their amino groups (Fig. 1). This method has found wide application (Micheel and Ewers, 1949; Mitz and Summaria, 1961; Hornby *et al.*, 1966; Wharton *et al.*, 1968a,b; Brümmer *et al.*, 1972). CM-cellulose azide has been converted to the isocyanate by treatment with acid (Brown *et al.*, 1968c).

The O-carboxymethyl derivatives of cellulose and cross-linked dextran have also been used directly for the coupling of proteins by activating the carboxyl groups on these polymers with N-ethyl-5-phenylisoxazolium 3'-sulfonate (Patel *et al.*, 1967, 1969) (see Fig. 4), with carbodiimides (Weliky and Weetall, 1965; Weliky *et al.*, 1969) (Fig. 3), and recently with N-ethoxycarbonyl-2-ethoxy-1,2-dihydroquinoline (Sundaram, 1974) (Fig. 5).

Various arylamino derivatives of polysaccharides have found wide application (see Table IV). Such polymers could be used, after diazotization, for the coupling of proteins mainly through tyrosyl or histidyl residues (Fig. 15); they could also be converted to the corresponding isothiocyanates by treatment with thiophosgene and used for coupling through lysyl ε-amino groups on the enzyme. The following ether derivatives of polysaccharides have been used as supports: O-(4-aminobenzyl) cellulose (p-aminobenzyl cellulose; Fig. 25) and O-(3-aminobenzyl) oxymethyl cellulose (Mitz and Summaria, 1961; Weliky and Weetall, 1965; Lilly *et al.*, 1965; Surinov and Manoilov, 1966; Barker *et al.*, 1968, 1969; Goldstein *et al.*, 1970; Datta *et al.*, 1973), the (3-amino-4-methoxyphenylsulfonyl)ethyl and the 3-(4-aminophenoxy)-2-hydroxypropyl ethers of cellulose (Surinov and Manoilov, 1966; Barker *et al.*, 1968, 1969; Li *et al.*, 1973) and the 3-(4-aminophenoxy)-2-hydroxypropyl ether of cross-linked dextran (Axén and Porath, 1966; Sundberg and Kristiansen, 1972). The O-3-(4-aminophenoxy)-2-hydroxypropyl derivatives of cellulose and dextran have also been converted to the corresponding isothiocyanates (Barker *et al.*, 1968; Axén and Porath, 1966). A different type of diazotizable polysaccharide supports containing or —NH—C_6H_4—C_6H_4—NH_2 or —NH—C_6H_4—CH_2—C_6H_4—NH_2 groups, was prepared by treating oxidized starch with benzidine or with p,p'-diaminodiphenylmethane followed by reduction of the polymeric Schiff's base with sodium borohydride (Goldstein *et al.*, 1970).

Several derivatives of polysaccharides, containing alkylating or arylating functional groups, have been reported. Bromoacetyl cellulose was prepared by treating cellulose with bromoacetic acid and bromoacetyl chloride in dioxan (Jagendorf *et al.*, 1963; Robbins *et al.*, 1967; Shaltiel *et al.*, 1970). The chloro- and iodoacetates of cellulose have also been used (Sato *et al.*, 1971; Maeda and Suzuki, 1972a,b).

Comparative studies on the binding of aminoacylase to the various haloacetate esters of cellulose have suggested that the recovery of enzymic activity in the insoluble derivative was considerably higher in the case of iodoacetates (Sato *et al.*, 1971). The haloacetate esters of cellulose suffer from an intrinsic drawback, the relative instability of the ester bond even at neutral pH values.

The activation of hydroxyl-containing polymers with cyanuric chloride (2,4,6-trichloro-*s*-triazine) and its derivatives was described in Section V,A,2 (see Fig. 7). Supports containing monochloro- or dichloro-*s*-triazinyl groups have been prepared by treating various polysaccharides (cellulose, agarose, cross-linked dextran, etc.) with 2,4,6-trichloro-*s*-triazine or with dichloro-*s*-triazine derivatives such as 2-amino-4,6-dichloro-*s*-triazine or one of the Procion dyes in which one of the chlorine atoms of cyanuric chloride is replaced by a chromophore containing anionic solubilizing groups (Surinov and Manoilov, 1966; Kay and Crook, 1967; Kay *et al.*, 1968; Wilson *et al.*, 1968a,b; Self *et al.*, 1969; Kay and Lilly, 1970; Stasiw *et al.*, 1970, 1972; Lilly, 1971; Wykes *et al.*, 1971; Wilson and Lilly, 1969). In all these derivatized polymers the halotriazinyl functional group is attached to the polysaccharide backbone through stable ether bonds (see Fig. 7).

As can be seen from Table IV, a variety of the polysaccharide halo-triazinyl supports including water-soluble materials derived from linear dextran, DEAE-dextran, and CM-cellulose have been employed. The cyanuric chloride activation method allows considerable flexibility as to the charge characteristics to the functionalized polysaccharide support. The overall charge of the matrix can be predetermined by using as starting material one of the commercially available polysaccharide ion-exchangers, such as CM-cellulose or CM-Sephadex, DEAE-cellulose, DEAE-Sephadex (Kay and Lilly, 1970; Wykes *et al.*, 1971; Stasiw *et al.*, 1970, 1972); the charge on the matrix can also be altered by treating a dichloro-*s*-triazinyl polysaccharide derivative with a charged-low molecular-weight nucleophile, such as 3-diethanolaminopropylamine or 3-dimethylaminopropylamine, before the addition of enzyme to the reaction mixture (Kay *et al.*, 1968).

Dialdehyde derivatives of polysaccharides, prepared by controlled periodate or dimethyl sulfoxide oxidation of vicinal hydroxyl groups have been used as supports (Axén *et al.*, 1971b; Goldstein *et al.*, 1970; Vretblad and Axén, 1973a; Flemming *et al.*, 1973a,b,c; Weakley and Mehltretter, 1973; Van Leemputten and Horisberger, 1974a; Royer *et al.*, 1975). Partially oxidized cellulose was used to bind enzymes, presumably through the formation of Schiff's base linkages with the protein amino groups (Flemming *et al.*, 1973a,b,c; Van Leemputten and Horisberger, 1974a). Dialdehyde starch has been used in a similar fashion (Weakley and Mehltretter, 1973). As mentioned earlier, the use of polymeric aldehydes in this manner poses some problems stemming from the uncertainty regarding the stability of the aldimine bond (see Section V,A,6). The stabilization of the Schiff's base, —CH=N— bonds, by reduction with sodium borohydride under conditions mild enough to avoid inactivation of the enzyme has been explored (Royer *et al.*, 1975).

The carbonyl function of partially oxidized polysaccharides has been utilized to bind enzymes through their amine or carboxyl groups via four-component condensation reactions in the presence of a water-soluble isocyanide and carboxylate or amine, respectively (Vretblad and Axén, 1973a) (see Section V,A,10 and Fig. 17).

Dialdehyde derivatives of polysaccharides could serve in principle as starting materials for the preparation of supports containing a variety of functional groups. This approach is illustrated in the synthesis of arylamino derivatives of starch prepared by allowing oxidized starch to react with a bifunctional amine, such as benzidine or *p,p'*-diaminodiphenylmethane, followed by sodium borohydride reduction (Goldstein *et al.*, 1970), mentioned earlier in this section.

A novel method of immobilization of proteins on polysaccharide and other macromolecular supports involving chelation has been investigated (Barker *et al.*, 1971b; Kennedy and Doyle, 1973). By this method the polymer, e.g., cellulose, is activated by treatment with a transition metal salt, most commonly titanium chloride ($TiCl_4$), and then brought into contact with the protein solution. The mechanism of activation and coupling is still unclear; it is believed that in the activation process water molecules in the octahedral, hexahydrated, titanium (IV) ion are replaced by polysaccharide hydroxyl groups, additional water molecules being replaced in the second stage by amino, hydroxyl, or carboxyl groups on the enzyme (Barker *et al.*, 1971b; Kennedy, 1974b).

One of the most extensively used procedures of immobilization involving polysaccharide supports has been the cyanogen bromide activation method (Porath *et al.*, 1967, 1973; Axén *et al.*, 1967, 1969, 1971a; Axén and Ernback, 1971; Axén and Vretblad, 1971a,b; Kristiansen *et al.*, 1969; Cuatrecasas, 1970, 1972b; Gabel *et al.*, 1970, 1971; Yunginger and Gleich, 1972; Jost *et al.*, 1974; Porath, 1974; Wilchek, 1974; Wilchek and Miron, 1974a; Wilchek *et al.*, 1975; March *et al.*, 1974). This method, carried out at low temperature and leading to conjugates containing relatively high amounts of bound protein (Axén and Ernback, 1971; Yunginger and Gleich, 1972; Lasch *et al.*, 1972; David *et al.*, 1974), has been of particular importance for the chemical modification of agarose gels, which are disrupted even at moderate temperatures, thus precluding the application of many of the methods commonly used for the derivatization of polysaccharides. The chemistry of cyanogen bromide activation of polysaccharides was discussed in some detail in Section V,A,3. It should be recollected that the CNBr activation reaction was considered to proceed via the 2,3-*trans*-imidocarbonate intermediate, leading to the formation of three possible types of bonds with protein amino groups: N-substituted imidocarbonates, N-substituted carbamates, and N-substituted isoureas (Axén and Ernback, 1971) (see Fig. 8). The isourea structure was demonstrated to be sensitive to attack by nucleophiles as well as being positively charged at physiological pH values (Svensson, 1973; Jost *et al.*, 1974; Wilchek *et al.*, 1975; Topper *et al.*, 1975) (see Fig. 9). The cyanogen bromide method, despite its numerous advantages, thus poses some problems regarding the nature and particularly the stability of the bonds formed between the cyclic imidocarbonate intermediate and the amino groups on the protein. These problems, most pertinent in the case of agarose when used as support for affinity-chromatography systems (with low-molecular-weight ligands attached by a single bond), have been partly resolved by coupling the CNBr-activated polysaccharide to multivalent macromolecules such as polylysine or polymeric hydrazides, e.g., polyglutamic acid hydrazide and linear polyacrylic acid hydrazide (Wilchek, 1974; Wilchek and Miron, 1974b). The high stability of these agarose derivatives resulted from the multipoint attachment of the polymers; it should be recollected that the stability of protein conjugates was attributed to the same cause. The polylysine and polyhydrazide-agarose conjugates could serve as starting materials for further substitution with different functional groups in a variety of ways (Wilchek, 1974; Wilchek and Miron, 1974a,b).

A method involving cyclic *trans*-2,3-carbonate derivatives of polysaccharides has been recently described (Barker *et al.*, 1971a,b; Kennedy and Zamir, 1973; Kennedy *et al.*, 1972, 1973; Kennedy and Rosevear, 1974; Kennedy, 1974a). The *trans*-2,3-carbonate group is analogous to the cyclic *trans*-2,3-imidocarbonate structure assumed to be the reactive intermediate in cyanogen bromide activation (see Fig. 8). Nucleophilic attack on the cyclic carbonate by an amino group hence leads to ring opening and the formation of N-substituted carbamate (Kennedy, 1974a,b) [compare Fig. 8, structure (II)]. This method has been applied mainly to the chemical modification of cellulose, since the conditions for the introduction of cyclic carbonate functional groups, treatment of a polysaccharide with ethyl chloroformate in anhydrous organic solvents, are relatively harsh and hence unsuitable for agarose gels (Barker *et al.*, 1971a,b; Kennedy, 1974a). Improved methods for the coupling of proteins to macroporous cellulose *trans*-2,3-carbonate, preswollen with dimethyl sulfoxide, have been reported (Kennedy, 1974a; Kennedy and Rosevear, 1974).

A new and extremely simple two-step procedure for the covalent attachment of proteins to polysaccharide supports, via activation of the carrier, agarose or cross-linked dextran with *p*-benzoquinone, has been recently published (Brandt *et al.*, 1975). In the first step, the polysaccharide was treated with benzoquinone and thereby converted to a reactive intermediate. In the second step, the protein was coupled to the activated support. On the basis of the available data this method seems to offer several advantages: the amounts of bound protein were high, with good retention of enzymic activity (80–100 mg of protein per gram of conjugate, with 70–90% retention of activity, determined with chymotrypsin and ribonuclease bound to agarose beads, Sepharose 4B); the bonds formed between the benzoquinone-activated polysaccharide support and the protein appeared to be stable; moreover, the coupling reaction could be carried out over a rather broad pH range (pH 3–10). The chemistry of the reaction of benzoquinone-activated polysaccharides with a protein is still unclear. Studies on the reaction of benzoquinone with various amino acid derivatives have suggested that amino and sulfhydryl groups as well as the phenolic hydroxyls of tyrosine were involved. The products of the reaction of benzoquinone with those functional groups were thought to be formed by successive addition–oxidation reactions to give 2,5-substituted hydroquinones (Morrison *et al.*, 1969). On the basis of these results, a probable mechanism for the coupling of proteins to polysaccharide supports by means of benzoquinone has been suggested. During the activation process nucleophilic attack of a matrix-hydroxyl group on a benzoquinone molecule would result in a 2-substituted hydroquinone molecule fixed to the matrix. Hydrogen could then be eliminated by reaction with a second molecule of benzoquinone to give a 2-substituted matrix-bound quinone derivative. A nucleophilic group on the protein would then react with the matrix-bound quinone, forming a covalent bond with the support through a 2,5-substituted hydroquinone structure.

3. Polyamides

The family of synthetic polyamides known as nylons are condensation polymers of ω-aminocarboxylic acids or α,ω-dicarboxylic acids or acid dichlorides and α,ω-diamines. The common types of nylon, designated according to the number of carbon atoms in the repeating monomeric unit, are: nylon-6, polycaprolactam, the condensation product of aminocaproic acid (5-aminohexanoic acid); nylon-6,6 prepared by the condensation of 1,6-diaminohexane and adipic dichloride; nylon 6,10 synthesized from sebacic dichloride and 1,6-diaminohexane (see Sorenson and Campbell, 1968; Sweeny and Zimmermann, 1969; Kohan, 1973). Nylons, available commercially in a variety of physical forms, such as membranes, powders, tubes, hollow fibers, and spun-bonded films, are mechanically strong and nonbiodegradable. The nylons of shorter methylene chains, e.g., nylon-6 and nylon-6,6 are relatively hydrophilic, as judged by their moisture regain, and thus offer supports that in principle would be suitable for the immobilization of enzymes. Several derivatives have been described in which a protein was adsorbed on the surface of a nylon structure and then fixed by cross-linking. For example, various antigens were immobilized by adsorption onto nylon filaments followed by cross-linking with a carbodiimide (Edelman *et al.*, 1971); by a similar approach enzymes were adsorbed on nylon floc, pellicular nylon, or nylon membranes and then cross-linked with bisimidates or glutaraldehyde (Reynolds, 1974; Horvath, 1974; Inman and Hornby, 1972). Nylon suffers, however, from an intrinsic drawback when considered as support for covalent immobilization of enzymes—the chemical inertness of the polyamide backbone leaves only the terminal carboxyls and amines as possible reactive functional groups. To increase the binding capacity of nylon, three basic approaches have been adopted: (a) controlled cleavage of amide bonds to increase the number of amine and carboxyl groups; (b) introduction of reactive centers via 0-alkylation of the backbone peptide bonds; and (c) introduction of reactive side chains via N-alkylation of backbone peptide bonds.

Controlled cleavage of peptide bonds was effected in the majority of cases by mild acid hydrolysis (Sundaram and Hornby, 1970; Inman and Hornby, 1972). The carboxyl and amino groups generated by this procedure could then be activated by several methods: carboxyl groups could be coupled to benzidine or hydrazine with a carbodiimide and then activated to the corresponding aryldiazonium salts or acylazide derivatives (Hornby and Filippusson, 1970); coupling of enzymes through amino groups on the nylon surface could be carried out through activation of the latter with glutaraldehyde or cyanuric chloride or through their transformation into isocyanate groups with phosgene (Sundaram and Hornby, 1970; Horvath and Solomon, 1972; Allison *et al.*, 1972; Inman and Hornby, 1972; Bunting and Laidler, 1974). The controlled cleavage of peptide bonds has also been accomplished with nucleophiles such as *N,N*-dimethylaminopropyl-amine in an organic solvent; such treatment blocks the carboxyl group with a cationic *N,N*-dimethylaminopropylamido residue, leaving only the complementary amino group free for further chemical modification (Hornby *et al.*, 1972; Inman and Hornby, 1974) (see Fig. 27).

Methods based on the controlled cleavage of peptide bonds on the surface of a polyamide structure, although successful in attaining their primary objective, i.e., increasing the binding capacity of nylon, are not entirely satisfactory since the mechanical strength of the support might be impaired by such treatments. To avoid this, techniques have been developed for the grafting of side chains containing reactive functional groups, by O-alkylation or N-alkylation of the backbone peptide bonds.

Secondary amides, of the general structure R—CO—NH—R', can be alkylated by powerful alkylating reagents to yield the corresponding imidate salts (Benson and Cairns, 1951). Treatment of nylon with two such reagents, dimethyl sulfate and triethyloxonium tetrafluoroborate, has been used for the O-alkylation of some of the peptide bonds on the

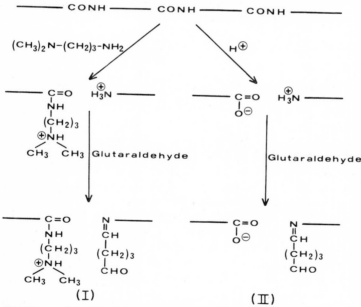

Fig. 27. Chemical modification of nylon by controlled cleavage of peptide bonds followed by treatment with glutaraldehyde (Sundaram and Hornby, 1970; Hornby *et al.*, 1972).

polymer surface, thereby producing the corresponding imidate salts (Campbell *et al.*, 1975; Morris *et al.*, 1975). Imidate salts and their free bases, imido esters, would react with nucleophiles such as amines to yield positively charged, substituted amidines (see Fig. 28). The imidate salt of nylon shown in Fig. 28 represents a versatile intermediate that affords several routes to enzyme immobilization (Hornby *et al.*, 1974; Campbell *et al.*, 1975; Morris *et al.*, 1975; Hornby, 1976): (a) An enzyme could be allowed to react directly through its free amino groups with the imidate salt of nylon [Fig. 28; structure (II)]; it has been noted, however, that coupling in this manner does not give the most active enzyme derivatives (Campbell *et al.*, 1975). (b) The imidate salts of the support could be allowed to react in a nonaqueous medium with an acid-dihydrazide, e.g., adipic acid dihydrazide, to give stable hydrazide-substituted nylon [Fig. 28, structure (I)]. The hydrazide derivatives of nylon could be used for enzyme coupling in a number of ways; the free acid-hydrazide group could be allowed to react with a bifunctional reagent, such as glutaraldehyde or a bisimidate, and the enzyme be coupled in turn to the new functional group introduced via the bifunctional reagent. Alternatively, the free acid

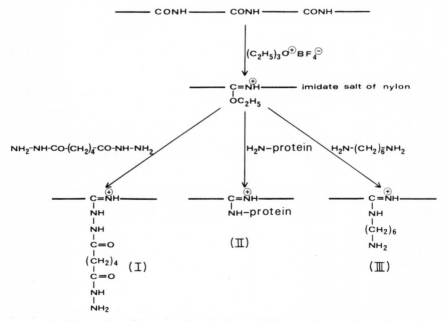

Fig. 28. Chemical modification of nylon by O-alkylation of peptide bonds to form imidate salts (Morris *et al.*, 1975).

hydrazide group could be converted into the acylazide and then reacted with an enzyme. (c) The O-alkylated nylon could be allowed to react with a diamine, such as 1,6-diaminohexane, to yield amine-substituted nylon [Fig. 28, structure (III)], which could be activated by one of the conventional procedures (Hornby *et al.*, 1974; Campbell *et al.*, 1975; Morris *et al.*, 1975; Hornby, 1976). It should be kept in mind that amine-substituted nylons carry a residue of positive charge on their surface in the form of protonated amidine groups (for a survey of amidination reactions and properties of amidines, the reader is referred to Section V,A,5).

N-Alkylation of backbone peptide bonds in nylon has been carried out by a two-step procedure: (a) mild acid hydrolysis to generate COOH . . . NH_2 pairs on the surface of a nylon structure and (b) resealing of the newly formed COOH . . . NH_2 pairs by a four-component condensation reaction involving the neighboring carboxyl and amino groups on the nylon backbone, an aldehyde, and an isocyanide (Goldstein *et al.*, 1974a,b). It should be recalled that in four-component condensation (4CC) reactions between carboxyl, amine, aldehyde, and isocyanide (Fig. 17) the carboxyl and amine components (R^1 and R^2) combine to form an N-substituted amide, the aldehyde and isocyanide components (R^3 and R^4) appearing as the side chain on the amide nitrogen (Ugi, 1962, 1971; Axén *et al.*, 1971b; Vretblad and Axén, 1971, 1973a,b; Goldstein *et al.*, 1974a,b) (see Section V,A,10). Four-component condensation reactions allow in principle for considerable versatility, since by proper choice of aldehyde and isocyanide various functional groups can be introduced on the N-alkyl side chains of the re-formed amide groups of nylon; moreover the modified polyamide backbone carries no residual charged groups resulting from the modification reaction. The procedure most commonly employed, utilizing acetaldehyde or isobutyral as aldehyde component and a bifunctional isocyanide, 1,6-diisocyanohexane, leads to nylon derivatives containing isocyanide (isonitrile) functional groups (Fig. 29) (see Goldstein *et al.*, 1974a,b). As mentioned earlier, enzymes can be coupled to the —N≡C group of polyisonitrile-nylon again by a 4CC reaction, carried out in aqueous buffer at neutral pH, in the presence of a water-soluble aldehyde, e.g., acetaldehyde; here the protein supplies either the amino or carboxyl component, the isocyanide group on the support being steered toward only one type of functional group on the protein by addition of the missing fourth component, carboxylate or amine, in excess, to the reaction medium (Goldstein *et al.*, 1974a,b).

The isocyanide groups on nylon could be easily modified to other

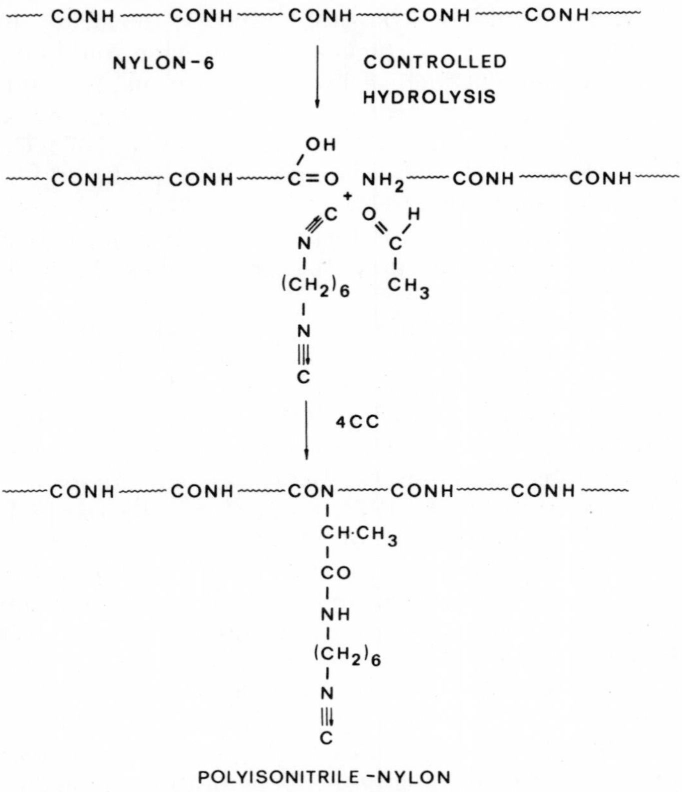

~~ CONH ~~~ CONH ~~~ CONH ~~~ CONH ~~~ CONH ~~~

NYLON-6 | CONTROLLED
|
↓ HYDROLYSIS

OH
/
~~ CONH ~~~ CONH ~~~ C=O   NH$_2$ ~~~ CONH ~~~ CONH ~~~
+
C   O   H
‖   ‖ /
N   C
| |
(CH$_2$)$_6$ CH$_3$
|
N
‖
C

| 4CC
↓

~~ CONH ~~~ CONH ~~~ CON ~~~ CONH ~~~ CONH ~~~
|
CH·CH$_3$
|
CO
|
NH
|
(CH$_2$)$_6$
|
N
‖
C

POLYISONITRILE -NYLON

Fig. 29. Chemical modification of nylon by N-alkylation of peptide bonds. Synthesis of polyisonitrile-nylon (Goldstein *et al.*, 1974a,b) (compare Fig. 17).

functional groups; the conversion of polyisonitrile-nylon to a diazotizable arylamino derivative, polyaminoaryl nylon, by 4CC with a bifunctional aromatic amine, *p,p′*-diaminodiphenylmethane, in the presence of acetaldehyde and acetic acid, is shown in Fig. 30. The conversion of the isocyanide groups of polyisonitrile-nylon to other functional groups should be useful in cases where coupling via 4CC reactions is undesirable owing to the sensitivity of the enzyme to aldehyde.

4. Inorganic Supports

Inorganic carriers offer in principle several advantages: high mechanical strength, resistance to solvents and microbial attack, and regenerability; moreover, inorganic supports do not change in structure

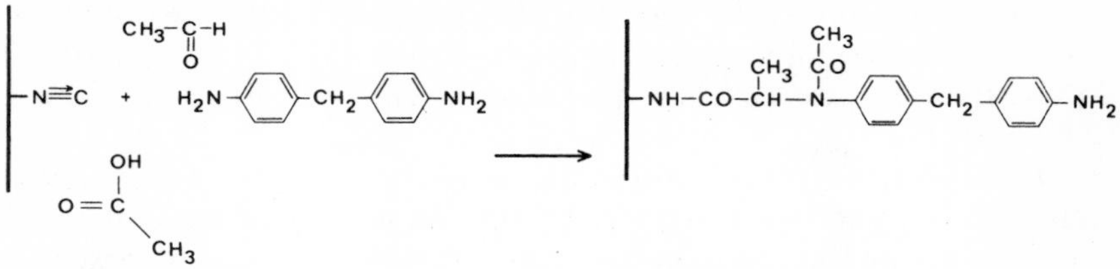

POLYISONITRILE-NYLON   POLYAMINOARYL-NYLON

Fig. 30. Synthesis of polyaminoaryl-nylon (Goldstein *et al.*, 1974a).

over wide ranges of pH, pressure, and temperature. Inorganic materials that have been used as supports include aluminas, bentonites, glass, nickel oxide, silicas, titanias, zirconias, and magnetic iron oxide powders (Weetall, 1969a,b, 1970; Weetall and Hersh, 1970; Weetall and Filbert, 1974; Robinson *et al.*, 1971, 1973; Herring *et al.*, 1972; Traher and Kittrell, 1974; Monsan and Durand, 1971; Van Leemputten and Horisberger, 1974b; Dunnill and Lilly, 1974; Eaton, 1974; Messing, 1974a,b, 1975; Filbert, 1975; Weetall, 1975; Baum and Lynn, 1975) (see also Table IV).

The immobilization of enzymes on controlled-pore glass or ceramics and more recently on magnetic iron oxide has been much more thoroughly investigated and will be dealt with in this section. Other inorganic supports, which have found only limited application, are listed in Table IV.

The technology of controlled-pore glass is based on the finding that certain borosilicate glass compositions can, after heat treatment (500°–700°), be leached to form a porous-glass framework (Nordberg, 1944). During heat treatment, the base glass separates into two intermingled and continuous phases. One phase, rich in boric acid, is soluble in acids; the other phase is rich in silica and insoluble in acids. The boric acid-rich phase may be leached out of the glass leaving a porous structure of very high silica content. Porous diameters for those glasses are in the range of 30 to 60 Å and the pore volume is about 28% of the total sample volume. Glasses of larger pore size can be prepared from the same sodium borosilicate glass compositions, by following the heat and acid-leaching steps with mild caustic treatment to enlarge the pore diameters by removing siliceous residue from pore interiors. By controlling the various physical and chemical parameters, glasses ranging in pore diameter from about 30 to 3000 Å and of rather narrow pore-size distributions are produced (Weetall and Filbert, 1974; Filbert, 1975; Messing, 1974a,b, 1975; Baum and Lynn, 1975). More recently, a technology for the preparation of controlled-pore ceramics has been developed by Messing (1974a,b, 1975). In most of the work on enzyme immobilization, controlled-pore glass, particles of 550 Å pore diameter and a surface area of approximately 40 m²/gm were used (Filbert, 1975).

The methods most commonly employed for the introduction of chemically reactive functional groups on a glass or ceramics surface are shown in Fig. 31. The inorganic support material is first treated with a trialkoxysilane derivative containing an organic functional group, γ-aminopropyltriethoxysilane (Fig. 31A). Coupling to the carrier takes place presumably through the surface silanol or oxide groups, polymerization most likely occurring between adjacent silanes. The product of the reaction, alkylamine glass [Fig. 31, structure (I)] is hence an inorganic carrier bearing organic functional groups (Weetall, 1969a,b, 1970; Weetall and Hersh, 1969; Weetall and Baum, 1970; Weetall and Filbert, 1974; Robinson *et al.*, 1971, 1973; Dixon *et al.*, 1973; Baum and Lynn, 1975). Improved methods whereby repeated silanization leads to a more uniform silane coat and higher functional group content have been recently described (Robinson *et al.*, 1971; Weibel *et al.*, 1973). Alkylamine glass can be converted to the arylamino derivative [Fig. 31B, structure (II)] by treatment with *p*-nitrobenzoylchloride followed by reduction (Weetall, 1969a,b;

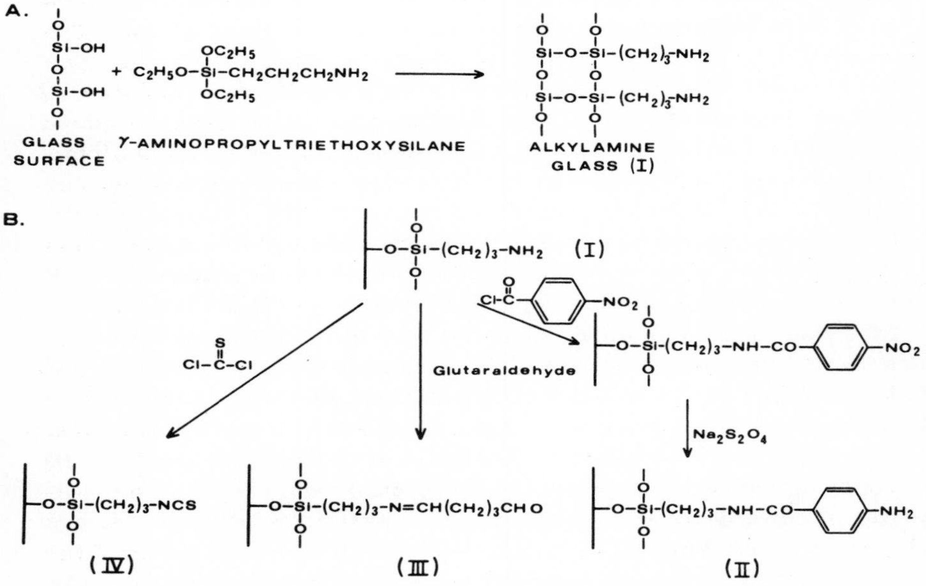

Fig. 31. Chemical modification of glass (Weetall, 1969a,b; Weetall and Filbert, 1974).

Weetall and Hersh, 1969; Weetall and Baum, 1970; Weetall and Filbert, 1974); alkylamine glass can also be converted to the isothiocyanate derivative [Fig. 31B, structure (IV)] with thiophosgene (Weetall, 1970; Weetall and Filbert, 1974), or activated directly with glutaraldehyde [Fig. 31B, structure (III)] (Robinson *et al.*, 1971; Dixon *et al.*, 1973). The covalent attachment of proteins to porous glass directly activated with cyanogen bromide has been recently reported (Weetall and Detar, 1975). Other methods are listed in Table IV (see also Baum and Lynn, 1975; Weetall and Filbert, 1974). The binding capacity of porous glass is about 10–25 mg of protein per gram of support (Table IV). The thermal and storage stabilities of enzymes bound to glass have been claimed to be higher than with most common organic polymeric supports (Weetall, 1970; Baum *et al.*, 1971). Controlled-pore ceramics have been utilized in a similar manner (Eaton, 1974; Messing, 1974a,b, 1975).

Enzymes bound to porous glass usually have shown highly perturbed kinetics due to the rather severe diffusional resistances prevalent within the pores. In the extreme case of a very fast enzyme, glucose oxidase coupled to diazotized arylamine glass, it was shown that only 6% of the active immobilized enzyme could be detected by a rate assay, owing to local depletion of substrate (Weibel and Bright, 1971).

Because of its high surface area, controlled-pore glass exhibits relatively high solubility, considering that it is essentially pure silica. The degree of solubility or corrosion rate of controlled-pore glass was found to be strongly dependent on the pH of the ambient solution as well as on parameters such as temperature, particle size, and surface area and on the dynamic flow characteristics of the system (Eaton, 1974; Filbert, 1975; Messing, 1975). As a general rule controlled-pore glass was most stable under acid conditions, the rate of chemical attrition increasing with increasing pH and rate of flow of liquid in the case of packed-bed column reactors; the increase in corrosion rate with increasing particle size could be related to changes in the void volume within a column, leading to higher liquid throughputs at a given linear velocity (Eaton, 1974; Filbert, 1975). Controlled-pore glass which had been treated with a hydrophilic silane (e.g., γ-aminopropyltriethoxysilane) showed improved durability, its solubility rate in aqueous solution decreasing substantially (Weetall and Filbert, 1974). Chemical attrition of the support could not,however, be completely eliminated by silanization presumably because of solvent penetration through "cracks" in the silane coat (see, e.g., Weetall *et al.*, 1974a,b). Improved dynamic durability at neutral and mildly alkaline pH values could be attained by coating the porous-glass particles with zirconium oxide (Messing, 1974a,b, 1975; Eaton, 1974). Such materials are rather costly, however, when considered for large-scale application (see Weetall *et al.*, 1975). Porous ceramics are being explored at present for their potential as cheap inorganic supports of improved chemical and dynamic durability at elevated pHs levels and temperatures (Eaton, 1974; Messing, 1974a,b, 1975; Weetall *et al.*, 1975; Benoit and Kohler, 1975). These materials appear to be of particular interest in processes aimed at large-volume production of relatively inexpensive end products (Weetall *et al.*, 1975).

A novel type of inorganic support materials based on magnetic iron oxide (magnetite, Fe_2O_3) is being explored (Robinson *et al.*, 1973; Dunnill and Lilly, 1974; Van Leemputten and Horisberger, 1974b). Such supports allow the retrieval of immobilized enzyme from reaction mixtures containing colloidal material or undissolved solids, where the conventional separation methods are not practicable; moreover, they make possible, in principle, the large-scale use of mixtures of immobilized enzymes, of different operational stabilities, since magnet-bound enzyme could be easily separated from an enzyme or enzymes immobilized on conventional nonmagnetic supports.

Several methods utilizing magnetic materials have been described. Magnetic iron oxide powder could be silanized with γ-aminopropyltriethoxysilane and then activated with glutaraldehyde (Robinson *et al.*, 1973); the amino groups on the support were claimed to be converted to the isocyanate by treatment with phosgene (Van Leemputten and Horisberger, 1974b). Alternatively, magnetite powders have been coated with cellulose and then activated with cyanogen bromide (Robinson *et al.*, 1973). The cellulose-coated magnetic particles were prepared by suspending the Fe_2O_3 powder in a solution of cellulose in ammoniacal copper hydroxide and extruding the suspension under pressure into a precipitant (dilute acid). Recently, "magnetic polyacrylamide gel" was prepared by copolymerizing acrylamide in the presence of cross-linker and magnetic iron oxide powder (Dunnill and Lilly, 1974). The polyacrylamide-coated particles were converted to the acyl hydrazide by treatment with hydrazine (compare Fig. 21) and then activated to the corresponding acylazide with nitrous acid.

These studies suggest a more general approach to enzyme immobilization whereby grafting of organic polymers onto inorganic materials may lead to a combination of the superior mechanical properties of an inorganic support with the larger protein binding capacities and the closer contact of an immobilized enzyme with the bulk solution usually encountered with macroporous organic polymeric structures and cross-linked gels. Several recent publications may serve to illustrate this approach (Royer *et al.*, 1973, 1975; Gray *et al.*, 1974; Horvath, 1974; Wan and Horvath, 1975).

Polyacrylic acid was coupled to alkylamine glass by means of a water-soluble carbodiimide; the polycarboxylate support could be used to bind proteins again by carbodiimide activation (Royer *et al.*, 1973). Polyethyleneimine was similarly bound to alkylamine glass via glutaraldehyde; the polyethyleneimine–glass complex could then be modified further by several procedures: (a) nitrobenzoylation of the amino groups on the support, followed by reduction and diazotization; (b) condensation with *p*-nitrobenzaldehyde followed by reduction and diazotization; and (c) alkylation with the bifunctional reagent 4,4'difluoro-3,3'-dinitrophenylsulfone (Royer *et al.*, 1973). Porous aluminas, porous glass, and other silicates could also be coated with polymeric acid anhydrides by allowing the alkylamine derivatives of these materials to react with maleic anhydride–methylvinyl ether copolymers in various organic solvents (Royer *et al.*, 1973). A method utilizing commercially available dextran-coated controlled-pore glass particles was recently employed to bind enzymes via reductive alkylation of protein lysyl ε-amino groups (Royer *et al.*, 1975). The support was oxidized with periodate to generate aldehyde groups on the polysaccharide chains, and then brought into contact with an enzyme solution, in the presence of sodium borohydride; under these conditions the reversible aldimine bonds formed between the aldehyde groups on the support and the amino groups on the protein were reduced to stable —CH₂—NH— groups.

A method for which considerable generality is claimed is based on the diazotization of *m*-diaminobenzene in the presence of a solid support and the exposure of the resulting material to a solution of the enzyme (Gray *et al.*, 1974). When *m*-diaminobenzene in acid solution is treated with NaNO₂ a red-brown precipitate known as Bismarck Brown is formed. This solid presumably a polymer produced by coupling of diazotized and undiazotized *m*-diaminobenzene molecules, contains free diazonium groups. If the diazotization reaction is carried out in the presence of a particulate support material the red solid adsorbs on its surface, forming a uniform and stable coat; upon bringing Bismark Brown-coated particles into contact with a protein solution, covalent binding, presumably through azo bonds, takes place (Gray *et al.*, 1974); possible deleterious effects on the stability of bound enzymes, due to the highly hydrophobic character of the support, cannot be eliminated at this stage.

Recently, the preparation and properties of pellicular immobilized enzymes, which consists of a solid, fluid impervious core supporting a spherical annulus of an enzymically active porous medium, have been investigated (Horvath and Engasser, 1973; Horvath, 1974; Wan and Horvath, 1975). This approach to enzyme immobilization was an outgrowth of developments in high-pressure liquid chromatography, where a variety of shell-structured sorbents and an inert support, using glass beads as the core material, have been employed (Horvath, 1973). Polyanionic pellicular enzyme resins were prepared by coreticulating the protein with a copolymer of maleic anhydride and vinylmethyl ether *in situ* on the surface of glass beads. Carbon, nylon, silica, and alumina in pellicular form were employed in a similar manner for the preparation of immobilized enzymes by cross-linking with glutaraldehyde (Horvath, 1974). The activity of the products compared favorably with that obtained with enzymes immobilized on porous glass. Pellicular enzyme particles were shown to offer significant advantages for use in packed beds because of their mechanical stability and favorable mass-transfer properties.

VI. CONCLUDING REMARKS

The large number of reactive polymers and immobilization procedures allows in principle the fixation of any biologically active macromolecule. The emergence of an "ideal" support of universal applicability cannot, however, be anticipated due to the compositional and structural diversity of proteins. Hence the immobilization of an enzyme aimed at a specific application still requires an empirical, essentially trial and error approach. Here our experience in devising coupling techniques and tailored polymeric materials would be put to use in the search for an optimized solution in terms of mechanical properties and catalytic activity. The available information is, however, inadequate due to the widely varied conditions under which immobilized enzyme preparations have been tested and characterized.

The standardization of the procedures for the characterization of immobilized enzyme derivatives seems in the authors' opinion to be a prerequisite for a fuller exploitation of the potential of immobilized enzymes in industry as well as in pure research.

ACKNOWLEDGMENT

The authors wish to thank Miss Joanne Dobry for her invaluable help in processing the manuscript of this article.

REFERENCES

Alexander, B., Rimon, A., and Katchalski, E. (1965). *Fed. Proc., Fed. Am. Soc. Exp. Biol.* **24**, 804.

Allison, J. P., Davidson, A., Gutierrez-Hartman, A., and Kitto, G. B. (1972). *Biochem. Biophys. Res. Commun.* **47**, 66.

Amsterdam, A., Er-El, Z., and Shaltiel, S. (1974). *Isr. J. Med. Sci.* **10**, 1580.

Amsterdam, A., Er-El, Z., and Shaltiel, S. (1975). *Arch. Biochem. Biophys.* **171**, 673.

Andrews, P. (1962). *Nature (London)* **196**, 36.

Araki, C. (1956). *Bull. Soc. Chem. Jpn.* **29**, 543.

Araki, C., and Arai, K. (1967). *Bull. Chem. Soc. Jpn.* **40**, 1452.

Arnott, S., Fulmer, A., Scott, W. E., Dea, I. C. M., Moorhouse, R., and Rees, D. A. (1974). *J. Mol. Biol.* **90**, 269.

Arthur, J. C., Jr. (1970). *Adv. Macromol. Chem.* **2**, 1.

Avrameas, S. (1969). *Immunochemistry* **6**, 43.

Avrameas, S., and Guilbert, B. (1971). *Biochimie* **53**, 603.

Avrameas, S., and Ternynck, T. (1967). *J. Biol. Chem.* **242**, 1651.

Avrameas, S., and Ternynck, T. (1969). *Immunochemistry* **6**, 53.

Axén, R., and Ernback, S. (1971). *Eur. J. Biochem.* **18**, 351.

Axén, R., and Porath, J. (1966). *Nature (London)* **210**, 367.

Axén, R., and Vretblad, P. (1971a). *Protides Biol. Fluids, Proc. Colloq.* **18**, 383.

Axén, R., and Vretblad, P. (1971b). *Acta Chem. Scand.* **25**, 2711.

Axén, R., Porath, J., and Ernback, S. (1967). *Nature (London)* **214**, 1302.

Axén, R., Heilbronn, E., and Winter, A. (1969). *Biochim. Biophys. Acta* **191**, 478.

Axén, R., Myrin, P., and Janson, J. (1970). *Biopolymers* **9**, 401.

Axén, R., Carlsson, J., Janson, J. C., and Porath, J. (1971a). *Enzymologia* **41**, 359.

Axén, R., Vretblad, P., and Porath, J. (1971b). *Acta Chem. Scand.* **25**, 1129.

Bachler, M. J., Strandberg, G. W., and Smiley, K. L. (1970). *Biotechnol. Bioeng.* **12**, 85.

Bar-Eli, A., and Katchalski, E. (1960). *Nature (London)* **188**, 856.

Bar-Eli, A., and Katchalski, E. (1963). *J. Biol. Chem.* **238**, 1960.

Barker, S. A., and Epton, R. (1970). *Process Biochem.* **5**, 14.

Barker, S. A., and Fleetwood, J. G. (1957). *J. Chem. Soc.* p. 4857.

Barker, S. A., Somers, P. J., and Epton, R. (1968). *Carbohydr. Res.* **8**, 491.

Barker, S. A., Somers, P. J., and Epton, R. (1969). *Carbohydr. Res.* **9**, 257.

Barker, S. A., Somers, P. J., and Epton, R. (1970a). *Carbohydr. Res.* **14**, 323.

Barker, S. A., Somers, P. G., Epton, R., and McLaren, J. V. (1970b). *Carbohydr. Res.* **14**, 287.

Barker, S. A., Doss, S. H., Gray, C. J., Kennedy, J. F., Stacey, M., and Yeo, T. H. (1971a). *Carbohydr. Res.* **20**, 1.

Barker, S. A., Emery, A. N., and Novais, J. M. (1971b). *Process Biochem.* **6**, 11.

Barnett, L. B., and Bull, H. B. (1959). *Biochim. Biophys. Acta* **36**, 244.

Barth, T., and Maskova, H. (1971). *Collect. Czech. Chem. Commun.* **36**, 2398.

Bartling, G. J., Brown, H. D., Forrester, L. J., Koes, M. T., Mather, A. N., and Stasiw, R. O. (1972). *Biotechnol. Bioeng.* **14**, 1039.

Bartling, G. J., Brown, H. D., and Chattopadhyay, S. K. (1974a). *Biotechnol. Bioeng.* **16**, 361.

Bartling, G. J., Chattopadhyay, S. K., Brown, H. D., Barker, C. W., and Vincent, J. K. (1974b). *Biotechnol. Bioeng.* **16**, 1425.

Baum, G. (1975). *Biotechnol. Bioeng.* **17**, 253.

Baum, G., and Lynn, M. (1975). *Process Biochem.* **10**(2), 14.

Baum, G., Ward, F. B., and Weetall, H. H. (1971). *Biochim. Biophys. Acta* **268**, 411.

Bauman, E. K., Goodson, L. H., Guilbault, G. G., and Kramer, D. N. (1965). *Anal. Chem.* **37**, 1378.

Bauman, E. K., Goodson, L. H., and Thomson, J. R. (1967). *Anal. Biochem.* **19**, 587.

Becker, W., and Pfeil, E. (1966). *J. Am. Chem. Soc.* **88**, 4299.

Bengtsson, S., and Philipson, L. (1964). *Biochim. Biophys. Acta* **79**, 399.

Benoit, M. R., and Kohler, J. T. (1975). *Biotechnol. Bioeng.* **17**, 1617.

Benson, R. E., and Cairns, T. L. (1951). *Org. Synth.* **31**, 72.

Bernfeld, P., and Bieber, R. E. (1969). *Arch. Biochem. Biophys.* **131**, 587.

Bernfeld, P., and Wan, J. (1963). *Science* **142**, 678.

Bernfeld, P., Bieber, R. E., and McDonnell, P. C. (1968). *Arch. Biochem. Biophys.* **127**, 779.

Bernfeld, P., Bieber, R. E., and Watson, D. M. (1969). *Biochim. Biophys. Acta* **191**, 587.

Bessmertnaya, L. Y., and Antonov, V. K. (1973). *Khim. Proteoliticheskikh Fermentov, Mater. Vses. Simp.* p. 43. [*Chem. Abstr.* **83**, 3666w (1975).]

Bigelow, C. C. (1967). *J. Theor. Biol.* **16**, 187.

Bikales, M. M., and Segal, L., eds. (1971). "Cellulose and Cellulose Derivatives," 2nd Ed., High Polymers, Vol. 5, Parts IV and V. Wiley (Interscience), New York.

Bishop, W. H., and Richards, F. M. (1968). *J. Mol. Biol.* **33**, 415.

Bishop, W. H., Quiocho, F. A., and Richards, F. M. (1966). *Biochemistry* **5**, 4077.

Bodlaender, P., Feinstein, G., and Shaw, E. (1969). *Biochemistry* **8**, 4941.

Bogulaski, R. C., Smith, R. S., and Mhatre, N. S. (1972). *Curr. Top. Microbiol. Immunol.* **58**, 1.

Boyer, P. D., ed. (1970) "Enzyme Structure and Control," The Enzymes, 3rd Ed., Vol. 1. Academic Press, New York.

Brandenberger, H. (1955). *Angew. Chem.* **67**, 661.

Brandt, J., Andersson, L.-O., and Porath, J. (1975). *Biochim. Biophys. Acta* **386**, 196.

Brocklehurst, K., Carlsson, J., Kierstan, M. P. J., and Crook, E. M. (1973). *Biochem. J.* **133**, 573.

Broun, G., Selegny, E., Avrameas, S., and Thomas, D. (1969). *Biochim. Biophys. Acta* **185**, 260.

Broun, G., Thomas, D., Gellf, G., Domurado, D., Berjonneau, A. M., and Guillon, C. (1973). *Biotechnol. Bioeng.* **15**, 359.

Broun, G. (1976). *Proc. Int. Symp. Anal. Control Immobilized Enzyme Syst., Compiegne, Fr. 1975* (in press).

Brown, E., and Racois, A. (1971a). *Bull. Soc. Chim. Fr.* p. 4357.

Brown, E., and Racois, A. (1971b). *Bull. Soc. Chim. Fr.* No. 12, p. 4351.

Brown, E., and Racois, A. (1972). *Tetrahedron Lett.* p. 5077.

Brown, E., and Racois, A. (1974a). *Tetrahedron* **30**, 675.

Brown, E., and Racois, A. (1974b). *Tetrahedron* **30**, 683.

Brown, E., Racois, A., and Gueniffey, H. (1970). *Tetrahedron Lett.* No. 25, 2139.

Brown, E., Racois, A., and Gueniffey, H. (1971). *Bull. Soc. Chim. Fr.* No. 12, p. 4341.

Brown, H. D., and Hasselberger, F. X. (1971). *In* "Chemistry of the Cell Interface" (H. D. Brown, ed.), Part B, p. 185. Academic Press, New York.

Brown, H. D., Chattopadhyay, S. K., and Patel, A. (1966). *Biochem. Biophys. Res. Commun.* **25**, 304.

Brown, H. D., Patel, A. B., and Chattopadhyay, S. K. (1968a). *J. Chromatogr.* **35**, 103.

Brown, H. D., Patel, A. B., and Chattopadhyay, S. K. (1968b). *J. Biomed. Mater. Res.* **2**, 231.

Brown, H. D., Patel, A. B., Chattopadhyay, S. K., and Pennington, S. N. (1968c). *Enzymologia* **35**, 215.

Brümmer, W., Hennrich, N., Klockow, M., Lang, H., and Orth, H. D. (1972). *Eur. J. Biochem.* **25**, 129.

Bunting, P. S., and Laidler, K. J. (1974). *Biotechnol. Bioeng.* **16**, 119.

Butterworth, P. H. W., Baum, H., and Porter, J. W. (1967). *Arch. Biochem. Biophys.* **118,** 716.

Calam, D. H., and Thomas, H. J. (1972). *Biochim. Biophys. Acta* **276,** 328.

Campbell, D. H., and Weliky, N. (1967). *Methods Immunol. Immunochem.* **1,** 365.

Campbell, D. H., Luescher, E., and Lerman, L. S. (1951). *Proc. Natl. Acad. Sci. U.S.A.* **37,** 575.

Campbell, J., Hornby, W. E., and Morris, D. L. (1975). *Biochim. Biophys. Acta* **384,** 307.

Carlsson, J., Axén, R., Brocklehurst, K., and Crook, E. M. (1974). *Eur. J. Biochem.* **44,** 189.

Carpenter, F. H., and Harriston, K. T. (1972). *J. Biol. Chem.* **247,** 5580.

Cavins, J. F., and Friedman, M. (1967). *Biochemistry* **6,** 3766.

Cebra, J. J., Givol, D., Silman, H. I., and Katchalski, E. (1961). *J. Biol. Chem.* **236,** 1720.

Chang, T. M. S. (1964). *Science* **146,** 524.

Chang, T. M. S. (1966). *Trans. Am. Soc. Artif. Int. Organs* **12,** 13.

Chang, T. M. S. (1967). *Sci. J.* **3,** 62.

Chang, T. M. S. (1969). *Sci. Tools* **16,** 35.

Chang, T. M. S. (1971a). *Biochem. Biophys. Res. Commun.* **44,** 1531.

Chang, T. M. S. (1971b). *Nature (London)* **229,** 117.

Chang, T. M. S. (1972a). *Proc. Eur. Dial. Transplant Assoc.* **9,** 568.

Chang, T. M. S. (1972b). *J. Dent. Res.* **51,** 319.

Chang, T. M. S. (1972c). "Artificial Cells." Thomas, Springfield, Illinois.

Chang, T. M. S. (1974). *In* "Insolubilized Enzymes" (M. Salmona, C. Saronio, and S. Garattini, eds.), p. 15. Raven, New York.

Chang, T. M. S., and Poznansky, M. J. (1968). *J. Biomed. Mater. Res.* **2,** 187.

Chang, T. M. S., McIntosh, T. C., and Mason, S. G. (1966). *Can. J. Physiol. Pharmacol.* **44,** 115.

Chalres, M., Coughlin, R. W., Paruchuri, E. K., Allen, B. R., Hasselberger, and F. X. (1975). *Biotechnol. Bioeng.* **17,** 203.

Cho, I. C., and Swaisgood, H. E. (1972). *Biochim. Biophys. Acta* **258,** 675.

Chukhrai, E. S., and Poltorak, O. M. (1973). *Vestn. Mosk. Univ., Khim.* **14,** 271. [*Chem. Abstr.* **80,** 1139r (1974).]

Cohen, L. A. (1968). *Annu. Rev. Biochem.* **37,** 695.

Cohen, L. A. (1974). *In* "Affinity Techniques: Enzyme Purification," Part B (W. B. Jakoby and M. Wilchek, eds.), Methods in Enzymology, Vol. 34, p. 102. Academic Press, New York.

Constantinides, A., Vieth, W. R., and Fernandes, P. M. (1973). *Mol. Cell. Biochem.* **1,** 127.

Conte, A., and Lehmann, K. (1971). *Hoppe-Seyler's Z. Physiol. Chem.* **352,** 533.

Cordes, E. H., and Jencks, W. P. (1962). *J. Am. Chem. Soc.* **84,** 826.

Coughlan, M. P., and Johnson, D. B. (1973). *Biochim. Biophys. Acta* **302,** 200.

Coulet, P. R., Julliard, J. H., and Gautheron, D. C. (1974). *Biotechnol. Bioeng.* **16,** 1055.

Čoupek, J., Křiváková, M., and Pokorny, S. (1973). *J. Polym. Sci., Polym. Symp. No. 42,* p. 185.

Cremonesi, P., Mazzola, G., Focher, B., and Vechio, G. (1975). *Angew. Makromol. Chem.* **48,** 17.

Crook, E. M. (1970). *Metab. Regul. Enzyme Action, FEBS Proc., 6th Meet.* **19,** 297.

Crook, E. M., Brocklehurst, K., and Wharton, C. W. (1970). *In* "Proteolytic Enzymes" (G. Perlmann and L. Lorand, eds.), Methods in Enzymology, Vol. 19, p. 963. Academic Press, New York.

Cuatrecasas, P. (1970). *J. Biol. Chem.* **245,** 3059.

Cuatrecasas, P. (1972a). *Proc. Natl. Acad. Sci. U.S.A.* **69**, 1277.

Cuatrecasas, P. (1972b). *Adv. Enzymol. Relat. Areas Mol. Biol.* **36**, 29.

Cuatrecasas, P., and Parikh, I. (1972). *Biochemistry* **11**, 2291.

Darlington, W. A., and Keay, L. (1965). *Can. J. Biochem.* **43**, 1171.

Datta, R., Armiger, W., and Ollis, D. F. (1973). *Biotechnol. Bioeng.* **15**, 993.

David, G. S., Chino, T. H., and Reisfeld, R. A. (1974). *FEBS Lett.* **43**, 264.

Davies, G. E., and Stark, G. R. (1970). *Proc. Natl. Acad. Sci. U.S.A.* **66**, 651.

Dea, I. C. M., McKinnon, A. A., and Rees, D. A. (1972). *J. Mol. Biol.* **68**, 153.

Degani, Y., and Miron, T. (1970). *Biochim. Biophys. Acta* **212**, 362.

Denti, E. (1974). *Biomater., Med. Devices, Artif. Organs* **2**, 293.

Desnuelle, P., Neurath, H., and Ottesen, M., eds. (1970). "Structure-Function Relationships of Protelytic Enzymes." Munksgaard, Copenhagen; Academic Press, New York.

Determann, H. (1968). "Gel Chromatography." Springer-Verlag, Berlin and New York.

Dinelli, D. (1972). *Process Biochem.* **7**, 9.

Dinelli, D., Marconi, W., and Morisi, F. (1975). *In* "Immobilized Enzymes, Antigens, Antibodies and Peptides. Preparation and Characterization" (H. H. Weetall, ed.), p. 171. Dekker, New York.

Dixon, J. E., Stolzenbach, F. E., Berenson, J. A., and Kaplan, N. O. (1973). *Biochem. Biophys. Res. Commun.* **52**, 905.

Dobo, J. (1970). *Acta Chim. Acad. Sci. Hung.* **63**, 453.

Duckworth, M., and Yaphe, W. (1971). *Carbohydr. Res.* **16**, 189.

Duckworth, M., Hong, K. C., and Yaphe, W. (1971). *Carbohydr. Res.* **18**, 1.

Dunnill, P., and Lilly, M. D. (1974). *Biotechnol. Bioeng.* **16**, 987.

Dutton, A., Adams, M., and Singer, S. J. (1966). *Biochem. Biophys. Res. Commun.* **23**, 730.

Eaton, D. L. (1974). *In* "Immobilized Biochemicals and Affinity Chromatography" (R. B. Dunlop, ed.), p. 241. Plenum, New York.

Edelman, G. M., Rutishauer, U., and Millette, C. F. (1971). *Proc. Natl. Acad. Sci. U.S.A.* **68**, 2153.

Ellman, G. L. (1959). *Arch. Biochem. Biophys.* **82**, 70.

Engasser, J. M., and Horvath, C. (1975). *Biochem. J.* **145**, 431.

Engvall, E., and Perlmann, P. (1971). *J. Immunol.* **109**, 129.

Epstein, C. J., and Anfinsen, C. B. (1962). *J. Biol. Chem.* **237**, 2175.

Epton, R. (1973). *In* "Reactions on Polymers" (J. A. Moore, ed.), p. 286. Reidel Publ., Dordrecht, Netherlands.

Epton, R., McLaren, J. V., and Thomas, T. H. (1972). *Carbohydr. Res.* **22**, 301.

Er-El, Z., Zaidenzaig, Y., and Shaltiel, S. (1972). *Biochem. Biophys. Res. Commun.* **49**, 383.

Erlanger, B. G., Isambert, M. F., and Michelson, A. M. (1970). *Biochem. Biophys. Res. Commun.* **40**, 70.

Fasold, H. (1964). *Biochem. Z.* **339**, 482.

Fasold, H. (1965). *Biochem. Z.* **342**, 288.

Fasold, H., Groschel-Steward, U., and Turba, F. (1963). *Biochem. Z.* **337**, 425.

Fasold, H., Klappenberger, J., Mayer, C., and Remold, H. (1971). *Angew. Chem., Int. Ed. Engl.* **10**, 795.

Fasold, H., Mayer, C., and Steinkopff, G. (1973). *Eur. J. Biochem.* **32**, 63.

Fawcett, J. S., and Morris, C. J. O. R. (1966). *Sep. Sci.* **1**, 9.

Feinstein, G., Bodlaender, P., and Shaw, E. (1969). *Biochemistry* **8**, 4949.

Filbert, A. M. (1975). *In* "Immobilized Enzymes for Industrial Reactors" (R. A. Messing, ed.), p. 39. Academic Press, New York.

Filippusson, H., and Hornby, W. E. (1970). *Biochem. J.* **120**, 215.

Filippusson, H., Hornby, W. E., and McDonald, A. (1972). *FEBS Lett.* **20**, 291.

Fink, D. J. (1973). *Diss. Abstr. Int. B* **34**, 1489.

Flemming, C., Gabert, A., and Roth, P. (1973a). *Acta Biol. Med. Ger.* **30**, 177.

Flemming, C., Gabert, A., and Roth, P. (1973b). *Acta Biol. Med. Ger.* **31**, 365.

Flemming, C., Gabert, A., Roth, P., and Wand, H. (1973c). *Acta Biol. Med. Ger.* **31**, 449.

Fletcher, G. L., and Okada, S. (1955). *Nature (London)* **176**, 882.

Fletcher, G. L., and Okada, S. (1959). *Radiat. Res.* **11**, 291.

Flodin, P. (1961). *J. Chromatogr.* **5**, 103.

Flodin, P., and Porath, J. (1961). *In* "Chromatography" (E. Heftmann, ed.), p. 328. Reinhold, New York.

Freedberg, W. B., and Hardman, J. K. (1971). *J. Biol. Chem.* **246**, 1439.

Fritz, H., Trautschold, I., Haendle, H., and Werle, E. (1968a). *Ann. N.Y. Acad. Sci.* **146**, 400.

Fritz, H., Hochstrasser, K., Werle, E., Brey, B., and Gebhardt, M. (1968b). *Z. Anal. Chem.* **243**, 452.

Fritz, H., Gebhardt, M., Fink, E., Schramm, W., and Werle, E. (1969a). *Hoppe-Seyler's Z. Physiol. Chem.* **350**, 129.

Fritz, H., Brey, B., Schmal, A., and Werle, E. (1969b). *Hoppe-Seyler's Z. Physiol. Chem.* **350**, 617.

Gabel, D., Vretblad, P., Axén, R., and Porath, J. (1970). *Biochim. Biophys. Acta* **214**, 561.

Gabel, D., Steinberg, I. Z., and Katchalski, E. (1971). *Biochemistry* **10**, 4661.

Garnett, J. L., Kenyon, R. S., and Liddy, M. J. (1974). *J. Chem. Soc., Chem. Commun.* p. 735.

Gelotte, B. (1960). *J. Chromatogr.* **3**, 330.

Gelotte, B., and Porath, J. (1967). *In* "Chromatography" (E. Heftmann, ed.), 2nd ed., Reinhold, New York.

Ghetie, V., and Schell, H. D. (1971). *Experientia* **27**, 1384.

Giovenco, S., Morisi, F., and Pansolli, P. (1973). *FEBS Lett.* **36**, 57.

Glassmeyer, C. K., and Ogle, J. D. (1971). *Biochemistry* **10**, 786.

Goldfeld, M. G., Vorobeva, E. S., and Poltorak, O. M. (1966). *Zh. Fiz. Khim.* **40**, 2594. [*Chem. Abstr.* **66**, 16713h (1967).]

Goldman, R., and Lenhoff, H. N. (1971). *Biochim. Biophys. Acta* **242**, 514.

Goldman, R., Silman, H. I., Caplan, S. R., Kedem, O., and Katchalski, E. (1965). *Science* **150**, 758.

Goldman, R., Kedem, O., Silman, H. I., Caplan, S. R., and Katchalski, E. (1968). *Biochemistry* **7**, 486.

Goldman, R., Kedem, O., and Katchalski, E. (1971a). *Biochemistry* **10**, 165.

Goldman, R., Goldstein, L., and Katchalski, E. (1971b). *In* "Biochemical Aspects of Reactions on Solid Supports" (G. R. Stark, ed.), p. 1. Academic Press, New York.

Goldstein, L. (1968). *In* "Fermentation Advances" (D. Perlman, ed.), p. 391. Academic Press, New York.

Goldstein, L. (1970). *In* "Proteolytic Enzymes" (G. Perlmann and L. Lorand, eds.), Methods in Enzymology, Vol. 19, p. 935. Academic Press, New York.

Goldstein, L. (1972a). *Anal. Biochem.* **50**, 40.

Goldstein, L. (1972b). *Biochemistry* **11**, 4072.

Goldstein, L. (1973a). *Biochim. Biophys. Acta* **315**, 1.

Goldstein, L. (1973b). *Biochim. Biophys. Acta* **327**, 132.

Goldstein, L. (1976). *In* "Immobilized Enzymes" (K. Mosbach, ed.), Methods in Enzymology, Vol. 44. Academic Press, New York.

Goldstein, L., and Katchalski, E. (1968). *Z. Anal. Chem.* **243**, 375.

Goldstein, L., Levin, Y., and Katchalski, E. (1964). *Biochemistry* **3**, 1913.

Goldstein, L., Pecht, M., Blumberg, S., Atlas, D., and Levin, Y. (1970). *Biochemistry* **9**, 2322.

Goldstein, L., Lifshitz, A., and Sokolovsky, M. (1971). *Int. J. Biochem.* **2**, 448.

Goldstein, L., Freeman, A., and Sokolovsky, M. (1974a). *Biochem. J.* **143**, 497.

Goldstein, L., Freeman, A., and Sokolovsky, M. (1974b). *In* "Enzyme Engineering" (E. K. Pye and L. B. Wingard, eds.), Vol. 2, p. 97. Plenum, New York.

Gorbunov, M. J. (1971). *Biochemistry* **10**, 250.

Gorbunov, M. J. (1970). *Arch. Biochem. Biophys.* **138**, 684.

Gorecki, M., Wilchek, M., and Patchornik, A. (1971). *Biochim. Biophys. Acta* **229**, 590.

Gounaris, A. D., and Perlmann, G. E. (1967). *J. Biol. Chem.* **242**, 2739.

Gray, C. J., Livingstone, C. M., Jones, C. M., and Barker, S. A. (1974). *Biochim. Biophys. Acta* **341**, 457.

Green, M. L., and Crutchfield, G. (1969). *Biochem. J.* **115**, 183.

Gregoriadis, G. (1974). *In* "Insolubilized Enzymes" (M. Salmona, C. Saronio, and S. Garattini, eds.), p. 165. Raven, New York.

Gregoriadis, G., and Buckland, R. A. (1973). *Nature (London)* **244**, 170.

Gregoriadis, G., and Ryman, B. E. (1972a). *Biochem. J.* **129**, 123.

Gregoriadis, G., and Ryman, B. E. (1972b). *Eur. J. Biochem.* **24**, 485.

Gregoriadis, G., Leathwood, P. D., and Ryman, B. E. (1971). *FEBS Lett.* **14**. 95.

Grove, M. J., Strandberg, G. W., and Smiley, K. L. (1971). *Biotechnol. Bioeng.* **13**, 709.

Grubhofer, N., and Schleith, L. (1953). *Naturwissenschaften* **40**, 508.

Grubhofer, N., and Schleith, L. (1954). *Hoppe-Seyler's Z. Physiol. Chem.* **297**, 108.

Guilbault, G. G., and Das, J. (1970). *Anal. Biochem.* **33**, 341.

Guilbault, G. G., and Kramer, D. N. (1965). *Anal. Chem.* **37**, 1675.

Guilbault, G. G., and Montalvo, J. G. (1970). *J. Am. Chem. Soc.* **92**, 2533.

Gutcho, J. J. (1974). "Immobilized Enzymes. Preparations and Engineering Techniques." Noyes Data Corp., Park Ridge, New Jersey.

Gyenes, L., and Sehon, A. H. (1960). *Can. J. Biochem. Physiol.* **38**, 1235.

Habeeb, A. F. S. A. (1967). *Arch. Biochem. Biophys.* **119**, 264.

Habeeb, A. F. S. A., and Hiramoto, R. (1968). *Arch. Biochem. Biophys.* **126**, 16.

Hand, E., and Jencks, W. P. (1962). *J. Am. Chem. Soc.* **84**, 3505.

Handschumacher, R. E., and Gaumond, C. (1972). *Mol. Pharmacol.* **8**, 59.

Hardy, P. M., Nicholls, A. C., and Rydon, H. N. (1969). *J. Chem. Soc., Chem. Commun.* p. 565.

Hartley, B. S. (1970). *Phil. Trans. R. Soc. London, Ser. B* **257**, 77.

Hartman, F. C., and Wold, F. (1966). *J. Am. Chem. Soc.* **88**, 3890.

Hartman, F. C., and Wold, F. (1967). *Biochemistry* **6**, 2439.

Hasselberger, F. X., Allen, B., Paruchuri, E. K., Charles, M., and Coughlin, R. W. (1974). *Biochem. Biophys. Res. Commun.* **57**, 1054.

Haynes, R., and Walsh, K. A. (1969). *Biochem. Biophys. Res. Commun.* **36**, 235.

Herring, W. M., Laurence, R. L., and Kittrell, J. R. (1972). *Biotechnol. Bioeng.* **14**, 975.

Hicks, G. P., and Updike, S. J. (1966). *Anal. Chem.* **38**, 726.

Higgins, H. G., and Harrington, K. J. (1959). *Arch. Biochem. Biophys.* **85**, 409.

Hirs, C. H. W., ed. (1967). "Enzyme Structure," Methods in Enzymology, Vol. 11. Academic Press, New York.

Hirs, C. H. W., and Timasheff, S. N., eds. (1972). "Enzyme Structure," Part B, Methods in Enzymology, Vol. 25. Academic Press, New York.

Hjerten, S. (1961). *Biochim. Biophys. Acta* **53**, 514.

Hjerten, S. (1962a). *Arch. Biochem. Biophys.* **99**, 466.

Hjerten, S. (1962b). *Biochim. Biophys. Acta* **62**, 445.
Hjerten, S. (1962c). *Biochim. Biophys. Acta* Suppl. 1, p. 147.
Hjerten, S. (1963). *J. Chromatogr.* **12**, 510.
Hjerten, S. (1964). *Biochim. Biophys. Acta* **79**, 393.
Hjerten, S., and Mosbach, R. (1962). *Anal. Biochem.* **3**, 109.
Hjerten, S., and Porath, J. (1962). *Methods Biochem. Anal.* **9**, 193.
Hoare, D. G., and Koshland, D. E. (1966). *J. Am. Chem. Soc.* **88**, 2057.
Hoare, D. G., and Koshland, D. E. (1967). *J. Biol. Chem.* **242**, 2447.
Hofstee, B. H. J. (1973). *Anal. Biochem.* **52**, 430.
Hofstee, B. H. J., and Otillio, N. F. (1973). *Biochem. Biohys. Res. Commun.* **53**, 1137.
Holmgren, J., and Svennerholm, A. M. (1973). *Infect. Immun.* **7**, 759.
Hopwood, D. (1972). *Histochem. J.* **4**, 267.
Hornby, W. E. (1976). *In* "Enzyme Engineering" (H. H. Weetall and K. E. Pye, eds.),
  Vol. 3. Plenum, New York. In press.
Hornby, W. E., and Filippusson, H. (1970). *Biochim. Biophys. Acta* **220**, 343.
Hornby, W. E., Lilly, M. D., and Crook, E. M. (1966). *Biochem. J.* **98**, 420.
Hornby, W. E., Inman, D. J., and McDonald, A. (1972). *FEBS Lett.* **23**, 114.
Hornby, W. E., Campbell, J., Inman, D. J., and Morris, D. L. (1974). *In* "Enzyme
  Engineering" (E. K. Pye and L. B. Wingard, eds.), Vol. 2, p. 401. Plenum, New York.
Horvath, C. (1973). *Methods Biochem. Anal.* **21**, 79.
Horvath, C. (1974). *Biochim. Biophys. Acta* **358**, 164.
Horvath, C., and Engasser, J. M. (1973). *Ind. Eng. Chem., Fund.* **12**, 229.
Horvath, C., and Solomon, B. A. (1972). *Biotechnol. Bioeng.* **14**, 885.
Howard, A. N., and Wild, F. (1957). *Biochem. J.* **65**, 651.
Hunter, M. J., and Ludwig, M. L. (1962). *J. Am. Chem. Soc.* **84**, 3491.
Hunter, M. J., and Ludwig, M. L. (1972). *In* "Enzyme Structure," Part B (D. H. W. Hirs
  and S. N. Timasheff, eds.). Methods in Enzymology, Vol. 25, p. 585. Academic
  Press, New York.
Inman, D. J., and Hornby, W. E. (1972). *Biochem. J.* **129**, 255.
Inman, D. J., and Hornby, W. E. (1974). *Biochem. J.* **137**, 25.
Inman, J. K. (1974). *In* "Affinity Techniques: Enzyme Purification," Part B (W. B. Jakoby
  and M. Wilchek, eds.). Methods in Enzymology, Vol. 34, p. 30. Academic Press, New
  York.
Inman, J. K., and Dintzis, H. M. (1969). *Biochemistry* **8**, 4074.
Izumi, K. (1971). *Carbohydr. Res.* **17**, 227.
Jagendorf, A. T., Patchornik, A., and Sela, M. (1963). *Biochim. Biophys. Acta* **78**, 516.
Jahn, E. C., ed. (1971). *J. Polym. Sci., Polym. Symp., Part C* No. 36.
James, L. K., and Augenstein, L. (1966). *Adv. Enzymol. Relat. Areas Mol. Biol.* **28**, 1.
Jansen, E. F., and Olson, A. C. (1969). *Arch. Biochem. Biophys.* **129**, 221.
Jansen, E. F., Tomimatsu, Y., and Olson, A. C. (1971). *Arch. Biochem. Biophys.* **144**, 394.
Jaworek, D. (1974). *In* "Insolubilized Enzymes" (M. Salmona, C. Saronio, and
  S. Garattini, eds.), p. 65. Raven, New York.
Jencks, W. P. (1964). *Prog. Phys. Org. Chem.* **2**, 63.
Jencks, W. P. (1969). "Catalysis in Chemistry and Enzymology." McGraw-Hill, New York.
Johansson, A. C., and Mosbach, K. (1974a). *Biochim. Biophys. Acta* **370**, 339.
Johansson, A. C., and Mosbach, K. (1974b). *Biochim. Biophys. Acta* **370**, 348.
Johnson, P., and Whateley, T. L. (1971). *J. Colloid Interface Sci.* **37**, 557.
Jones, D. M. (1964). *Adv. Carbohydr. Chem.* **19**, 219.
Josephs, R., Eisenberg, H., and Reisler, E. (1973). *Biochemistry* **12**, 4060.
Jost, R., Miron, T., and Wilchek, M. (1974). *Biochim. Biophys. Acta* **362**, 75.

Joustra, M. K. (1969). *In* "Modern Separation Methods of Macromolecules and Particles" (T. Gerritsen, ed.), p. 183. Wiley (Interscience), New York.
Julliard, J. H., Godinot, C., and Gautheron, D. C. (1971). *FEBS Lett.* **14**, 185.
Kågedal, L., and Åkerström, S. (1970). *Acta Chem. Scand.* **24**, 1601.
Karube, I., and Suzuki, S. (1972a). *Biochem. Biophys. Res. Commun.* **47**, 51.
Karube, I., and Suzuki, S. (1972b). *Biochem. Biophys. Res. Commun.* **48**, 320.
Kasche, V., and Bergwall, M. (1974). *In* "Insolubilized Enzymes" (M. Salmona, C. Saronio, and S. Garattini, eds.), p. 77. Raven, New York.
Katchalski, E., Silman, I. H., and Goldman, R. (1971). *Adv. Enzymol. Relat. Areas Mol. Biol.* **34**, 445.
Kay, G., and Crook, E. M. (1967). *Nature (London)* **216**, 514.
Kay, G., and Lilly, M. D. (1970). *Biochim. Biophys. Acta* **198**, 276.
Kay, G., Lilly, M. D., Sharp, A. K., and Wilson, R. J. H. (1968). *Nature (London)* **217**, 641.
Kennedy, J. F. (1974a). *In* "Insolubilized Enzymes" (M. Salmona, C. Saronio and S. Garattini, eds.), p. 29. Raven, New York.
Kennedy, J. F. (1974b). *Adv. Carbohydr. Chem. Biochem.* **29**, 305.
Kennedy, J. F., and Doyle, C. E. (1973). *Carbohydr. Res.* **28**, 89.
Kennedy, J. F., and Epton, J. (1973). *Carbohydr. Res.* **27**, 11.
Kennedy, J. F., and Rosevear, A. (1974). *J. Chem. Soc., Perkin Trans.* 1 **7**, 757.
Kennedy, J. F., and Zamir, A. (1973). *Carbohydr. Res.* **29**, 497.
Kennedy, J. F., Barker, S. A., and Rosevear, A. (1972). *J. Chem. Soc., Perkin Trans.* 1 2568.
Kennedy, J. F., Barker, S. A., and Rosevear, A. (1973). *J. Chem. Soc., Perkin. Trans.* 1 2293.
Khorana, H. G. (1953). *Chem. Rev.* **53**, 145.
Khoricova, E. S., Chukhrai, E. S., and Poltorak, O. M. (1973). *Vestn. Mosk. Univ., Khim.* **14**, 474. [*Chem. Abstr.* **80**, 11696j (1974).]
Kitajima, M., Miyano, S., and Kondo, A. (1969). *Kogyo Kagaku Zasshi* **72**, 493. [*Chem. Abstr.* **70**, 118067a (1969).]
Knights, R. J., and Light, A. (1974). *Arch. Biochem. Biophys.* **160**, 377.
Kobayashi, T., and Laidler, K. J. (1973). *Biochim. Biophys. Acta* **302**, 1.
Koelsch, R., Lasch, J., and Hanson, H. (1970). *Acta Biol. Med. Ger.* **24**, 833.
Kohan, M. I., ed. (1973). "Nylon Plastics." Wiley, New York.
Korn, A. H., Feairheller, S. H., and Filachione, E. M. (1972). *J. Mol. Biol.* **65**, 525.
Korus, R., and O'Driscoll, K. F. (1974). *Can. J. Chem. Eng.* **52**, 775.
Krämer, D. M., Lehmann, K., Plainer, H., Reisner, W., and Sprossler, B. G. (1974). *J. Polym. Sci., Polym. Symp.* No. 47, p. 77.
Krieger, M., Kay, L. M., and Stroud, R. M. (1974). *J. Mol. Biol.* **83**, 209.
Kristiansen, T., Sundberg, L., and Porath, J. (1969). *Biochim. Biophys. Acta* **184**, 93.
Krug, F., Desbuquois, B., and Cuatrecasas, P. (1971). *Nature (London), New Biol.* **234**, 268.
Kurzer, F., and Douraghi-Zadeh, K. (1967). *Chem. Rev.* **67**, 107.
Lasch, J., Iwig, M., and Hanson, H. (1972). *Eur. J. Biochem.* **27**, 431.
Lasch, J., Iwig, M., Koelsch, R., David, H., and Marx, I. (1975). *Eur. J. Biochem.* **60**, 163.
Laurent, T. C. (1967). *Biochim. Biophys. Acta* **136**, 199.
Ledingham, W. M., and Ferreira, M. S. S. (1973). *Carbohydr. Res.* **30**, 196.
Ledingham, W. M., and Hornby, W. E. (1969). *FEBS Lett.* **5**, 118.
Lerman, L. S. (1953). *Proc. Natl. Acad. Sci. U.S.A.* **39**, 232.
Levin, Y., Pecht, M., Goldstein, L., and Katchalski, E. (1964). *Biochemistry* **3**, 1905.
Li, K. H., Chang, S. K., Sun, W. R., Ku, S. F., Yang, L. W., Li, S. F., and Yang, K. Y. (1973). *Wei Sheng Wu Hseuh Pao* **13**, 31. [*Chem. Abstr.* **80**, 24312f (1974).]

Liddy, M. J., Garnett, J. L., and Kenyon, R. S. (1975). *J. Polym. Sci., Polym. Symp., Part C* **49**, 109.
Lilly, M. D. (1971). *Biotechnol. Bioeng.* **13**, 589.
Lilly, M. D., Money, C., Hornby, W. E., and Crook, E. M. (1965). *Biochem. J.* **95**, 45p.
Lin, L. J., and Foster, J. F. (1975). *Anal. Biochem.* **63**, 485.
Lindsey, A. S. (1969). *J. Macromol. Sci., Rev. Macromol. Chem.* **3**, 1.
Line, W. F., Kwong, A., and Weetall, H. H. (1971). *Biochim. Biophys. Acta* **242**, 194.
Liu, C. C., Lahoda, E. J., Galasco, R. T., and Wingard, L. B., Jr. (1975). *Biotechnol. Bioeng.* **17**, 1695.
Ludwig, M. L., and Byrne, R. (1962). *J. Am. Chem. Soc.* **84**, 4160.
McLaren, A. D. (1960). *Enzymologia* **21**, 356.
McLaren, A. D., and Estermann, E. F. (1956). *Arch. Biochem. Biophys.* **61**, 158.
McLaren, A. D., and Estermann, E. F. (1957). *Arch. Biochem. Biophys.* **68**, 157.
McLaren, A. D., and Packer, L. (1970). *Adv. Enzymol. Relat. Areas Mol. Biol.* **33**, 245.
Maeda, H. (1975). *Biotechnol. Bioeng.* **17**, 1571.
Maeda, H., and Suzuki, H. (1972a). *Agric. Biol. Chem.* **36**, 1581.
Maeda, H., and Suzuki, H. (1972b). *Agric. Biol. Chem.* **36**, 1839.
Maeda, H., Yamauchi, A., and Suzuki, H. (1973a). *Biochim. Biophys. Acta* **315**, 18.
Maeda, H., Suzuki, H., and Yamauchi, A. (1973b). *Biotechnol. Bioeng.* **15**, 607.
Maeda, H., Suzuki, H., and Sakimae, A. (1973c). *Biotechnol. Bioeng.* **15**, 403.
Maeda, H., Suzuki, H., Yamauchi, A., and Sakimae, A. (1974). *Biotechnol. Bioeng.* **16**, 1517.
Maeda, H., Suzuki, H., Yamauchi, A., and Sakimae, A. (1975). *Biotechnol. Bioeng.* **17**, 119.
Manecke, G. (1962). *Pure Appl. Chem.* **4**, 507.
Manecke, G. (1964). *Naturwissenshcaften* **51**, 25.
Manecke, G. (1975). *Proc. Int. Symp. Macromol., Rio de Janeiro, 1974* p. 397.
Manecke, G., and Förster, H. J. (1966). *Makromol. Chem.* **91**, 136.
Manecke, G., and Gillert, K. E. (1955). *Naturwissenschaften* **42**, 212.
Manecke, G., and Günzel, G. (1962). *Makromol. Chem.* **51**, 199.
Manecke, G., and Günzel, G. (1967a). *Naturwissenschaften* **54**, 531.
Manecke, G., and Günzel, G. (1967b). *Naturwissenschaften* **54**, 647.
Manecke, G., and Singer, S. (1960a). *Makromol. Chem.* **37**, 119.
Manecke, G., and Singer, S. (1960b). *Makromol. Chem.* **39**, 13.
Manecke, G., and Vogt, H. G. (1976). *Makromol. Chem.* **177**, 725.
Manecke, G., Singer, S., and Gillert, K. E. (1958). *Naturwissenschaften* **45**, 440.
Manecke, G., Singer, S., and Gillert, K. E. (1960). *Naturwissenschaften* **47**, 63.
Manecke, G., Günzel, G., and Förster, H. J. (1970). *J. Poly. Sci., Polym. Symp., Part C* **30**, 607.
March, S. C., Parikh, I., and Cuatrecassas, P. (1974). *Anal. Biochem.* **60**, 149.
Marconi, W., Gulinelli, S., and Morisi, F. (1974a). *In* "Insolubilized Enzymes" (M. Salmona, C. Saronio, and S. Garattini, eds.), p. 51. Raven, New York.
Marconi, W., Gulinelli, S., and Morisi, F. (1974b). *Biotechnol. Bioeng.* **16**, 501.
Marfrey, P. S., and King, M. V. (1965). *Biochim. Biophys. Acta* **105**, 178.
Martensson, K., and Mosbach, K. (1972). *Biotechnol. Bioeng.* **14**, 715.
Martinek, K., Goldmacher, V. S., Klibanov, A. M., and Berezin, I. V. (1975). *FEBS Lett.* **51**, 152.
Mason, R. D., and Weetall, H. H. (1972). *Biotechnol. Bioeng.* **14**, 637.
Matthews, B. W., Sigler, P. B., Henderson, R., and Blow, D. M. (1967). *Nature (London)* **214**, 652.
May, S. W., and Li, N. N. (1972). *Biochem. Biophys. Res. Commun.* **47**, 1179.

Means, G. E., and Feeney, R. E. (1971). "Chemical Modification of Proteins." Holden-Day, San Francisco, California.

Meighen, E. A., and Schachman, H. K. (1970). *Biochemistry* 9, 1163.

Melrose, G. J. H. (1971). *Rev. Pure Appl. Chem.* 21, 83.

Messing, R. A. (1969). *J. Am. Chem. Soc.* 91, 2370.

Messing, R. A. (1970a). *Enzymologia* 39, 12.

Messing, R. A. (1970b). *Enzymologia* 38, 39.

Messing, R. A. (1970c). *Enzymologia* 38, 370.

Messing R. A. (1974a). *In* "Immobilized Enzymes in Food and Microbial Processes" (A. C. Olson and C. L. Cooney, eds.), p. 149. Plenum, New York.

Messing, R. A. (1974b). *Process Biochem.* 9(9), 26.

Messing, R. A., ed. (1975). "Immobilized Enzymes for Industrial Reactors." Academic Press, New York.

Micheel, F., and Ewers, J. (1949). *Makromol. Chem.* 3, 200.

Mitz, M. A., and Schlueter, R. (1959). *J. Am. Chem. Soc.* 81, 4024.

Mitz, M. A., and Summaria, L. J. (1961). *Nature (London)* 189, 576.

Miyamoto, K., Fujii, T., and Miura, Y. (1971) *Hakko Kogaku Zasshi* 49, 565. [*Chem. Abstr.* 75, 71845p (1971).]

Miyamoto, K., Fujii, T., Tamaoki, N., Okazaki, M., Miura, Y. (1973). *Hakko Kogaku Zasshi* 51, 556. [*Chem. Abstr.* 80, 24314h (1974).]

Miyamura, M., and Suzuki, S. (1972). *Nippon Kagaku Kaishi* 47, 1274. [*Chem. Abstr.* 77, 165224d (1972).]

Mohan, R. R., and Li, N. N. (1974). *Biotechnol. Bioeng.* 16, 513.

Monsan, P., and Durand, G. (1971). *FEBS Lett.* 16, 39.

Monsan, P., Puzo, G., and Mazarguil, H. (1975). *Biochimie* 57, 1281.

Moore, J. E., and Ward, W. H. (1956). *J. Am. Chem. Soc.* 78, 2414.

Moore, P. W. (1972). *Rev. Pure Appl. Chem.* 20, 139.

Mori, T., Sato, T., Tosa, T., and Chibata, I. (1972). *Enzymologia* 43, 213.

Morris, D. L., Campbell, J., and Hornby, W. E. (1975). *Biochem. J.* 147, 593.

Morrison, M., Steele, W., and Danner, D. J. (1969). *Arch. Biochem. Biophys.* 134, 515.

Mosbach, K. (1970). *Acta Chem. Scand.* 24, 2084.

Mosbach, K. (1971). *Sci. Am.* 224, 26.

Mosbach, K., and Mattiasson, B. (1970). *Acta Chem. Scand.* 24, 2093.

Mosbach, K., and Mosbach, R. (1966). *Acta Chem. Scand.* 20, 2807.

Munier, R. L. (1973). *In* "Experimental Methods in Biophysical Chemistry" (C. Nicolau, ed.), p. 209. Wiley (Interscience), New York.

Nadler, H. L., and Updike, S. J. (1974). *Enzyme* 18, 150.

Nagai, W., Kirk, K. L., and Cohen, L. (1973). *J. Org. Chem.* 38, 1971.

Nikolaev, A. Y. (1962). *Biokhimiya* 27, 843. [Engl. transl., *Biochemistry (USSR)* 27, 713.]

Nikolaev, A. Y., and Mardashev, S. R. (1961). *Biokhimiya* 26, 641. [Engl. transl., *Biochemistry (USSR)* 26, 565.]

Nikolaev, A. Y., Benko, E. M., and Vovchenko, G. D. (1973a). *Zh. Fiz. Khim.* 47, 1310. [Engl. transl., *Russ. J. Phys. Chem.* 47, 742.]

Nikolaev, A. Y., Benko, E. M., and Vovchenko, G. D. (1973b). *Zh. Fiz. Khim.* 47, 1311. [Engl. transl., *Russ. J. Phys. Chem.* 47, 743.]

Nilsson, H., Mosbach, R., and Mosbach, K. (1972). *Biochim. Biophys. Acta* 268, 253.

Nordberg, M. E. (1944). *J. Am. Ceram. Soc.* 27, 299.

O'Driscoll, F. K., Izu, M., and Korus, R. (1972). *Biotechnol. Bioeng.* 14, 847.

Ogata, K., Ottesen, M., and Svendsen, I. (1968). *Biochim. Biophys. Acta* 159, 403.

Ohmiya, K., Terao, C., and Shimizu, S. (1975). *Agric. Biol. Chem.* 39, 491.

Ohno, Y., and Stahmann, M. A. (1972a). *Immunochemistry* **9,** 1077.
Ohno, Y., and Stahmann, M. A. (1972b). *Immunochemistry* **9,** 1087.
Oka, T., and Topper, Y. J. (1974). *Proc. Natl. Acad. Sci. U.S.A.* **71,** 1630.
Ong, E. B., Tsang, Y., and Perlmann, G. E. (1966). *J. Biol. Chem.* **241,** 5661.
Orth, H. D., and Brümmer, W. (1972). *Angew. Chem., Int. Ed. Engl.* **11,** 249.
Ottesen, M., and Svensson, B. (1971). *C. R. Trav. Lab. Carlsberg* **38,** 171.
Ozawa, H. (1967a). *J. Biochem. (Tokyo)* **62,** 419.
Ozawa, H. (1967b). *J. Biochem. (Tokyo)* **62,** 531.
Parikh, I., March, S., and Cuatrecasas, P. (1974). *In* "Affinity Techniques: Enzyme
    Purification," Part B (W. B. Jakoby and M. Wilchek, eds.), Methods in Enzymology,
    Vol. 34, p. 77. Academic Press, New York.
Patel, R. P., and Price, S. (1967). *Biopolymers* **5,** 583.
Patel, R. P., Lopiekes, D. V., Brown, S. P., and Price, S. (1967). *Biopolymers* **5,** 577.
Patel, A. B., Pennington, S. N., and Brown, H. D. (1969). *Biochim. Biophys. Acta* **178,**
    626.
Patel, A. B., Stasiw, R., Brown, H. D., and Ghiron, C. (1972). *Biotechnol. Bioeng.* **14,**
    1031.
Pennington, S. N., Brown, H. D., Patel, A. B., and Chattopadhyay, S. K. (1968a). *J.
    Biomed. Mater. Res.* **2,** 443.
Pennington, S. N., Brown, H. D., Patel, A. B., and Knowles, C. O. (1968b). *Biochim.
    Biophys. Acta* **167,** 479.
Phillips, D. C., Blow, D. M., Hartley, B. S., and Lowe, G., eds. (1970). *Phil. Trans. R.
    Soc. London, Ser. B* **237**(813), 63–266.
Pigman, W., Horton, D., and Herp, A., eds. (1970). "The Carbohydrates, Chemistry and
    Biochemistry." Academic Press, New York.
Polson, A. (1961). *Biochim. Biophys. Acta* **50,** 565.
Porath, J. (1960). *Biochim. Biophys. Acta* **39,** 193.
Porath, J. (1962). *Adv. Protein Chem.* **17,** 209.
Porath, J. (1967). *In* "Gamma Globulins" (J. Killander, ed.), Nobel Symp. No. 3, p. 287.
    Wiley (Interscience), New York.
Porath, J. (1968). *Nature (London)* **218,** 834.
Porath, J. (1974). *In* "Affinity Techniques: Enzyme Purification," Part B (W. B. Jakoby
    and M. Wilchek, eds.), Methods in Enzymology, Vol. 34, p. 13. Academic Press,
    New York.
Porath, J., and Flodin, P. (1959). *Nature (London)* **183,** 1657.
Porath, J., and Kristiansen, T. (1975). *In* "The Proteins" (H. Neurath and R. L. Hill,
    eds.), 3rd Ed., Vol. 1, p. 95. Academic Press, New York.
Porath, J., and Sundberg, L. (1972). *Nature (London), New Biol.* **238,** 261.
Porath, J., Axén, R., and Ernback, S. (1967). *Nature (London)* **215,** 1491.
Porath, J., Janson, J.-C., and Låås, T. (1971). *J. Chromatogr.* **60,** 167.
Porath, J., Aspberg, K., Drevin, H., and Axén, R. (1973). *J. Chromatogr.* **86,** 53.
Poznansky, M. J., and Chang, T. M. S. (1974). *Biochim. Biophys. Acta* **334,** 103.
Quiocho, F. A. (1974). *In* "Insolubilized Enzymes" (M. Salmona, C. Saronio, and S.
    Garattini, eds.), p. 113. Raven, New York.
Quiocho, F. A., and Richards, F. M. (1964). *Proc. Natl. Acad. Sci. U.S.A.* **52,** 833.
Quiocho, F. A., and Richards, F. M. (1966). *Biochemistry* **5,** 4062.
Rao, S. S., Patki, V. M., and Kulkarni, A. D. (1970). *Indian J. Biochem.* **7,** 210.
Rees, D. A. (1969). *Advan. Carbohydr. Chem. Biochem.* **24,** 267.
Reiland, J. (1971). *In* "Enzyme Purification and Related Techniques" (W. B. Jakoby, ed.),
    Methods in Enzymology, Vol. 22, p. 287. Academic Press, New York.

Reiner, R., Siebeneick, H. U., Christensen, I., and Lukas, H. (1975). *J. Mol. Catal.* **1**, 3.

Reynolds, J. H. (1974). *Biotechnol. Bioeng.* **16**, 135.

Richards, F. M., and Knowles, J. R. (1968). *J. Mol. Biol.* **37**, 231.

Riordan, J. F., Wacker, W. E. C., and Vallee, B. L. (1965). *Biochemistry* **4**, 1758.

Robbins, J. B., Haimovich, J., and Sela, M. (1967). *Immunochemistry* **4**, 11.

Robinson, P. J., Dunnill, P., and Lilly, M. D. (1971). *Biochim. Biophys. Acta* **242**, 659.

Robinson, P. J., Dunnill, P., and Lilly, M. D. (1973). *Biotechnol. Bioeng.* **15**, 603.

Royer, G. P., and Andrews, J. P. (1973). *J. Biol. Chem.* **248**, 1807.

Royer, G. P., and Green, G. M. (1971). *Biochem. Biophys. Res. Commun.* **44**, 426.

Royer, G. P., and Uy, R. (1973). *J. Biol. Chem.* **248**, 2627.

Royer, G. P., Andrews, J. P., and Uy, R. (1973). *Enzyme Technol. Digest* **1**(3), 99.

Royer, G. P., Liberatore, F. A., and Green, G. M. (1975). *Biochem. Biophys. Res. Commun.* **64**, 478.

Ruaho, A., Bartlett, P. A., Dutton, A., and Singer, S. J. (1975). *Biochem. Biophys. Res. Commun.* **63**, 417.

Russell, B., Mead, T. H., and Polson, A. (1964). *Biochim. Biophys. Acta* **86**, 169.

Sabatini, D. D., Bensch, K., and Barnett, R. J. (1963). *J. Cell Biol.* **17**, 19.

Saini, R., Vieth, W. R., and Wang, S. S. (1972). *Trans. N.Y. Acad. Sci.* **34**, 664.

Salomaa, P. (1966). *In* "The Chemistry of the Carbonyl Group" (S. Patai, ed.), p. 177. Wiley (Interscience), New York.

Sato, T., Mori, T., Tosa, T., and Chibata, I. (1971). *Arch. Biochem. Biophys.* **147**, 788.

Schejter, A., and Bar-Eli, A. (1970). *Arch. Biochem. Biophys.* **136**, 325.

Schell, H. D., and Ghetie, V. (1968). *Rev. Roum. Biochim.* **5**, 295.

Schell, H. D., and Ghetie, V. (1971). *Rev. Roum. Biochim.* **8**, 251.

Schick, A. F., and Singer, S. J. (1961). *J. Biol. Chem.* **236**, 2477.

Schwabe, C. (1969). *Biochemistry* **8**, 795.

Self, D., Kay, G., Lilly, M. D., and Dunnill, P. (1969). *Biotechnol. Bioeng.* **11**, 337.

Shaltiel, S. (1967). *Biochem. Biophys. Res. Commun.* **29**, 178.

Shaltiel, S. (1974a). *In* "Affinity Techniques: Enzyme Purification," Part B (W. B. Jakoby and M. Wilchek, eds.), Methods in Enzymology, Vol. 34, p. 126. Academic Press, New York.

Shaltiel, S. (1974b). *Isr. J. Chem.* **12**, 403.

Shaltiel, S., and Er-El, Z. (1973). *Proc. Natl. Acad. Sci. U.S.A.* **70**, 778.

Shaltiel, S., and Soria, M. (1969). *Biochemistry* **8**, 4411.

Shaltiel, S., Mizrahi, S., Stupp, Y., and Sela, M. (1970). *Eur. J. Biochem.* **14**, 509.

Sharon, N., and Lis, H. (1972). *Science* **177**, 949.

Sharp, A. K., Kay, G., and Lilly, M. D. (1969). *Biotechnol. Bioeng.* **11**, 363.

Shaw, E. N., Hirs, C. H. W., Koenig, D. F., Popenoe, E. A., and Siegelman, H. W., eds. (1969). *Brookhaven Symp. Biol.* **21**.

Silman, I. H., and Katchalski, E. (1966). *Annu. Rev. Biochem.* **35**, 873.

Silman, I. H., Albu-Weissenberg, M., and Katchalski, E. (1966). *Biopolymers* **4**, 441.

Simon, S. R., and Konigsberg, W. H. (1966). *Proc. Natl. Acad. Sci. U.S.A.* **56**, 749.

Sluyterman, L. A. A., and deGraaf, M. J. M. (1969). *Biochim. Biophys. Acta* **171**, 277.

Sluyterman, L. A. A., and Wijdenes, J. (1972). *Biochim. Biophys. Acta* **263**, 329.

Smiley, K. L. (1971). *Biotechnol. Bioeng.* **13**, 309.

Smiley, K. L., and Strandberg, G. W. (1972). *Adv. Appl. Microbiol.* **14**, 13.

Sokolovsky, M., and Vallee, B. L. (1966). *Biochemistry* **5**, 3574.

Sokolovsky, M., and Vallee, B. L. (1967). *Biochemistry* **6**, 700.

Solomon, B., and Levin, Y. (1974a). *Biotechnol. Bioeng.* **16**, 1161.

Solomon, B., and Levin, Y. (1974b). *Bioetechnol. Bioeng.* **16**, 1393.

Sorenson, W. R., and Campbell, T. W. (1968). "Preparative Methods of Polymer Chemistry," 2nd Ed. Wiley (Interscience), New York.

Spande, T. F., and Glenner, G. G. (1973). *J. Am. Chem. Soc.* **95**, 3400.

Stanley, W. L., and Olson, A. C. (1974). *J. Food Sci.* **39**, 660.

Stanley, W. L., Watters, G. G., Chan, B., and Mercer, J. M. (1975). *Biotechnol. Bioeng.* **17**, 315.

Stark, G. R. (1965a). *Biochemistry* **4**, 1030.

Stark, G. R. (1965b). *Biochemistry* **4**, 2363.

Stark, G. R. (1970). *Adv. Protein Chem.* **24**, 261.

Stasiw, R. O., Brown, H. D., and Hasselberger, F. X. (1970). *Can. J. Biochem.* **48**, 1314.

Stasiw, R. O., Patel, A. B., and Brown, H. D. (1972). *Biotechnol. Bioeng.* **14**, 629.

Steers, E., Jr., Cuatrecasas, P., and Pollard, H. B. (1971). *J. Biol. Chem.* **246**, 196.

Strandberg, G. W., and Smiley, K. L. (1971). *Appl. Microbiol.* **21**, 588.

Stroud, R. M., Kay, L. M., and Dikerson, R. E. (1974). *J. Mol. Biol.* **83**, 185.

Sulkowski, E., and Laskowski, M. (1974). *Biochem. Biophys. Res. Commun.* **57**, 463.

Sundaram, P. V. (1973). *Biochim. Biophys. Acta* **321**, 319.

Sundaram, P. V. (1974). *Biochem. Biophys. Res. Commun.* **61**, 667.

Sundaram, P. V., and Hornby, W. E. (1970). *FEBS Lett.* **10**, 325.

Sundberg, L., and Kristiansen, T. (1972). *FEBS Lett.* **22**, 175.

Sundberg, L., and Porath, J. (1974). *J. Chromatogr.* **90**, 87.

Surinov, B. P., and Manoilov, S. E. (1966). *Biokhimiya* **31**, 387. [Engl. transl., *Biochemistry (USSR)* **31**, 387 (1966).]

Suzuki, H., Ozawa, Y., and Maeda, H. (1966). *Agric. Biol. Chem.* **30**, 807.

Suzuki, S., Karube, I., and Watanabe, Y. (1972). In "Fermentation Technology Today" (G. Terui, ed.), p. 375. Soc. Ferment. Technol., Tokyo.

Suzuki, S., Sonobe, N., Karube, I., and Aizawa, M. (1974). *Chem. Lett.* **1**, 9.

Suzuki, S., Karube, I., and Namba, K. (1976). *Proc. Int. Symp. Anal. Control Immobilized Enzymes Syst., Compiegne, Fr., 1975* (in press).

Švec, F., Hradil, J., Coupek, J., and Kálal, J. (1975). *Angew. Makromol. Chem.* **48**, 135.

Svensson, B. (1973). *FEBS Lett.* **29**, 167.

Sweeney, W., and Zimmermann, J. (1969). In "Encyclopedia of Polymer Science and Technology" (H. F. Mark and N. G. Gaylord, eds.), Vol. 10, p. 483. Wiley (Interscience), New York.

Tawney, P. O., Snyder, R. H., Conger, R. P., Leibbrand, K. A., Stileter, C. H., and Williams, A. R. (1961). *J. Org. Chem.* **26**, 15.

Ternynck, T., and Avrameas, S. (1972). *FEBS Lett.* **23**, 24.

Tesoro, G. C., and Willard, J. J. (1971). In "Cellulose and Cellulose Derivatives" (N. M. Bikales and L. Segal, eds.), 2nd Ed., High Polymers, Vol. 5, Part V, p. 835. Wiley (Interscience), New York.

Tesser, G. I., Fisch, H.-U., and Schwyzer, R. (1972). *FEBS Lett.* **23**, 56.

Topper, Y. J., Oka, T., Vonderhaar, B. K., and Wilchek, M. (1975). *Biochem. Biophys. Res. Commun.* **66**, 793.

Tosa, T., Mori, T., Fuse, N., and Chibata, I. (1966a). *Enzymologia* **31**, 214.

Tosa, T., Mori, T., Fuse, N., and Chibata, I. (1966b). *Enzymologia* **31**, 225.

Tosa, T., Mori, T., Fuse, N., and Chibata, I. (1967a). *Enzymologia* **32**, 153.

Tosa, T., Mori, T., Fuse, N., and Chibata, I. (1967b). *Biotechnol. Bioeng.* **9**, 603.

Tosa, T., Mori, T., and Chibata, I. (1969a). *Agric. Biol. Chem.* **33**, 1053.

Tosa, T., Mori, T., Fuse, N., and Chibata, I. (1969b). *Agric. Biol. Chem.* **33**, 1047.

Tosa, T., Mori, T., and Chibata, I. (1971). *Enzymologia* **40**, 49.

Tosa, T., Sato, T., Mori, T., Chibata, J. (1974). *Appl. Microbiol.* **27**, 886.

Traher, A. D., and Kittrell, J. R. (1974). *Biotechnol. Bioeng.* **16**, 413.
Traub, A., Kaufmann, E., and Teitz, Y. (1969). *Anal. Biochem.* **28**, 469.
Turkova, J. (1974). *Chem. Listy* **68**, 489.
Turkova, J., Hubalkova, O., Krivakova, M., and Coupek, J. (1973). *Biochim. Biophys. Acta* **322**, 1.
Tveritinova, E. A., Kirai, E., Chukhrai, E. S., and Poltorak, O. M. (1969). *Vestn. Mosk. Univ., Khim.* **24**, 16. [*Chem. Abstr.* **73**, 73325b (1970).]
Ugi, I. (1962). *Angew. Chem., Int. Ed. Engl.* **1**, 8.
Ugi, I., ed. (1971). "Isonitrile Chemistry." Academic Press, New York.
Usami, S., and Shirasaki, H. (1970). *Hakko Kagaku Zasshi* **48**, 506. [*Chem. Abstr.* **73**, 127178b (1970).]
Usami, S., and Taketomi, N. (1965). *Hakko Kyokaishi* **23**, 267. [*Chem. Abstr.* **63**, 11939f (1965).]
Usami, S., Yamada, T., and Kimura, A. (1967). *Hakko Kyokaishi* **25**, 513. [*Chem. Abstr.* **68**, 84499j (1968).]
Usami, S., Noda, J., and Goto, K. (1971). *Hakko Kagaku Zasshi* **49**, 598. [*Chem. Abstr.* **75**, 74889s 1971).]
Vallee, B. L., and Riordan, J. F. (1969). *Annu. Rev. Biochem.* **38**, 733.
Van Leemputten, E., and Horisberger, M. (1974a). *Biotechnol. Bioeng.* **16**, 997.
Van Leemputten, E., and Horisberger, M. (1974b). *Biotechnol. Bioeng.* **16**, 385.
Velikanov, L. L., Velikanov, N. L., and Zvyagintsev, D. G. (1971). *Pochvovedenie* No. 3, p. 62. [*Chem. Abstr.* **74**, 135651m (1971).]
Vieth, W. R., and Venkatasubramanian, K. (1974). *Chem. Technol.* **4**(1), 47.
Vieth, W. R., Wang, S. S., and Gilbert, S. G. (1972a). *Biotechnol. Bioeng. Symp.* No. 3, p. 285.
Vieth, W. R., Wang, S. S., Bernath, F. R., and Mogensen, A. O. (1972b). *Recent Dev. Sep. Sci.* **1**, 175.
Visser, J., and Strating, M. (1975). *FEBS Lett.* **57**, 183.
Vorobeva, E. S., and Poltorak, O. M. (1966). *Vestn. Mosk. Univ., Khim.* **21**, 17. [*Chem. Abstr.* **66**, 62194n (1967).]
Vretblad, P. (1974). *FEBS Lett.* **47**, 86.
Vretblad, P., and Axén, R. (1971). *FEBS Lett.* **18**, 254.
Vretblad, P., and Axén, R. (1973a). *Acta Chem. Scand.* **27**, 2769.
Vretblad, P., and Axén, R. (1973b). *Biotechnol. Bioeng.* **15**, 783.
Walsh, K. A., Hauston, L. L., and Kenner, R. A. (1970). *In* "Structure-Function Relationships of Protolytic Enzymes" (P. Desnuelle, H. Neurath, and M. Ottesen, eds.), p. 56. Academic Press, New York.
Walton, H. M., and Eastman, J. E. (1973). *Biotechnol. Bioeng.* **15**, 951.
Wan, H., and Horvath, C. (1975). *Biochim. Biophys. Acta* **410**, 135.
Wang, J. H. C., and Tu, J. I. (1969). *Biochemistry* **8**, 4403.
Wang, S. S., and Vieth, W. R. (1973). *Biotechnol. Bioeng.* **15**, 93.
Weakley, F. B., and Mehltretter, C. L. (1973). *Biotechnol. Bioeng.* **15**, 1189.
Weetall, H. H. (1969a). *Nature (London)* **233**, 959.
Weetall, H. H. (1969b). *Science* **166**, 615.
Weetall, H. H. (1970). *Biochim. Biophys. Acta* **212**, 1.
Weetall, H. H. (1971). *Res./Dev.* **22**, 18.
Weetall, H. H., ed. (1975). "Immobilized Enzymes, Antigens, Antibodies and Peptides; Preparation and Characterization." Dekker, New York.
Weetall, H. H., and Baum, G. (1970). *Biotechnol. Bioeng.* **12**, 399.
Weetall, H. H., and Detar, C. C. (1975). *Biotechnol. Bioeng.* **17**, 295.

Weetall, H. H., and Filbert, A. M. (1974). *In* "Affinity Techniques: Enzyme Purification," Part B (W. B. Jakoby and M. Wilchek, eds.), Methods in Enzymology, Vol. 34, p. 59. Academic Press, New York.

Weetall, H. H., and Hersh, L. S. (1969). *Biochim. Biophys. Acta* **185,** 464.

Weetall, H. H., and Hersh, L. S. (1970). *Biochim. Biophys. Acta* **206,** 54.

Weetall, H. H., and Mason, R. D. (1973). *Biotechnol. Bioeng.* **15,** 455.

Weetall, H. H., and Messing, R. A. (1972). *In* "The Chemistry of Biosurfaces" (M. L. Hair, ed.), Vol. 2, p. 563. Dekker, New York.

Weetall, H. H., and Weliky, N. (1966). *Anal. Biochem.* **14,** 160.

Weetall, H. H., Havewala, N. B., Garfinkel, H. M., Buehl, W. M., and Baum, G. (1974a). *Biotechnol. Bioeng.* **16,** 169.

Weetall, H. H., Havewala, N. B., Pitcher, W. H., Detar, C. C., Vann, W. P., and Yaverbaum, S. (1974b). *Biotechnol. Bioeng.* **16,** 295, 689.

Weetall, H. H., Vann, W. P., Pitcher, W. H., Lee, D. D., Lee, Y. Y., and Tsao, G. T. (1975). *In* "Immobilized Enzyme Technology" (H. H. Weetall and S. Suzuki, eds.), p. 269. Plenum, New York.

Weibel, M. K., and Bright, H. J. (1971). *Biochem. J.* **124,** 801.

Weibel, M. K., Dritschhilo, W., Bright, H. J., and Humphrey, A. E. (1973). *Anal. Biochem.* **52,** 402.

Weir, D. M., ed. (1973). "Handbook of Experimental Immunology." Blackwell, Oxford.

Weliky, N., and Weetall, H. H. (1965). *Immunochemistry* **2,** 293.

Weliky, N., Brown, F. S., and Dale, E. C. (1969). *Arch. Biochem. Biophys.* **131,** 1.

Westman, T. L. (1969). *Biochem. Biophys. Res. Commun.* **35,** 313.

Weston, P. D., and Avrameas, S. (1971). *Biochem. Biophys. Res. Commun.* **45,** 1574.

Wetz, K., Fasold, H., and Meyer, C. (1974). *Anal. Biochem.* **58,** 347.

Wharton, C. W., Crook, E. M., and Brocklehurst, K. (1968a). *Eur. J. Biochem.* **6,** 565.

Wharton, C. W., Crook, E. M., and Brocklehurst, K. (1968b). *Eur. J. Biochem.* **6,** 572.

Wheeler, K. P., Edwards, B. A., and Whittam, R. (1969). *Biochim. Biophys. Acta* **191,** 187.

Whipple, E. B., and Ruta, M. (1974). *J. Org. Chem.* **39,** 1666.

Wieland, T., Determann, H., and Bünning, K. (1966). *Z. Naturforsch., Teil B* **21,** 1003.

Wilchek, M. (1974). *In* "Immobilized Biochemicals and Affinity Chromatography" (R. B. Dunlop, ed.), p. 15. Plenum, New York.

Wilchek, M., and Miron, T. (1974a). *Mol. Cell. Biochem.* **4,** 181.

Wilchek, M., and Miron, T. (1974b). *In* "Affinity Techniques: Enzyme Purification," Part B (W. B. Jakoby and M. Wilchek, eds.), Methods in Enzymology, Vol. 34, p. 72. Academic Press, New York.

Wilchek, M., Oka, T., and Topper, Y. J. (1975). *Proc. Natl. Acad. Sci. U.S.A.* **72,** 1055.

Wilson, R. J. H., and Lilly, M. D. (1969). *Biotechnol. Bioeng.* **11,** 349.

Wilson, R. J. H., Kay, G., and Lilly, M. D. (1968a). *Biochem. J.* **108,** 845.

Wilson, R. J. H., Kay, G., and Lilly, M. D. (1968b). *Biochem. J.* **109,** 137.

Wold, F. (1961). *J. Biol. Chem.* **236,** 106.

Wold, F. (1972). *In* "Enzyme Structure," Part B (C. H. W. Hirs and S. N. Timasheff, eds.), Methods in Enzymology, Vol. 25, p. 623. Academic Press, New York.

Woodward, R. B., and Olofson, R. A. (1961). *J. Am. Chem. Soc.* **83,** 1007.

Woodward, R. B., Olofson, R. A., and Mayer, H. (1961). *J. Am. Chem. Soc.* **83,** 1010.

Woodward, R. B., Olofson, R. A., and Mayer, H. (1966). *Tetrahedron, Suppl.* **8,** 321.

Wykes, J. R., Dunnill, P., and Lilly, M. D. (1971). *Biochim. Biophys. Acta* **250,** 522.

Yamamoto, K., Sato, T., Tosa, T., and Chibata, J. (1974a). *Biotechnol. Bioeng.* **16,** 1589.

Yamamoto, K., Sato, T., Tosa, T., and Chibata, J. (1974b). *Biotechnol. Bioeng.* **16,** 1601.

Yarovaya, G. A., Gulyanskaya, T. N., Dotsenko, V. L., Bessmertnaya, L. Y., Kozlov, L. V., and Antonov, V. K. (1975). *Bioorg. Khim.* **1**, 646. [*Chem. Abstr.* **83**, 128206c (1975).]

Yunginger, J. W., and Gleich, G. J. (1972). *J. Allergy Clin. Immunol.* **50**, 109.

Zaborsky, O. R. (1973). "Immobilized Enzymes." CRC Press, Cleveland, Ohio.

Zaborsky, O. R. (1974a). *In* "Immobilized Enzymes in Food and Microbiol Process" (A. C. Olson and C. C. Cooney, eds.), p. 187. Plenum, New York.

Zaborsky, O. R. (1974b). *In* "Enzyme Engineering" (E. K. Pye and L. B. Wingard, eds.), Vol. 2, p. 115. Plenum, New York.

Zabriskie, D., Ollis, D. F., and Burger, M. M. (1973). *Biotechnol. Bioeng.* **15**, 981.

Zahn, H. (1955). *Angew. Chem.* **67**, 561.

Zahn, H., and Lumper, L. (1968). *Hoppe-Seyler's Z. Physiol. Chem.* **349**, 485.

Zhirkov, Y. A., Chukhrai, E. S., and Poltorak, O. M. (1971). *Vestn. Mosk. Univ., Khim.* **12**, 405. [*Chem. Abstr.* **76**, 43461n (1972).]

Zingaro, R. A., and Uziel, M. (1970). *Biochim. Biophys. Acta* **213**, 371.

Zittle, C. A. (1953). *Adv. Enzymol. Relat. Subj. Biochem.* **14**, 319.

Diffusion and Kinetics with Immobilized Enzymes[1]

Jean-Marc Engasser

Laboratoire des Sciences du Génie Chimique,
E.N.S.I.C., Nancy, France

and

Csaba Horvath

Biochemical Engineering Group, Department of
Engineering and Applied Science, Yale University,
New Haven, Connecticut

[1] The authors' research and the preparation of this chapter were supported by Grants No. GM 20993 and CA 17245 from the National Institutes of Health, U.S. Public Health Service.

I. INTRODUCTION

This chapter is devoted to the analysis of the kinetic behavior of enzymes entrapped in or bound to membranes or other supports, as can be observed from macroscopic measurements. There are several reasons for dealing with the subject. The growing employment of immobilized enzymes (Zaborsky, 1973) in various technological applications requires an understanding of the overall kinetic properties in order to design systems in which the potential of enzymes can be fully exploited. Immobilized enzymes can also serve as experimental and theoretical models for bound enzymes in living systems (Goldstein and Katchalski, 1968; Katchalski *et al.*, 1971; Goldman *et al.*, 1971a) so that their study has a much broader scope than that of technological utility. It is increasingly recognized that in the cellular environment most enzymes are localized in various cell compartments (Greville, 1969) and the catalytic properties can be quite different from those of the same enzymes in free solution (Brown and Chattopadhyay, 1972). The allotopy of cellular enzymes (Racker, 1967) requires a departure from the conventional approach, which is based on the observation of chemical events in free solution, for the elucidation of metabolic processes. Consequently, phenomena related to the structure of the heterogeneous enzymic environment are receiving more and more attention in cellular physiology. In addition, soil chemistry and other areas of agricultural chemistry also provide a theater of great scientific and practical significance for the action of bound enzymes (McLaren and Peterson, 1967).

The step from homogeneous to heterogeneous enzyme catalysis is a major one, and the theoretical treatment of the subject can be extremely difficult. In order to attempt a description of the kinetic behavior of bound enzymes which reveals itself to the macroscopic probe of the observer, we have to assume that the conformation of the fixed enzyme can be different from that of the same enzyme in free solution. Additional assumptions are that the properties of the local environment provided by the matrix for the enzyme can be significantly different from that of the medium in which the reaction events are followed and that the slowness of diffusion of the species that participate in the reaction can play a major role in determining the overall system behavior. The system, which in the simplest case consists of a single enzyme attached to a surface and the contacting solution, can be quite complex when Michaelis–Menten kinetics, microenvironmental effects, diffusional resistances, and kinetic complications, such as inhibition by product or substrate, occur simultaneously. The treatment becomes more difficult when the enzyme is embedded in a porous medium so that the coupled effect of chemical reaction and internal diffusion has to be considered. The complexity greatly increases when the system consists of several bound enzymes that catalyze a series of reactions. Finally, in living systems the heterogeneous medium is anisotropic and transport processes other than passive molecular and convective diffusion may play an important role. The theoretical study of such situations has not been attempted yet and remains a challenging task for the future together with the transient analysis of even simple systems.

So far the interplay of enzymic reaction and diffusion has received the greatest attention, probably because our tools have been best developed to tackle this problem. Chemical engineers are often credited with the establishment of the fundamentals of diffusion–reaction interactions. The perusal of the literature clearly shows, however, that the first treatment of the subject was made by physiologists. The diffusion of oxygen, carbon dioxide, and lactic acid in tissues, where they are either consumed or produced by a zeroth- or first-order reaction, was examined by Krogh (1919), Warburg (1923), Hill (1928), and Jacobs (1935). Rigorous analytical solutions of these cases for different geometrical configurations were obtained by Roughton (1932) and Rashevsky (1948). Later the calculations were extended to Michaelis–Menten kinetics by Blum (1956) and Blum and Jenden

(1957), who presented their results in the form of very complex analytical expressions. Nevertheless chemical engineering can claim to have developed a comprehensive theory of diffusion and reaction since Damköhler (1937) and Thiele (1939) introduced dimensionless moduli to characterize the interplay of diffusion and reaction. Indeed, the wide use of powerful dimensionless numbers such as the so-called effectiveness factor, which express the utilization of the catalyst in the presence of diffusional effects, and the extensive investigations required for the design of heterogeneous reactors put chemical engineers in distinct advantage for coping with such problems. Important contributions have been made, among others, by Wheeler (1951), Weisz and Prater (1954), Aris (1957), Roberts and Satterfield (1965); and Carberry (1970).

The literature on the kinetics of heterogeneous biocatalysis is still relatively small in view of the importance of the subject, and only a few review articles have been published (Katchalski *et al.,* 1971; Wingard, 1972; Sundaram and Laidler, 1972; Vieth and Venkatasubramanian, 1974). It is expected, however, that the growing interest in transport phenomena on the part of the life scientists and the increasing involvement of chemical engineers in bio-oriented research (Carbonell and Kostin, 1972) will provide a fertile ground for future work. In our journey through the subject, we intend to give an overview of the most important results described in the literature and to establish a coherent framework for the somewhat fragmented pieces of knowledge in this field. Because of our active involvement in this kind of research we may have some biases in weighing the results obtained by others despite our effort to present as complete a picture as possible. Naturally we have omitted any discussion of matters related to biochemical reactors, as those are dealt with by Vieth *et al.* in this volume.

Our goal is to relate the rate of heterogeneous enzyme reactions as observed by measuring changes in the macroenvironment to the intrinsic kinetic parameters of the bound enzyme and the system parameters, such as diffusivity and fixed charges. Most of the results have been obtained by numerical calculations and are presented in graphical form by using conventional plots. Under certain circumstances, the results permit the evaluation of the microscopic properties of the bound enzyme and the system parameters from macroscopic measurements. We hope that the material brought together in this chapter will also be of help in both the theoretical and experimental modeling of physiological systems as well as in the design of immobilized enzyme reactors and analytical devices for practical applications.

II. FACTORS AFFECTING THE KINETICS OF BOUND ENZYMES

As mentioned in Section I, the behavior of a bound enzyme can differ significantly from that of the same enzyme in free solution. Thus far the differences have been most extensively investigated with artificially immobilized enzymes, and the results have been invaluable to explain the changes caused by confining the enzyme to a solid matrix. It is believed that many of the findings are applicable also to membrane-bound enzymes in physiological systems.

In view of the great variety of enzymes, reactions catalyzed by the same enzyme, solid matrices, and immobilization techniques investigated, it is not surprising that the results are often contradictory. It has been observed that a particular immobilization technique can impart different changes in properties to different enzymes and the behavior of the same enzyme can change with the immobilization method employed.

[Most investigations have been focused on the evaluation of the kinetic parameters, particularly the K_m value and the effect of immobilization on the pH activity profile and on the stability of the bound enzymes toward denaturing agents and elevated temperature.\ Some studies have addressed themselves to changes in the substrate specificity, particularly with respect to the molecular weight of the substrates, to the activation energy, and to the effect of inhibition and activators. In most instances the properties of the native enzyme—as measured in free solution—have been significantly altered by immobilization. Only a few studies have been carried out under conditions that permit a detailed analysis of the phenomena observed with the matrix bound enzyme.

In the following discussion, we assume that the enzyme is uniformly distributed on a surface or in a porous medium, and the system is isotropic. Under such conditions the changes in the enzymic behavior due to immobilization in a heterogeneous medium can be attributed to two major factors. The first involves changes in the enzyme molecule or in its immediate vicinity that arise directly from the attachment of the enzyme molecule to the carrier. These can manifest themselves in conformational changes in the protein structure and/or in restrictions on the accessibility of the active and allosteric sites. Direct interaction between the matrix with fixed functional groups and the bound enzyme molecules falls into this category. The second factor arises from the heterogeneous nature of the local enzymic environment in which the concentration of the substrate, product, and effectors may be different from that in the bulk solution, where the course of the enzymic reaction is followed.

A clear separation of these factors is usually very difficult in practice because in most cases the structure of the enzyme, the way of attachment, and the matrix properties are not known sufficiently well. Nevertheless, the distinction between the two major factors is necessary when a theoretical treatment of heterogeneous enzyme kinetics is attempted and can be very useful in the design of experiments and the interpretation of data.

A. The Enzyme Molecule: Conformational Changes, Matrix and Steric Effects

The effect of immobilization on the enzyme per se has been investigated mainly by comparing the activity of the bound enzyme with that of the native enzyme in free solution. In view of the wide variety of enzymes and binding procedures employed, it is natural that different results have been obtained. According to most studies the activity of immobilized enzymes has been lower than that of the same amount of soluble enzyme at a given concentration of substrate and effectors. In some cases, however, immobilization has not been accompanied by a loss of activity, and even an increase in activity has been reported (Weetall and Hersh, 1970; Sluyterman and De Graaf, 1969).

The decrease in enzyme activity on immobilization is often attributed to conformational changes in the enzyme structure or to steric hindrances in the immediate vicinity of the enzyme molecules. These two effects are schematically illustrated in Fig. 1. It is well established that the properties of the active and allosteric sites of an enzyme molecule depend strongly on the three-dimensional structure of the protein molecule. Thus, when an enzyme is adsorbed or covalently bound to a solid support, the interaction with the support likely results in a modification of the enzyme conformation. Covalent bonds between the enzyme and the matrix, for instance, can stretch the whole molecule and thus alter the three-dimensional structure at the active site. Gabel and Hofsten (1971), Cho and Swaisgood (1974), and Swaisgood and Horton (1974) have suggested that such conformational alterations may account for the observed changes in the enzyme activity upon immobilization.

Steric hindrances, on the other hand, are caused by the shielding effect of the matrix, which renders certain parts of the enzyme molecule less accessible to the substrate or effectors. Although it is often difficult to distinguish between the shielding of the active site and the reduced diffusivity of the substrate in the porous medium,

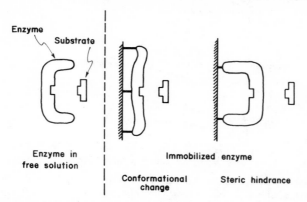

Fig. 1. Schematic illustration of conformational changes and steric hindrances that may affect the intrinsic kinetic behavior of an enzyme upon immobilization.

which is usually a gel, the shielding or steric effects have been proposed to explain the decrease in the activity of immobilized enzymes. For instance, trypsin and papain have been found to be less active in cross-linked dextran gel than in agarose. This effect has been attributed to the greater steric hindrances in the cross-linked dextran than may exist in the more open agarose matrix (Porath *et al.*, 1967; Axén and Ernback, 1971). A decline in the enzyme activity has also been observed with increasing degree of cross-linking of the carrier whereby the accessibility of the enzyme by the substrate has been reduced (Degani and Miron, 1970; Porath *et al.*, 1967). Similar steric effects have been used also to account for the observation that the degree of inhibition of insoluble trypsin has been inversely related to the molecular weight of the inhibitor (Haynes and Walsh, 1969; Glassmeyer and Ogle, 1971). For instance, trypsin bound to aminoethyl cellulose has been inhibited to the same extent as soluble trypsin by small inhibitor molecules, such as phenylmethyl sulfonyl fluoride. On the other hand, ovomucoid (MW 33,000), soybean trypsin inhibitor (MW 22,000), and lima bean inhibitor (MW 9000) have inhibited bound trypsin much less than the soluble enzyme under otherwise identical conditions. As shown in Fig. 2, the highest and lowest degrees of inhibition have been obtained with lima bean inhibitor and ovomucoid, respectively. In view of the size difference between the molecules these results suggest that whereas some trypsin molecules in the matrix are accessible to small inhibitors, they are inaccessible to the large-molecular-weight inhibitors.

In order to reduce the shielding of the active or allosteric sites, which may accompany the binding of an enzyme to a carrier, a

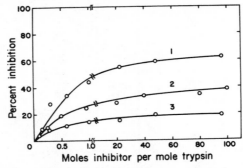

Fig. 2. Effect of the molecular dimensions of inhibitors on the inhibition of trypsin immobilized on aminoethyl cellulose in 0.1 N phosphate buffer, pH 8.0. The substrate is N-α-benzoyl-DL-arginine-p-nitrophenylanilide hydrochloride. Curve 1, chicken ovomucoid, MW 33,000; curve 2, soybean trypsin inhibitor, MW 22,000; curve 3, lima bean trypsin inhibitor, MW 9000. (From Glassmeyer and Ogle, 1971. Reprinted with permission from *Biochemistry*, **10**, 786–792. Copyright by The American Chemical Society.)

"spacer" can be used to keep the enzyme at a certain distance from the matrix. This approach has found wide application in affinity chromatography, where steric hindrances would otherwise impair the binding of large-molecular-weight substances to the matrix-bound moiety (Wilchek and Rotman, 1970; Mosbach *et al.*, 1974a; Steers *et al.*, 1971).

Additional interactions between the bound enzyme and the matrix have also been investigated. It is now well established that the properties of the support, such as its hydrophobic and hydrophilic nature, the dielectric constant of the medium, and the presence of fixed charges can significantly affect the mode of action of the biological catalyst (McLaren and Packer, 1970; Katchalski *et al.*, 1971). For instance, Goldstein (1972), Valenzuela and Bender (1971), and Gatfield and Stute (1972) attributed the changes in enzyme activity upon immobilization to interactions with fixed electrically charged groups on the matrix. On the other hand, the effect of hydrophobic interactions between the enzyme and carrier has been demonstrated by the behavior of yeast β-fructofuranosidase bound to cross-linked polystyrene (Filippusson and Hornby, 1970). As shown in Fig. 3, the width of the pH activity profile is narrower for the bound enzyme or for the soluble enzyme in the mixture of buffer and dioxan than that obtained with soluble β-fructofuranosidase in neat buffer (i.e., aqueous buffer without organic solvent). The narrow pH activity profile of the enzyme in the hydroorganic medium or on the polystyrene surface can be ex-

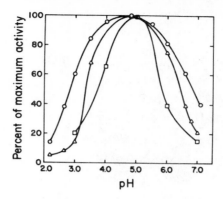

Fig. 3. Profiles of pH activity of soluble and immobilized β-fructofuranosidase for the hydrolysis of sucrose in citrate–phosphate–borate buffer. ○, Soluble enzyme in neat buffer (i.e., aqueous buffer without organic solvent); ▽, enzyme covalently bound to polystyrene beads; □, soluble enzyme in free solution containing 40% (v/v) dioxan. (From Filippusson and Hornby, 1970.)

plained by the low dielectric constant in the surroundings, according to the theory of Bass and McIlroy (1968).

Evidently, conformational changes and matrix interactions can modify not only the catalytic activity, but also the selectivity and stability of the bound enzyme with respect to the enzyme in free solution. In many cases the stability of the enzyme is increased by immobilization (Zaborsky, 1973; Vieth and Venkatasubramanian, 1974) because of the stabilization of the protein structure, the prevention of autolysis, or simply because the immobilized enzyme is less accessible to denaturing agents and microbial attack. Nevertheless, the widely held notion (Melrose, 1971) that the stability of an enzyme always is increased by immobilization is incorrect; and there is experimental evidence to the contrary (Suzuki *et al.*, 1966; Goldman *et al.*, 1968a). In some cases, no changes in the stability of the enzyme have been observed upon immobilization (Hicks and Updike, 1966).

B. Concentrations in the Surroundings: Partition Effects and Diffusional Resistances

1. Micro- and Macroenvironment of Bound Enzymes

When an enzymic reaction takes place in a well stirred homogeneous solution, the concentration of all species is uniform throughout the system. With bound enzymes, however, the interaction between the matrix and the substrate and/or effectors, as well as the presence of dif-

MICROENVIRONMENT

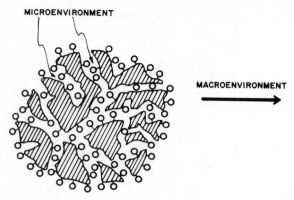

MACROENVIRONMENT

Fig. 4. Schematic illustration of a porous medium containing immobilized enzyme molecules, each of them having its own microenvironment, which is determined by partition effects as well as by diffusional resistances in both the exterior and interior of the enzymic medium.

fusional resistances, often results in concentration nonuniformities in the system. In the *microenvironment*, i.e., in the immediate vicinity of the bound enzyme, the concentrations of those species that influence the rate of reaction differ from those in the bulk solution, which is called the *macroenvironment*. This is illustrated in Fig. 4. In most cases, when the kinetics of immobilized enzymes are investigated, the solution in contact with the solid phase constitutes the macroenvironment, whereas the activity of the bound enzyme is determined by the local concentrations in the microenvironment. The experimental conditions permit the measurement of the respective bulk concentrations only in the macroenvironment. Thus, in view of the preceding discussion, the intrinsic kinetic behavior of a heterogeneous enzyme may not be evaluated by measuring only the macroenvironmental concentrations of the substrate and/or effectors.

2. *Partition Effects between the Micro- and Macroenvironment*

Hydrophobic, hydrophilic, and electrostatic interactions between the carrier and the substrate and/or effectors often produce an unequal distribution of these species between the micro- and macroenvironment. This is called the partition effect. For instance, a relatively nonpolar substance can be more soluble in a hydrophobic membrane than in the aqueous bathing solution. Thus, its concentration in an enzymic membrane can be greater than in the surrounding solution. Such a partition effect explained the observed change in the degree of inhibition of μ-fructofuranosidase by aniline and Tris when the enzyme was attached to a polystyrene surface (Filippusson and Hornby, 1970). The hydrophobic polystyrene was a good sorbent for aniline; therefore, the inhibitor concentration in the microenvironment of the bound fructofuranosidase was greater than that in the bulk solution. As a result, the inhibition constant measured with aniline was lower for the bound enzyme. On the other hand, the highly hydrophilic Tris was largely excluded from the relatively hydrophobic microenvironment of the enzyme. This phenomenon accounted for the 3-fold increase in the inhibition constant with Tris when the enzyme was immobilized on polystyrene relative to that obtained with the soluble enzyme.

The micro- and the macroenvironmental concentrations of the substrate and effectors can also differ when both the support and the species are electrically charged. Unequal concentration distribution due to electrostatic interactions has been frequently observed with enzymes bound to a polyelectrolytic matrix. A theoretical analysis of such electrostatic phenomena is presented in Section VIII, A.

3. Diffusional Resistances

When a single enzyme is bound to a carrier, the substrate diffuses from the bulk solution to the catalytic sites, and the products of the reaction usually diffuse back to the bulk solution. These processes can involve both molecular and convective diffusion. The low molecular diffusivities in aqueous solutions and gels and the high catalytic activity of enzymes often result in significant diffusional resistances. Consequently, concentration gradients are established in the surroundings of the bound enzyme so that concentrations of substrate and product both differ between the micro- and the macroenvironment. Figure 5 illustrates some possible concentration profiles for the substrate and the product for the case of the enzyme fixed in a porous membrane. The partition of these species between the matrix and the liquid phase due to electrostatic, hydrophilic, or hydrophobic interactions results in a steep concentration change at the interface. On the other hand, the effect of diffusional resistances is such that the concentration of the substrate or product gradually decreases or increases from the bulk solution toward the interior of the porous medium. Concentration gradients due to diffusional resistances and partition frequently occur together.

In the case of diffusional resistances, the concentration differences in the system are caused by the respective depletion and accumulation of the substrate and product as a result of the chemical reaction in the enzymic microenvironment. The extent of substrate depletion and product accumulation in the matrix usually depends on the size of the

BULK SOLUTION   ENZYMIC MATRIX

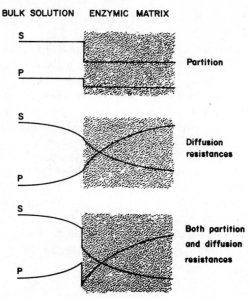

Fig. 5. Schematic illustration of the concentration profiles of the substrate, S, and the product, P, in a porous medium containing an immobilized enzyme and in the surrounding solution. The three cases shown are: the species are distributed between the two phases due to partition effects, but the reaction is kinetically controlled; diffusional resistances are present for both species without partitioning effects; and both diffusional limitations and partition occur.

species involved. Large molecules have a relatively small diffusivity in the porous medium so that they usually encounter significant diffusional resistances. Numerous experimental data demonstrate the effect of the size of the substrate molecules, which are related to the diffusivity, on the observed enzyme activity. The molecular weight and diffusivity in aqueous solution are shown for certain biological substances in Table I. As expected, the decrease in enzyme activity due to immobilization has been found to be greater with high- than with low-molecular-weight substrates (Silman *et al.*, 1966; Mosbach and Mosbach, 1966). The activity of immobilized Pronase normalized to the activity in free solution was linearly dependent on the logarithm of the substrate molecular weight (Cresswell and Sanderson, 1970). When the substrate molecules are larger than the pore size of the medium containing the enzyme, the substrate cannot diffuse to the catalytic sites. In the case of such molecular sieve effects, no reaction can take place even if the enzyme is fully active in the interior (Chang, 1964).

TABLE I
MOLECULAR WEIGHT AND DIFFUSION COEFFICIENT FOR
MOLECULES DIFFUSING THROUGH WATER AT 20°C

Species	MW	Diffusion coefficient $(10^6 \times cm^2/sec)$
Glucose	180	6.7
Sucrose	342	4.5
Inulin	5200	2.3
Ribonuclease	13,683	1.1
Serum albumin	66,500	0.6
Fibrinogen	330,000	0.2
Myosin	440,000	0.105
Deoxyribonucleic acid	6,000,000	0.013

III. HETEROGENEOUS ENZYME KINETICS

A. Local and Overall Reaction Rate

It was shown Section II,B,3 that enzymes embedded in a porous medium can be exposed to different local concentrations of substrate and product because of diffusional resistances. Therefore, in the presence of concentration gradients enzymes at different local positions exhibit different activities even if the enzyme molecules have the same intrinsic catalytic constants and K_m values. Consequently, the *local rate* of the reaction varies with the distance from the surface. In practice, however, the rate of reaction is determined by measuring concentration changes of the substrate and the product in the bulk solution. Therefore, the observed reaction rate is an *overall rate* that represents the sum of all local rates.

In this section we will also consider reactions that are catalyzed by enzymes attached to a flat surface. Under these conditions, the local substrate and product concentrations, and thus the local rate of reaction, will be assumed to be invariant on the surface even though there are concentration gradients in the surrounding solution.

As previously discussed, the local concentration of substrate, product, and other species that affect the enzyme activity may also differ from the measured macroenvironmental concentration because of unequal equilibrium distribution of these species between the enzymic matrix and the bulk solution. Thus, owing to diffusional limitations and partition effects, the observed dependence of the overall rate of reaction on the macroenvironmental concentration of substrate, product, or other effectors may often differ from the intrinsic dependence of the

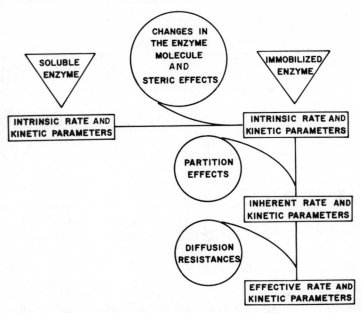

Fig. 6. Schematic illustration of the different rates and kinetic parameters and their interrelation.

enzyme activity on the local concentration of these species in the microenvironment.

B. Intrinsic, Inherent, and Effective Rates

We have seen that the observed kinetic behavior of a heterogeneous enzyme system of fixed intrinsic kinetic parameters can be significantly affected by partition effects and diffusional resistances. In order to facilitate the theoretical treatment of this phenomenon, we distinguish between intrinsic, inherent, and effective rates of reaction. Each is characterized by different values of the kinetic parameters. The relationship between these overall rates is schematically illustrated in Fig. 6.

1. Intrinsic Rate and Kinetic Parameters

The true kinetic behavior of a bound enzyme is characterized by the intrinsic kinetic parameters. Thus, the intrinsic kinetic behavior of a heterogeneous enzyme could be observed in the macroenvironment only if the concentration of the substrate, product, and effectors was the same in both the micro- and macroenvironment. According to this definition, the intrinsic kinetic parameters of an immobilized enzyme are not necessarily the same as those of the enzyme in the free solution because conformational changes, matrix interactions, and steric effects can change the intrinsic kinetic parameters of an enzyme upon immobilization. Most theoretical and experimental studies are aimed at the evaluation of the intrinsic kinetic behavior of the enzyme from the concentration of the different species and the pertinent concentration changes in the macroenvironment.

2. Inherent Rate and Kinetic Parameters

The inherent rate of reaction is defined as the rate that would be observed in the absence of any diffusional limitations, i.e., if the transport of substrate and product between the enzymic micro- and macroenvironment were infinitely fast. In practice, the inherent enzyme activity can be observed with relatively thin membranes, with low enzyme activity, and with sufficient stirring of the bulk solution. The inherent rate and the inherent kinetic parameters can be different from the intrinsic rate and parameters when the partitioning due to electrostatic and other interactions between the matrix and various soluble species yields different concentrations in the micro- and the macroenvironment.

3. Effective Rate and Kinetic Parameters

The effective rate of reaction and the effective kinetic parameters for bound enzymes are observed when diffusional limitations occur in the presence or in the absence of partition effects. Values can be determined from the overall rate as measured under usual experimental conditions.

Because of the relatively slow transport of biochemical substances in liquids and gels, diffusional resistances most often affect the activity of heterogeneous enzyme systems. Therefore, the interrelationship between the inherent and effective kinetic behavior has been at the center of most theoretical studies on heterogeneous enzyme systems. In the ensuing part of this chapter it will be shown quantitatively how the inherent kinetic behavior is altered by the slowness of diffusion at different concentrations in the macroenvironment to yield the effective kinetic behavior. Generally, Michaelis–Menten kinetics are no longer obeyed in the presence of diffusional limitations.

The above definitions of various rates and parameters are expected to facilitate both the theoretical and experimental analysis of immobilized enzyme kinetics by distinguishing between the different factors that affect the kinetics of the bound enzyme. The first step in an experimental investigation is the determination of the effective rate and kinetic parameters. Although these parameters adequately describe the observed enzymic behavior under certain experimental conditions they cannot be used to describe the system in general, in view of the many phenomena that are involved. Greater insight is gained by separating the diffusional effects in order to extract the inherent parameters of the system, and by further removing the partition effects in order to obtain the intrinsic parameters of the enzymic reaction. Then, and only then, is a comparison of the intrinsic parameters of the soluble and bound enzyme possible, along with an evaluation of the effects of immobilization on enzyme structure, the significance of enzyme–matrix interactions, and the presence of shielding effects. Comparison of the intrinsic and inherent parameters of the bound enzyme, on the other hand, can give information about the enzyme microenvironment, namely, the interaction between the matrix and the various species that play a role in determining the rate of reaction.

C. External and Internal Diffusion

The present review is mainly concerned with the effect of diffusional resistances on the kinetic behavior of heterogeneous enzyme systems, i.e., the relationship between inherent and effective rates. The kinetics of an enzyme entrapped in a porous medium can be affected by diffusional resistances for the *external transport* of the substrates and products between the bulk solution and the outer surface of the enzymic membrane or particle. On the other hand, diffusional resistances for the *internal transport* of these species inside the porous catalytic medium often play an even more significant role. External transport, which in the biochemical literature is often referred to as diffusion through unstirred layers, takes place by passive molecular diffusion and convection. Thus, when the term external diffusion is used hereafter, it will encompass both molecular and convective diffusion. In contrast, transport inside the porous matrix usually takes place by passive molecular diffusion only.

An additional distinction between external and internal diffusion is necessary when considering their interaction with the catalytic reaction. Internal diffusion proceeds simultaneously, or in parallel, with the chemical reaction so that the two events are coupled in the mathematical sense. On the other hand, external diffusion occurs before, i.e., in series with the actual reaction step process (Levenspiel, 1972). As a result, the theoretical approaches employed to analyze the interplay of the enzymic reaction with external and internal transport are different.

In the following treatment the effect of the two types of diffusional resistances will be discussed. The effect of external diffusion alone is examined for the enzyme bound to an impervious surface. The effect of internal diffusion is investigated when the enzyme is embedded in a porous matrix, and external diffusional resistances are negligible. The overall effect of the combined external and internal diffusional resistances are then discussed briefly in the light of the results obtained for the two cases.

D. Open and Closed Systems

In this study the kinetic behavior of immobilized enzymes will be examined only in an open system, i.e., one that can exchange substrates and products with its surroundings. The kinetics of enzymes in free solution are usually determined in a stirred vessel, which must be considered as a closed system since no substrate is added or product is removed over the course of the reaction. It is well established that in closed systems the Michaelis–Menten kinetic law adequately describes the classical two-step enzyme mechanism, provided the substrate concentration is much greater than the total enzyme concentration (Wong, 1965).

When enzymic membranes or particles are present in the stirred vessel, the container as a whole must still be considered as a closed system. The enzymic membranes or particles themselves, however, represent open systems, since the substrates and products can be, and in fact are, exchanged between the solid matrix and the surrounding solution. When bound enzymes are located in living cells or in a continuous-flow reactor, the entire system can be considered open.

An important consequence of the fact that bound enzymes constitute open systems is that at steady state the intrinsic Michaelis–Menten kinetic law holds even at relatively high enzyme concentrations inside the matrix (Aris, 1972; Engasser and Horvath, 1973). Therefore, the more complex kinetic laws that have been derived from relatively high enzyme concentrations in closed systems (Reiner, 1969; Cha, 1970) are not relevant to immobilized enzyme systems of practical interest.

E. Steady-State and Transient Kinetics of Bound Enzymes

In the following, only the steady-state behavior of immobilized enzymes will be discussed; that is, we shall assume that the enzymic reaction, the transport of substrate from the macro- to the microenvironment, and the transport of the product in the reverse direction all take place at the same rate which is invariant in time. First, most previous theoretical studies have been restricted to steady-state kinetics because of its greater simplicity. Second, the experimental investigations with bound enzymes have mostly been carried out at or close to steady state. Steady state is rigorously valid only when the concentration of various species in the system do not vary with time. Such conditions are easily fulfilled in an isothermal continuous-flow reactor when the inlet and outlet concentrations and the flow rate are constant. In a stirred vessel, on the other hand, the concentrations of the substrate and product in the bulk solution decrease and increase with time, respectively. Nevertheless, steady state for bound enzyme particles within a stirred vessel can still be regarded as a good approximation when the volume of the surrounding solution is sufficiently large so that the variations of bulk solution substrate and product concentrations with time can be neglected.

IV. EFFECT OF EXTERNAL DIFFUSIONAL LIMITATIONS

The influence of external diffusional limitations can be examined with enzymes immobilized in porous membranes and particles. Under such conditions, however, internal diffusional limitations also have to be taken into account (Goldman *et al.*, 1971a; Rony, 1971; Vieth *et al.*, 1973). Therefore, the theoretical and experimental study of the singular effect of external diffusion is preferably carried out with enzymes immobilized on fluid impervious solid surfaces (Hornby *et al.*, 1968; Shuler *et al.*, 1972; O'Neill, 1972; Kobayashi and Laidler, 1974; Horvath and Engasser, 1974). Microencapsulated enzymes have also been used as models for external diffusion with the assumption that diffusional resistances are only in the inert porous wall of the microcapsules (Sundaram, 1973; Mogensen and Vieth, 1973; Koch and Coffman, 1970).

Different mathematical approaches have been proposed to describe the effect of external transport on the kinetics of the enzymic reaction. When combined external and internal diffusion were examined, either the treatment was restricted to the limiting first-order kinetics (Vieth *et al.*, 1973; Rony, 1971; Mogensen and Vieth, 1973) or numerical calculations were performed with a given set of transport and kinetic parameters (Goldman *et al.*, 1971a). The interplay of external diffusion with Michaelis–Menten kinetics has been most conveniently analyzed by the use of dimensionless parameters. Thereby the number of pertinent parameters is reduced and the results can be generalized in order to represent a wide range of conditions. The various authors, however, have introduced different dimensionless groups to characterize the interaction of diffusion with chemical reaction. Some employed a parameter that depended on the substrate concentration (Sundaram, 1973; Shuler *et al.*, 1972; Kobayashi and Laidler, 1974), whereas others preferred to use a dimensionless group that was a function of the kinetic and transport parameters only (Horvath and Engasser, 1974; Engasser and Horvath, 1974b; Fink *et al.*, 1973).

In the following treatment the effect of external diffusion on an enzymic reaction at an impervious surface is discussed in terms of dimensionless quantities, which have been found most convenient for the interpretation and representation of the results. The most important parameter of the system is·the dimensionless modulus, which has been defined so that it is independent of the substrate concentration. This approach allows the illustration of diffusional effects on graphs commonly used in enzyme studies when the rate of reaction is measured as a function of the substrate concentration.

A. Inherent and Effective Rates with External Transport

When enzymes are attached to a solid surface that is in contact with the substrate solution, the overall heterogeneous reaction consists of three consecutive steps: (a) the transport of the substrate from the bulk solution to the surface; (b) the transformation of the substrate into the product at the surface; (c) the transport of the product from the surface to the bulk solution.

As the actual chemical reaction is localized at the surface, the substrate and product concentrations are lower and higher at the surface than in the bulk solution, respectively. The corresponding concentration profiles near the surface are schematically illustrated in Fig. 7. The magnitude of substrate depletion and product accumulation de-

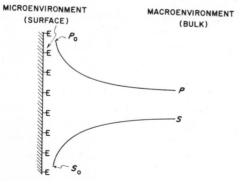

Fig. 7. Concentration profiles of the substrate and product for a reaction catalyzed by surface-bound enzymes. The substrate and product concentrations in the bulk solution are S and P, and at the enzymic surface S_0 and P_0, respectively.

pend both on the catalytic activity at the surface and the transport phenomena in the solution.

A rigorous treatment of the effect of external diffusion on heterogeneous enzyme kinetics would require a precise knowledge of the hydrodynamic conditions in the liquid and the integration of the differential equation, which expresses the conservation of the substrate and product, with the appropriate boundary conditions. In many practical cases, however, the treatment can be greatly simplified by the use of an external transport coefficient. In equiaccessible systems, where the transport of the substrate and product are identical over the whole surface (Frank-Kamenetskii, 1969), the rate of transport of the substrate, J_S, and the product, J_P (moles/sec) to and from the surface can simply be expressed by the product of a transport coefficient and the corresponding driving force, which is the concentration difference between the surface and the bulk. Thus,

$$J_S = h_S(S - S_0) \tag{1}$$
$$J_P = h_P(P_0 - P) \tag{2}$$

where h_S and h_P (cm³/sec) are the transport coefficients for the substrate and product, S and P are the macroenvironmental and S_0 and P_0 (moles/cm³) are the microenvironmental concentrations of the substrate and product, respectively.

In theoretical studies the transport coefficient, h, is usually expressed per unit area and is sometimes replaced by an effective boundary layer thickness, δ, which satisfies the following relation:

$$h = D/\delta \tag{3}$$

where D is the diffusivity of the substance under consideration. It is important to note that δ is only a fictitious distance, and its use by no means implies that the transport of the substrate takes place only by molecular diffusion through a stagnant liquid layer of thickness δ adjacent to the surface (Levich, 1962). Indeed, experimental measurements have shown that liquid motion may occur at distances from the surface much smaller than δ. Thus, the frequently used treatment of external diffusion in kinetic studies with immobilized enzymes and biological systems by assuming molecular diffusion through a so-called unstirred layer or Nernst film, though convenient, cannot be always reconciled with the physical reality. The assumption of an unstirred layer and molecular diffusion as the sole transport mechanism can be highly misleading when the simultaneous transport of several species is investigated. It is because δ depends not only on the hydrodynamic conditions, but also on the diffusivity, so that in general each species has a particular value of δ.

Nevertheless, if the transport of a given species takes place by molecular diffusion only—for instance, through an inert membrane—then h is accurately determined by Eq. (3), and in this case δ and D correspond to the membrane thickness and the diffusivity through the membrane, respectively. In physiology, h is generally referred to as the permeability coefficient, and in chemical engineering as the mass-transfer coefficient. For many experimental systems, the value of the transport coefficient can be obtained from the literature on mass and heat transfer. The results of mass- and heat-transfer experiments are available for a wide variety of geometrical and flow configurations and are usually expressed in dimensionless form, for instance, by the Nusselt number, Nu (Bird *et al.*, 1960). The mass transport coefficient for a given species can then be calculated from that Nusselt number which corresponds to the pertinent experimental conditions by the following relationship:

$$h = \text{Nu}\, D/d \qquad (4)$$

where d is the characteristic length of the system, for example, the particle or the tube diameter. Table II shows some typical values of the external-transport coefficient per unit area of enzymic surface.

Let us first consider the effect of diffusional limitations when the rate of the intrinsic enzymic reaction depends on the substrate concentration only, and follows the Michaelis–Menten kinetic expression. In the case of a surface reaction, the transport of the substrate from the macro-to the microenvironment and the consumption of the substrate in the reaction catalyzed by a surface-bound enzyme take place consecu-

TABLE II

TYPICAL VALUES OF THE EXTERNAL TRANSPORT COEFFICIENT PER UNIT AREA OF
ENZYMICALLY ACTIVE SURFACE IN DIFFERENT IMMOBILIZED ENZYME SYSTEMS

Immobilized enzyme system	External transport coefficient, h (cm/sec)	References
Open tubular reactor with laminar flow	5×10^{-4}	Horvath *et al.* (1973b)
Open tubular reactor with turbulent flow	2×10^{-3}	Horvath *et al.* (1973b)
Enzymic membranes	1×10^{-3}	Goldman *et al.* (1971b)
Packed-bed reactor	2×10^{-2}	Rovito and Kittrell (1973)

tively and, at steady state, proceed at the same rate. This effective rate of the reaction, V, depends both on the mass transport coefficient for the substrate, h_S, and the kinetic parameters of the reaction, V_{max} and K_m. The effective rate is usually more strongly influenced by the parameters of one process than by those of the other. In order to interpret this dependence, it is convenient to define two virtual maximal rates characteristic of the two steps involved (Horvath and Engasser, 1974). (i) The virtual maximum rate of the reaction, V_{kin}, is defined as the inherent reaction rate which is solely determined by the kinetic parameters of the enzymic reaction and the concentration of the substrate in the macroenvironment, S, and would be obtained if the diffusion of the substrate were infinitely fast. Thus, for Michaelis–Menten kinetics

$$V_{kin} = V_{max}S/(K_m + S) \tag{5}$$

where V_{max} is the saturation rate at the surface and K_m is the Michaelis constant. (ii) The virtual maximum rate of combined molecular and convective diffusion of the substrate, V_{diff}, is defined as the rate that is solely determined by the transport coefficient and the macroenvironmental concentration and would be obtained if the enzyme activity were infinitely high. Thus,

$$V_{diff} = h_S S \tag{6}$$

In the following, V_{diff} will also be considered to be the inherent rate of substrate diffusion.

The effective rate of the surface reaction is more influenced by that process which has the lower virtual maximum rate. In the limit, when one of the virtual maximum rates is much smaller than the other, the effective rate is practically equal to the lower virtual maximum rate. Under such conditions, it is convenient to distinguish between kinetically controlled reaction, which proceeds with the inherent rate equal to V_{kin}, and diffusion-controlled reaction, which proceeds with a rate equal to V_{diff}.

Figure 8 illustrates the dependence of V_{kin}, V_{diff}, and the effective rate, V, on the macroenvironmental concentration for an arbitrarily chosen set of the kinetic and transport parameters. By definition, V_{kin} and V_{diff} both represent an upper limit for V. Figure 8 clearly shows that at sufficiently high substrate concentrations V is always equal to V_{max}, the saturation value of V_{kin}, and the reaction is kinetically controlled. At low concentrations, on the other hand, either V_{diff} or V_{kin} plays the predominant role in determining the effective rate, depending on the relative magnitude of the initial slopes of V_{diff} and V_{kin}.

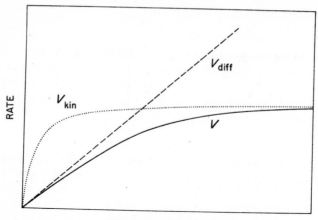

Fig. 8. Overall rate of a reaction, V, catalyzed by a surface-bound enzyme against the substrate concentration in the bulk solution. In this case, V is determined by both the inherent rate of the enzymic reaction at the surface, V_{kin}, and the maximum possible rate of substrate diffusion to the surface, V_{diff}, which are also illustrated. (From Horvath and Engasser, 1974.)

B. Substrate Modulus and Effectiveness Factor

In order to determine the steady-state dependence of the effective reaction rate, V, on the macroenvironmental concentration, S, it is necessary to calculate the substrate concentration at the surface, S_0, for any given S, h_S, V_{max}, and K_m. At steady state the rate of substrate transport to the surface is equal to the rate of substrate consumption by the reaction

$$h_S(S - S_0) = V_{max}S_0/(K_m + S_0) \tag{7}$$

The interpretation and graphical representation of the results are greatly facilitated by introducing a dimensionless substrate concentration, β, defined as

$$\beta = S/K_m \tag{8}$$

and a dimensionless substrate modulus, μ, given by

$$\mu = V_{max}/h_S K_m \tag{9}$$

This substrate modulus, which is defined as the ratio of a rate of reaction and a rate of transport characteristic of the system, will be shown later to conveniently express the magnitude of external diffusional resistances for heterogeneous enzymes.

With the above-defined quantities, Eq. (7) can be written as

$$\beta - \beta_0 = \mu[\beta_0/(1 + \beta_0)] \tag{10}$$

Thus, the microenvironmental concentration, β_0, at the surface can be determined for any given value of the macroenvironmental concentration β, and modulus, μ, from Eq. (10). The effective rate of reaction, V, is determined by the value of β_0. Thus, it can be expressed, after normalizing to V_{max}, simply by

$$V/V_{max} = \beta_0/(1 + \beta_0) \tag{11}$$

From Fig. 9, which shows the dependence of V/V_{max} on β at different values of μ, it can be seen that the substrate modulus, μ, is a measure of the importance of external diffusional resistances in the heterogeneous system under consideration. At small values of μ, $\mu < 0.1$, the inherent enzymic activity at the surface is sufficiently low, or the inherent rate of substrate transport is sufficiently fast, to prevent any significant depletion of substrate at the surface. Under these conditions, the substrate concentration is practically the same in the micro-

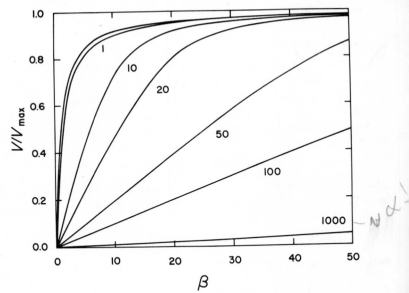

Fig. 9. Effective rate of an enzyme-catalyzed surface reaction, V, normalized to V_{max}, against the dimensionless bulk concentration, β, at different values of the substrate modulus, μ. The effect of diffusional limitations increases with increasing values of μ and results in a decrease of the rate of reaction with respect to the kinetically controlled rate, V_{kin}/V_{max}, which is obtained at about $\mu \leqslant 0.1$. (From Horvath and Engasser, 1974.)

and macroenvironment. Consequently, the heterogeneous reaction is kinetically controlled and V is equal to V_{kin}. On the other hand, when, at a given β, μ is sufficiently large, the surface is so active and diffusion of the substrate is so slow that all substrate molecules that arrive at the surface are converted into product. Under these conditions, the surface concentration is practically zero and V is equal to V_{diff}, the maximum rate of the transport to the surface. Then, for all practical purposes the activity of the bound enzyme is independent of the kinetic parameters of the reaction. At intermediate values of the modulus, however, the rate of the reaction is influenced by both the transport and reaction processes.

Figure 9 clearly shows that the rate of a heterogeneous enzymic reaction is attenuated by the relative slowness of the substrate diffusion to the enzyme. By analogy to the well established chemical action of inhibitors, this phenomenon can be referred to as "diffusional inhibition." In the chemical engineering literature the effect of diffusional limitations on the catalyst activity is quantitatively expressed by the effectiveness factor, η, which measures the departure of V from V_{kin}, and is defined by the following relationship:

$$V = \eta V_{kin} \qquad \eta = \frac{actual\ rate}{rate\ if\ no\ diffusion\ limitation} \qquad (12)$$

Thus, the magnitude of diffusional inhibition is conveniently expressed by the reciprocal of the effectiveness factor, which is a complex function of the kinetic parameters, transport coefficient, and substrate concentration.

Figure 10 shows the dependence of η on β and μ. The value of η is unity when the reaction is kinetically controlled, but decreases with increasing diffusional limitations. The straight lines obtained at high values of μ represent the diffusion-controlled reaction domain. Under such conditions the magnitude of diffusional inhibition is essentially proportional to μ. At small substrate concentrations, such as $\beta \leqslant 0.1$, when the rate of reaction is first order, η approaches a limiting value, ϵ, which can be shown to obey the following dependence on μ:

$$\epsilon = 1/(1 + \mu) \qquad (13)$$

As indicated in Fig. 9, V is a complex function of β at intermediate concentrations. In the two limiting cases of high and low concentrations, however, the relation is greatly simplified. At a sufficiently high β both V and V_{kin} have the same limiting value, V_{max}, the saturation rate of the enzymic reaction. On the other hand, at low β values, V follows a first-order dependence on β with a rate constant V_{max}/κ,

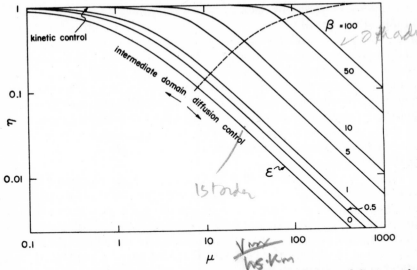

Fig. 10. External effectiveness factor, η, as a function of the substrate modulus, μ, with the dimensionless bulk concentration, β, as the parameter, for an enzymic surface reaction. At sufficiently small concentrations, the limiting first-order effectiveness factor ϵ is reached. (From Horvath and Engasser, 1974.)

where κ is defined by

$$\kappa = K_m/\epsilon = K_m(1 + \mu) \tag{14}$$

This similarity in the limiting behaviors of V and V_{kin} does not mean that when diffusional limitations occur the functional relationship between V and β can be illustrated by a rectangular hyperbola characteristic of the Michaelis–Menten kinetic law. Figure 11 shows plots of the normalized reaction rate, V/V_{max}, against $V/(V_{max}\beta)$. This type of plot, like the so-called Eadie–Hofstee plot used in kinetic studies with soluble enzymes, yields straight lines when V obeys the Michaelis–Menten law, as seen for $\mu \to 0$. With increasing diffusional limitations, however, the curves depart significantly from straight lines, particularly when a wide concentration range is examined. Experimental evidence for external diffusional limitations has been provided by many authors (Lilly *et al.*, 1966; Wilson *et al.*, 1968; Sharp *et al.*, 1969; Goldman *et al.*, 1971b; Horvath and Solomon, 1972; Taylor and Swaisgood, 1972; Mogensen and Vieth, 1973; Sundaram, 1973; Kobayashi and Moo-Young, 1973; Rovito and Kittrell, 1973; Horvath *et al.*, 1973b; Brams and McLaren, 1974). They found that the rate of

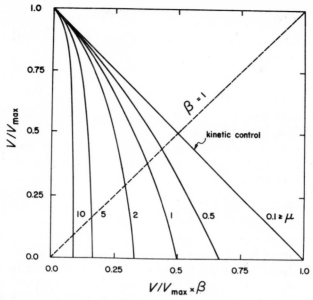

Fig. 11. Departure from Michaelis–Menten kinetics due to external transport limitations as illustrated by Eadie–Hofstee-type plots at different values of the substrate modulus, μ. The effective surface reaction rate, V, obtained at different dimensionless bulk concentrations, β, is normalized to V_{max}.

reaction was dependent on the stirring and flow rate in batch and continuous-flow reactors, respectively. It was also observed that the substrate concentration that yielded half of the saturation rate was larger than the intrinsic K_m value of the enzyme; or, alternatively, the limiting first-order rate constant for the bound enzyme was smaller than V_{max}/K_m for the soluble enzyme.

C. Interplay of Diffusional and Chemical Inhibition

The inhibition of soluble enzymes by various substances has been extensively treated in the literature (Webb, 1963). With bound enzymes, however, the simultaneous presence of diffusional inhibition causes the chemical inhibition to manifest itself in a different way than in free solution. In the following section the interplay of chemical and external diffusional inhibition is examined first in the general case, when the inhibitor is neither a substrate nor a product, then when it is a product, and finally when it is a substrate.

1. General Inhibition

In the case of noncompetitive inhibition the rate of the enzymic reaction is expressed by

$$V = \frac{V_{\text{max}}S_0}{[1 + (I/K_I)][K_m + S_0]} \tag{15}$$

where I is the inhibitor concentration assumed to be uniform in the solution, and K_I is the inhibition constant. Thus, at steady state, for a given bulk concentration, β, and substrate modulus, μ, the dimensionless substrate concentration at the surface, β_0, is obtained from the following equation

$$\beta - \beta_0 = \frac{\mu\beta_0}{[1 + (I/K_I)][1 + \beta_0]} \tag{16}$$

The enzymic activity V is readily determined from the value of β_0 by using Eq. (15).

The combined effect of chemical and diffusional inhibition is conveniently expressed by the efficiency factor for inhibition, η_I, which is defined as the ratio of the effective enzymic activity, V, to the activity of the enzyme in the absence of both kinds of inhibition (Engasser and Horvath, 1974b). Then the following relationship holds:

$$V = \eta_I[V_{\text{max}}S/(K_m + S)] \tag{17}$$

In the absence of diffusional inhibition the efficiency factor is simply given by $[1 + (I/K_I)]^{-1}$, as seen from Eq. (15), and the magnitude of the chemical inhibition is expressed by the factor $[1 + (I/K_I)]$. On the other hand, in the absence of chemical inhibition η_I is equal to the effectiveness factor η, defined in the preceding section.

When both chemical and diffusional inhibition occur, the efficiency factor depends on both μ and I/K_I as shown in Fig. 12. The variation of η_I also represents the variation of the enzymic activity, since at a constant β the rate of reaction is proportional to η_I. The curve for $\mu \leqslant 0.1$ shows the intrinsic dependence of the enzymic activity on the dimensionless inhibitor concentration, I/K_I, because at such low values of μ diffusional inhibition is absent. At higher values of μ that is, with increasing diffusional inhibition, the overall activity decreases less with increasing inhibitor concentration, and at $\mu = 50$ the activity becomes practically independent of I/K_I.

In Fig. 12 the curve corresponding to $\mu \leqslant 0.1$ illustrates the effect of chemical inhibition alone without diffusional effects. As seen, the decrease in enzymic activity is solely due to the increasing inhibitor

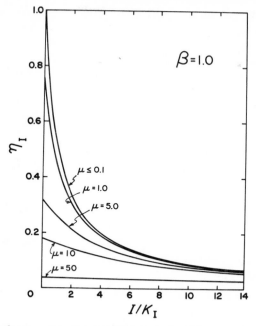

Fig. 12. Effect of a noncompetitive inhibitor on the activity of bound enzymes with external diffusion of the substrate. The efficiency factor η_I, which expresses the combined effect of chemical and diffusional inhibition, is plotted as a function of the normalized inhibitor concentration I/K_I, at different values of the substrate modulus, μ, i.e., at different magnitudes of the external diffusional resistance. (From Engasser and Horvath, 1974b. Reprinted with permission from *Biochemistry* 13, 3845–3849. Copyright by the American Chemical Society.)

concentration. On the other hand, at sufficiently large values of the substrate modulus, $\mu > 0.1$, diffusional inhibition already occurs in the absence of the inhibitor as indicated by the intercept of the curves with the ordinate. The antienergistic interaction of chemical and diffusional inhibition (Engasser and Horvath, 1974b) manifests itself in the fact that when diffusional resistances are present at $\mu > 0.1$ at any given inhibitor concentration the enzymic activity is greater than it would be if the rate decreasing effect of pure chemical inhibition, as shown at $\mu \leqslant 0.1$, and that of pure diffusional inhibition, as shown by the ordinate intercepts, had been exerted independently. Consequently, the relative decrease in enzymic activity due to chemical inhibition is smaller in the presence than in the absence of diffusional resistances, as shown in Fig. 12.

The antienergism between chemical and diffusional inhibition can be qualitatively explained as follows. Substrate depletion at the enzymic surface occurs when the inherent diffusion rate is slow with respect to the inherent enzymic activity. An enzyme inhibitor that decreases the inherent enzymic activity also reduces the relative slowness of substrate transport. Thus, chemical inhibition in heterogeneous enzyme systems can be characterized by two antagonistic effects acting simultaneously on the rate of reaction: the decrease of the inherent enzyme activity and the reduction of the degree of diffusional inhibition.

Equation 9 defines the substrate modulus, μ, as the ratio of the saturation rate, V_{max}, to the characteristic transport rate, $h_S K_m$, to express the magnitude of diffusional resistances in the absence of an inhibitor. When an enzyme inhibitor is present in the system, however, another substrate modulus, μ_I, is used to quantify the effect of diffusional resistances on the chemically inhibited reaction. As the maximum rate of the enzymic reaction is reduced by the factor $[1 + I/K_I]$ in the presence of a noncompetitive inhibitor, the substrate modulus μ_I is conveniently defined as

$$\mu_I = \mu/[1 + (I/K_I)] \tag{18}$$

Equation 18 shows that an enzyme inhibitor reduces the substrate modulus and consequently the magnitude of diffusional inhibition in the system. Thus Eq. (18) is a quantitative representation of the antienergism between diffusional and chemical inhibition.

It was previously shown in Fig. 9 that the depletion of substrate in the enzymic microenvironment is negligible when the substrate modulus is smaller than 0.1. In view of Eq. (18), diffusional inhibition due to substrate depletion can become insignificant even in systems characterized by values of μ larger than 0.1 when the inhibitor concentration is sufficiently high, so that $\mu_I \leq 0.1$. Equation 18 also explains that at high values of μ, $\mu \leq 50$, the effective enzyme activity is unaffected by the presence of an inhibitor. As long as μ_I is larger than 10 the magnitude of diffusional inhibition is proportional to μ_I, therefore the addition of a noncompetitive inhibitor reduces diffusional inhibition by a factor $[1 + I/K_I]$. Since chemical inhibition reduces the inherent enzyme activity by the same factor, the effective rate of the enzymic reaction, as observed from the macroenvironment, remains unchanged. The antienergism described above for noncompetitive inhibition is applicable to all other kinds of inhibitions, such as competitive and uncompetitive inhibition, with the exception of substrate inhibition.

2. Product Inhibition

Many enzymes are inhibited by the products of their reactions. Since the degree of inhibition depends on the product concentration in the microenvironment, both substrate depletion and product accumulation must be taken into account in the quantitative treatment of product inhibition. At steady state the transport of substrate from the bulk solution to the surface, the enzymic reaction, and the transport of product from the surface to the solution proceed at the same rate. Thus, for competitive product inhibition:

$$h_S(S - S_0) = \frac{V_{max}S_0}{K_m[1 + (P_0/K_P)] + S_0} = h_P(P_0 - P) \tag{19}$$

and for noncompetitive inhibition

$$h_S(S - S_0) = \frac{V_{max}S_0}{[1 + (P_0/K_P)][K_m + S_0]} = h_P(P_0 - p) \tag{20}$$

where K_p is the product inhibition constant.

From Eqs. (19) and (20), S_0, P_0, and, thus, the effective rate V, are readily determined for given values of S, P, K_m, K_p, h_S, and h_p. In addition to the previously introduced substrate modulus, μ, the heterogeneous reaction is also characterized by the external accumulation factor, which is defined as

$$\alpha_e = h_S K_m / h_P K_P \tag{21}$$

and expresses the competition between the substrate and the product at the enzymic surface (Engasser and Horvath, 1974c).

Figure 13 shows the effect of the dimensionless product concentration P/K_p, on the value of the efficiency factor, η_I, thus, on the enzyme activity when the product is a competitive inhibitor. Owing to the negligible diffusional resistance for the substrate at $\mu = 0.1$, there is no substrate depletion under the conditions illustrated in Fig. 13, and the value of η_I is solely determined by product inhibition. The dependence of the dimensionless product concentration at the surface, P_0/K_p, on P/K_p is also illustrated. As long as $\alpha_e < 1$, the concentration of the product is essentially the same in the micro- and macroenvironment of the bound enzyme, which is then inhibited as if it were in free solution. When α_e is larger than unity, however, product accumulates at the enzymic surface as shown by the dotted lines for α_e equal to 10 and 100. Consequently, the value of the efficiency factor is lower, and the magnitude of product inhibition is greater, than that obtained when $\alpha_e < 1$ with the same value of P/K_p.

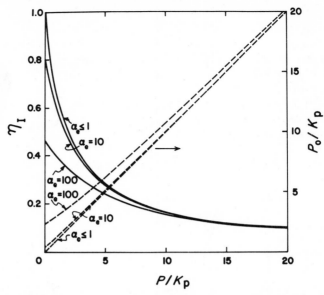

Fig. 13. Decrease of enzymic activity as expressed by the decrease of the efficiency factor, η_I, which represents the combined effect of diffusional and product inhibition, with increasing values of the dimensionless product concentration in the macroenvironment, P/K_P. In addition, the dashed lines show the dimensionless product concentration in the microenvironment, P_0/K_P, as a function of P/K_P. The external accumulation factor, α_e, is used as the parameter. The data have been calculated for competitive product inhibition with external diffusional resistances when the dimensionless substrate concentration, β, and the substrate modulus, μ, are 1 and 0.1, respectively. (From Engasser and Horvath, 1974c. Reprinted with permission from *Biochemistry* **13**, 3849–3854. Copyright by the American Chemical Society.)

The concurrent effect of substrate depletion as well as product accumulation and inhibition is shown in Fig. 14 by the double logarithmic plot of η_I against P/K_P. At $\mu = 10$ and $\alpha_e = 100$ the relatively low values of η_I are attributed to three different effects; each reduces the value of η_I by an increment Δ. These are as follows: Δ_1 represents the effect of pure chemical inhibition in the absence of any significant diffusional resistance, that is, when $\mu = 0.1$ and $\alpha_e = 0.01$; Δ_2 is due to diffusional inhibition, which arises from substrate depletion, when μ increases from 0.1 to 10 at $\alpha_e = 0.01$; Δ_3 is the result of product accumulation when α_e increases from 0.01 to 100 at $\mu = 10$ and manifests itself in an enhanced chemical inhibition.

As shown in Fig. 14, the decreases in enzyme activity due to substrate depletion, Δ_2, and product accumulation, Δ_3, are greatest at low product concentrations, that is, when the activity of the enzyme is the

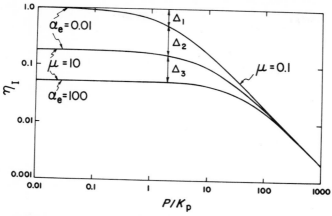

Fig. 14. Double logarithmic plot of the efficiency factor, η_{I_i} against the dimensionless product concentration in the macroenvironment, P/K_p, for different values of the external accumulation factor, α_e, and the substrate modulus, μ. The dimensionless substrate concentration, μ, is unity. The graph shows that when competitive product inhibition and external diffusional inhibition occur simultaneously at $\alpha_e = 100$ and $\mu = 10$, the decrease in enzyme activity as shown by the decrease in the efficiency factor is, at a given value of P/K_p, the result of three effects, each of which decreases η_I by an incremental value: Δ_1 is the contribution of pure chemical inhibition without diffusional resistances, Δ_2 is due to diffusional inhibition caused by substrate depletion, and Δ_3 is the contribution of chemical inhibition due to produce accumulation. (From Engasser and Horvath, 1974c. Reprinted with permission from *Biochemistry* **13**, 3849–3854. Copyright by the American Chemical Society.)

highest for a given system. Since the enzyme activity, and as a result the magnitude, of diffusional resistances are reduced with increasing product concentration, both substrate depletion and product accumulation vanish at a sufficiently high value of P/K_p. This antienergistic interaction between diffusional and chemical inhibition again results in a relative insensitivity of the bound enzyme activity to changes in the macroenvironmental concentration of the product in the presence of diffusional limitations. The results are much the same also if the product is a noncompetitive inhibitor.

When the substrate concentration varies, however, the combined effect of diffusional and product inhibition is different for competitive and noncompetitive inhibitors, especially at high substrate concentrations. In the case of competitive inhibition, only the "apparent" K_m of the enzyme increases, therefore the saturation rate, V_{max}, is always reached at sufficiently high substrate concentrations where both diffusional and product inhibition vanish. This variation of the enzyme activity on β is illustrated in Fig. 15 for different values of α_e. With

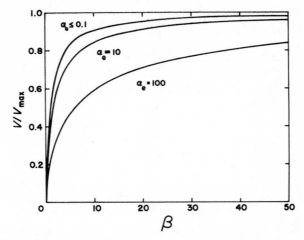

Fig. 15. Effect of the accumulation factor, α_e, on the normalized rate of enzymic surface reaction, V/V_{max}, at varying substrate concentration, β, when competitive product inhibition occurs. The substrate modulus, μ, is equal to 0.1, and the dimensionless product concentration in the bulk solution, P/K_p, has a value of 0.01.

competitive product inhibition Eadie–Hofstee plots have also been found to be highly sensitive to diffusional limitations, as shown in Fig. 16 for $\mu = 0.1$ and $\mu = 10$, i.e., at relatively mild and strong diffusional

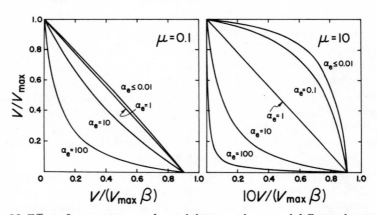

Fig. 16. Effect of competitive product inhibition with external diffusional resistances by plots of the dimensionless rate of reaction, V/V_{max}, against $V/V_{max}\beta$, where β is the dimensionless substrate concentration in the macroenvironment, for different values of the external accumulation factor, α_e, and at two values of the substrate modulus, μ. The dimensionless product concentration in the macroenvironment, P/K_p is 0.01. (From Engasser and Horvath, 1974c. Reprinted with permission from *Biochemistry* **13**, 3849–3854. Copyright by the American Chemical Society.)

inhibition, respectively. In both graphs the convex curves for $\alpha_e = 0.01$ represent the sole effect of substrate depletion, since at such a low value of the accumulation factor no product accumulation occurs. As long as $\alpha_e < 1$, the curves remain convex, which suggests that substrate depletion is the preponderant effect. When product accumulation becomes important at $\alpha_e > 1$, however, concave curves are obtained, and their curvature increases with diffusional limitations. It is important to note that when the product is a competitive inhibitor all curves representing various values of α_e converge at limiting low and high β values because the effect of product accumulation disappears at sufficiently low and high substrate concentrations.

On the other hand, both graphs in Fig. 16 show a straight line for $\alpha_e = 1$, indicating that the dependence of the reaction rate on the bulk concentration of the substrate also follows a Michaelis–Menten scheme when the accumulation factor is unity. This finding is easily verified analytically because for $\alpha_e = 1$ the reaction rate, V, can be expressed as a function of β by the Michaelis–Menten equation with a K_m^* value given by

$$K_m^* = K_m[1 + \mu + (P/K_P)] \tag{22}$$

Thus, an accumulation factor of unity with a competitive product inhibitor represents a unique situation, where the Michaelis–Menten kinetic law is obeyed in the presence of diffusional limitations.

When the product is a noncompetitive inhibitor, the saturation activity, V_{max}, is not reached at high substrate concentrations. Instead, lower saturation rates are obtained with increasing values of the accumulation factor, α_e, as demonstrated in Fig. 17. Consequently, the Eadie–Hofstee plots shown in Fig. 18 intercept the ordinate at different points, depending on the values of μ and α_e. In addition, comparing Figs. 16 and 18 we can see that at large values of α_e noncompetitive inhibition results in plots having very small curvature in comparison to those obtained for competitive product inhibition.

3. Substrate Inhibition

When substrate inhibition occurs, the rate of the enzymic reaction is generally expressed by Eq. 23.

$$V = \frac{V_{max}}{1 + (K_1/S_0) + (S_0/K_2)} \tag{23}$$

where K_1 and K_2 are two kinetic parameters of the system. If we define the dimensionless substrate concentration, β, as

$$\beta = S/K_1 \tag{24}$$

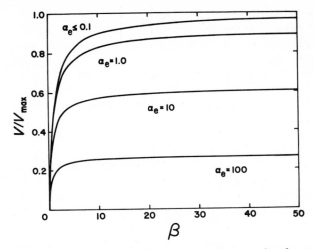

Fig. 17. Effect of the accumulation factor, α_e, on the normalized rate of enzymic surface reaction, V/V_{max}, at varying substrate concentration, β, when noncompetitive product inhibition occurs. The values of the substrate modulus, μ, and the dimensionless product concentration in the bulk solution, P/K_P, are chosen as 0.1 and 0.01, respectively.

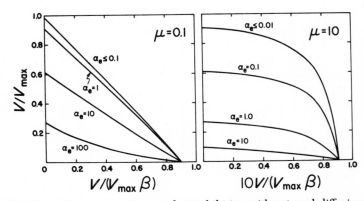

Fig. 18. Effect of noncompetitive product inhibition with external diffusional resistances by plots of the dimensionless rate of reaction, V/V_{max}, against $V/V_{max}\beta$, where β is the dimensionless substrate concentration in the macroenvironment, for different values of the external accumulation factor, α_e, and at two values of the substrate modulus, μ. The dimensionless product concentration in the macroenvironment, P/K_P, is 0.01. (From Engasser and Horvath, 1974c. Reprinted with permission from *Biochemistry* **13**, 3849–3854. Copyright by the American Chemical Society.)

the substrate modulus, μ, as

$$\mu = V_{max}/h_S K_1 \tag{25}$$

and the dimensionless inhibition constant, K, which expresses the degree of inhibition by the substrate, as

$$K = K_1/K_2 \tag{26}$$

then the substrate concentration at the surface, β_0, is expressed by the following third-order equation:

$$K\beta_0^3 + (1 + K\beta)\beta_0^2 + (1 + \mu - \beta)\beta_0 - \beta = 0 \tag{27}$$

In the acceptable domain of $0 < \beta_0 < \beta$ three different β_0 values have been found to satisfy Eq. (27) at certain values of β when μ is sufficiently high. As a result, the kinetic behavior of the bound enzyme can show multiple steady states in a particular range of substrate concentrations, provided its diffusion to the enzyme is sufficiently slow (Engasser and Horvath, 1974d).

This phenomenon is illustrated in Fig. 19, which shows the dependence of the enzyme activity on the substrate concentration, β, for $K = 1$. When $\mu < 1$, the overall reaction is kinetically controlled and

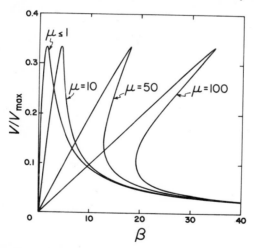

Fig. 19. The effect of substrate inhibition in heterogeneous systems is illustrated by plots of the normalized activity, V/V_{max}, against the dimensionless substrate concentration in the macroenvironment, β. The curves represent different values of the substrate modulus, μ, which expresses the magnitude of the diffusional resistance for the substrate. At high values of μ, that is, with high diffusional resistances, three steady-state reaction rates are obtained at a given β in a particular concentration domain. The value of the inhibition constant, K, is unity. (From Engasser and Horvath, 1974d. Reprinted with permission from *Biochemistry* **13**, 3855–3859. Copyright by the American Chemical Society.)

transport phenomena do not affect the enzyme behavior. For $\mu = 10$, substrate depletion occurs only at low β values and the substrate concentration that yields maximum enzyme activity is shifted toward higher values. With higher diffusion resistances, e.g., when μ is equal to 50 and 100, the reaction rate is essentially diffusion and kinetically controlled at low and high substrate concentrations, respectively. The straight line at low concentrations represents pseudo first-order dependence, which is characteristic for the diffusion controlled rate. At high β values, all curves converge and the reaction becomes kinetically controlled.

At certain intermediate substrate concentrations, three different steady-state rates can be obtained. The possibility of multiple steady states due to the interplay of substrate inhibition with substrate diffusion is illustrated in Fig. 20. Both the rate of surface reaction at a given set of the kinetic parameters and the transport rate of the substrate at fixed h_S and β are plotted against the surface concentration of the substrate, β_0. Since at steady state the two rates are equal, the intersections of the two curves yield the appropriate concentrations in the microenvironment at the surface. Under the conditions selected in Fig. 20, there are three possible surface concentrations, β_{0I}, β_{0II}, and β_{0III}, which give rise to three possible steady-state rates. It can be shown, however, that only the highest and the lowest rates are stable and may occur in practice. The actual rate of reaction is determined by

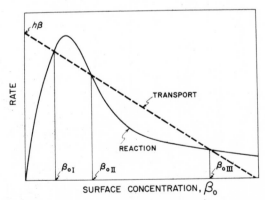

Fig. 20. Schematic illustration of the rate of a substrate inhibited enzymic reaction and of substrate transport as a function of the microenvironmental substrate concentration at the surface, when three steady-state rates of the overall reaction are possible at β_{0I}, β_{0II}, and β_{0III}. The graph can be used to show that only the rates obtained at β_{0I} and β_{0III} are stable. (From Engasser and Horvath, 1974d. Reprinted with permission from *Biochemistry* 13, 3855–3859. Copyright by the American Chemical Society.)

the direction of the change in the substrate concentration, as illustrated in Fig. 21. Beginning at low substrate concentrations the rate increases linearly with the concentration on the diffusion-controlled branch, then drops suddenly and decreases further on the kinetically controlled branch. On the other hand, when an originally high substrate concentration decreases, the rate first increases slowly then jumps to a higher value on the diffusion controlled branch and subsequently decreases. The graph illustrates that the interaction of transport processes with substrate inhibition results in a hysteresis phenomenon because the effective enzyme activity depends on the direction of variation in the substrate concentration.

It is important to note that with substrate inhibition at relatively high substrate concentrations the rate of reaction is higher in the presence than in the absence of diffusional limitations. Therefore, in this particular case of inhibition diffusional resistances can augment the rate of reaction in contradistinction to the previous cases where diffusional resistances for the substrate and product brought about a decrease in the rate of reaction. Another example for the enhancement of

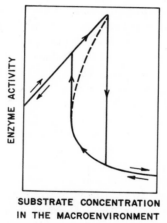

SUBSTRATE CONCENTRATION
IN THE MACROENVIRONMENT

Fig. 21. The variation of the effective enzyme activity with the substrate concentration when the substrate is an inhibitor and encounters diffusional resistances in a heterogeneous enzyme system. The solid lines represent the variation of the allowable enzymic activity. The arrows indicate the direction of change for the concentration or the activity. It is seen that in each case small changes in the concentration can bring about a dramatic decrease or increase in the enzymic activity as a result of sudden jumps from one allowed branch of the activity curve to the other. The broken lines indicate the theoretically possible but unstable, therefore not allowable, enzyme activities. (From Engasser and Horvath, 1974d. Reprinted with permission from *Biochemistry* 13, 3855–3859 Copyright by the American Chemical Society.)

the activity of bound enzymes is afforded when the product of the reaction is an activator for the enzyme (Caplan *et al.*, 1973; Engasser and Horvath, 1974a).

V. EFFECT OF INTERNAL DIFFUSIONAL LIMITATIONS

Whereas external diffusional limitations are often negligibly small owing to efficient mixing of the bulk solution, the effect of internal diffusion on the kinetic behavior of enzymes embedded in a porous matrix often can be very significant. As a result, the interplay of reaction and internal diffusion has received more theoretical attention in spite of the greater mathematical complexity. The immobilized enzyme models that have been used to quantitatively describe the interaction of reaction with internal diffusion usually have consisted of a porous membrane or spherical particle containing a uniformly distributed enzyme. Michaelis–Menten kinetics usually have been assumed. Yet, even for this relatively simple model a wide range of mathematical approaches have been employed.

Closed analytical solutions have been obtained in the limiting first- and zeroth-order kinetic domains only. The numerous expressions for the overall rate of reaction at low and high substrate concentrations are essentially the same as those derived in the physiological and chemical engineering literature (Doscher and Richards, 1963; Van Duijn *et al.*, 1967; Sluyterman and De Graaf, 1969; Sundaram *et al.*, 1970; Rony, 1971; Blaedel *et al.*, 1972; Lasch, 1973; Vieth *et al.*, 1973). For the full range of Michaelis–Menten type kinetics, analytical expressions in forms of series expansion have also been proposed (Blum, 1956; Blum and Jenden, 1957; Selegny *et al.*, 1971a; Thomas *et al.*, 1972), but these results are of limited practical value.

Most often the appropriate differential equations have been solved by numerical calculations, the results presented in terms of dimensionless quantities and the interplay of internal diffusion with chemical reaction characterized by a dimensionless modulus. Some authors have defined this modulus so that it was a function of the substrate concentration (Moo-Young and Kobayashi, 1972; Kobayashi *et al.*, 1973; Horvath and Engasser, 1973). In other treatments of the subject the modulus has been dependent only on transport and kinetic parameters (Roberts and Satterfield, 1965; Goldman *et al.*, 1968b; Engasser and Horvath, 1973; Thomas and Broun, 1973; Fink *et al.*, 1973; Marsh *et al.*, 1973; Engasser and Horvath, 1974c; Hamilton *et al.*, 1974; Thomas *et al.*, 1974).

In the following, a dimensionless modulus independent of the substrate concentration will be used to present the results of numerical calculations, since it greatly facilitates the illustration of diffusional effects, using graphs of interest to both the biochemist and the chemical engineer.

A. Effect of Internal Diffusional Resistances on Michaelis–Menten Kinetics

When enzymes are embedded in a porous medium, the diffusion of the substrate and the product both in the external solution and in the porous medium has to be taken into account. The simplest model consists of a membrane containing uniformly distributed bound enzyme molecules. The membrane is exposed to the substrate solution at one side and sealed at the other. Alternatively, both sides can be open and in contact with the same substrate solution.

The overall reaction behavior of such a heterogeneous system shows again the two limiting cases. At sufficiently low enzyme activity, the concentration of both the substrate and product are essentially the same inside and outside the membrane. Under these conditions, the reaction rate is kinetically controlled and unaffected by transport phenomena. At sufficiently high enzyme activity, on the other hand, all substrate molecules are converted at the external surface of the membrane before diffusing toward the interior. The reaction rate is then equal to the maximum rate of transport in the bulk solution.

In general both external and internal transport limitations influence the rate of the enzymic reaction. As illustrated in Fig. 5 the substrate and product concentration, respectively, decreases and increases from the bulk solution to the interior of the membrane. The relative importance of external and internal diffusional limitations can readily be assessed by considering the conservation of mass for the substrate at the external surface of the membrane. At the interface of the membrane the transport of substrate usually takes place by molecular diffusion only. Thus, the conservation of mass can be expressed by

$$-D_S(dS/dx)_e = -D_S^{\text{eff}}(dS/dx)_i \tag{28}$$

where the derivatives express the concentration gradients at the interface toward the external solution and the inside of the membrane, respectively. D_S^{eff} is the effective diffusivity of the substrate inside the membrane, which can be related to the substrate diffusivity in the bulk solution by the following expression (Satterfield, 1970):

$$D_S^{\text{eff}} = \epsilon D_S/\tau \tag{29}$$

In the above equation, ϵ is the void fraction in the porous membrane and τ is a tortuosity factor that takes into account the pore geometry and by definition is larger than unity. Since the diffusivity inside the matrix is smaller than in the bulk solution, the gradient labeled i must be larger than the gradient labeled e in view of Eq. (28). Then it follows that for relatively thick membranes substrate depletion is more significant inside than outside the membrane. Indeed, in many experimental situations, external transport can be sufficiently enhanced so that the surface concentration is practically equal to the bulk concentration. Under these conditions, the enzymic reaction is affected by internal diffusional resistances only. Therefore, in the present section the kinetics of enzymes entrapped in a porous matrix will be examined in this simplified but nevertheless realistic situation of negligible external diffusional limitations. At the end, the combined effect of internal and external diffusion will be discussed briefly.

At steady state, the decrease of the substrate concentration from the membrane surface toward the inside of the membrane is described by the following differential equation:

$$D_S^{\text{eff}}\left(\frac{d^2 S_0}{dx^2}\right) = \frac{V_{\text{max}}''' S_0}{K_m + S_0} \tag{30}$$

where V_{max}''' is the saturation rate per unit volume of the membrane and x is the distance from the external surface.

It is convenient to write Eq. (30) in dimensionless form as

$$\frac{d^2 \beta_0}{dz^2} = \phi^2 \left(\frac{\beta_0}{1 + \beta_0}\right) \tag{31}$$

for the calculation of the local values of the dimensionless substrate concentration, β_0, in the membrane. In Eq. (31) z is the position in the membrane given by

$$z = x/l \tag{32}$$

where l is the thickness of the membrane. On the other hand, ϕ is the substrate modulus for internal diffusion defined by

$$\phi = l(V_{\text{max}}'''/K_m D_S^{\text{eff}})^{1/2} \tag{33}$$

Numerical integration of Eq. (31) with the appropriate boundary conditions

$$\beta_0 = \beta \qquad \text{at} \quad z = 0 \tag{34}$$

and

$$d\beta_0/d_z = 0 \qquad \text{at} \quad z = 1 \tag{35}$$

yields the substrate concentration profile in the membrane, which is shown in Fig. 22 for $\beta = 1$ and at different values of ϕ. This graph illustrates that the substrate modulus is a suitable expression for showing the importance of diffusional resistances inside the membrane as the concentration gradients, and thus the magnitude of substrate depletion, increase with increasing values of ϕ.

In Fig. 23 the effective rate of reaction in the membrane, V, normalized to the saturation rate, V_{max}, is plotted as a function of the external concentration β with the modulus, ϕ, as the parameter. As seen, for $\phi \leq 1$ the reaction is essentially kinetically controlled, but at higher values of ϕ the rate of reaction is lower owing to internal substrate depletion. It should be noted that the shape of curves describing the bound enzyme activity against the substrate concentration resembles that of a rectangular hyperbola even at large moduli and does not display sigmoidal character (Selegny *et al.*, 1971a; Thomas *et al.*, 1972).

The effect of diffusional limitations is conveniently characterized by the effectiveness factor, η, defined in Eq. (12) as the ratio of the actual reaction rate to that rate which would be obtained if all enzyme

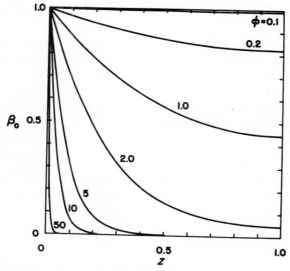

Fig. 22. Concentration profiles of the substrate in a membrane containing a bound enzyme uniformly distributed. The dimensionless substrate concentration in the membrane, β_0, is plotted against the dimensionless distance from the membrane surface, z, with the modulus, ϕ, as the parameter. The concentration of the substrate at the surface, β, is unity.

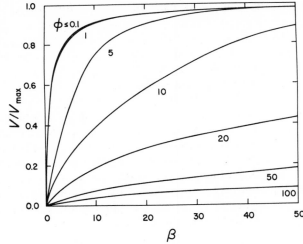

Fig. 23. Plots of the overall rate of reaction in an enzymic membrane, V, normalized to V_{max}, against the dimensionless concentration of the substrate in the macroenvironment β, at different values of the modulus, ϕ. The effect of diffusional limitations increases with increasing values of ϕ and results in a decrease of the overall rate of reaction with respect to the kinetically controlled rate obtained at $\phi \leqslant 0.1$. (From Horvath and Engasser, 1974.)

molecules in the membrane were exposed to the macroenvironmental concentration β. As in the case of external diffusion, the magnitude of internal diffusional inhibition can be expressed by the inverse of the effectiveness factor. In general, the effectiveness factor, η, is a function of ϕ and β (Engasser and Horvath, 1973). For $\beta < 0.1$, however, η becomes independent of β and approaches a limiting value, ϵ, which is related to ϕ by

$$\epsilon = \tanh \phi / \phi \qquad (36)$$

The dependence of V on the substrate concentration is usually complex, as seen in Fig. 23. At small concentrations, such as $\beta < 0.1$, however, the overall reaction rate is first order with the rate constant V_{max}/κ, where κ is defined by

$$\kappa = K_m/\epsilon = K_m\phi/\tanh \phi \qquad (37)$$

On the other hand, at sufficiently high values of ϕ, internal diffusion has no effect on the overall rate, as V is equal to the saturation rate, V_{max}. Thus, the observed rate shows a limiting behavior similar to that of the Michaelis–Menten scheme. Yet, its functional dependence on the surface concentration is different as illustrated by the Eadie–

Hofstee plots in Fig. 24. Such a plot yields a straight line only when diffusional limitations are negligible, i.e., for $\phi < 0.1$. Otherwise, with increasing values of ϕ sigmoidal curves are obtained. Comparison of Figs. 11 and 24 demonstrates that Eadie–Hofstee types of plots represent valuable diagnostic tools that enable one to distinguish between external and internal diffusion because internal, unlike external, diffusional resistances yield sigmoidal curves on such graphs.

The results of the numerical calculations have been verified by many experimental studies with immobilized enzymes. The influence of internal diffusion has been demonstrated by varying the thickness of polyacrylamide film (Van Duijn *et al.*, 1967; Bunting and Laidler, 1972) and collodion membranes (Goldman *et al.*, 1968a) or the diameter of porous-glass particles (Marsh *et al.*, 1973), Sephadex (Axén *et al.*, 1970) and ion exchange resin beads (Kobayashi and Moo-Young, 1973), which have been used as supports for the enzymes. In agreement with the theory, an increase in the characteristic length was

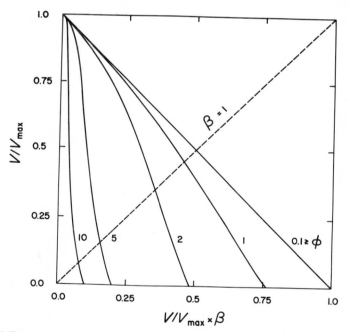

Fig. 24. Departure of the effective rate from Michaelis–Menten kinetics in an enzymic membrane due to diffusional limitations as illustrated on an Eadie–Hofstee type of plot at different values of the modulus, ϕ. The effective reaction rate, V, obtained at different macroenvironmental concentrations β, is normalized to V_{max}. As seen, diffusional limitations at $\phi > 0.1$ result in sigmoid curves. (From Horvath and Engasser, 1974.)

associated with a corresponding decrease in the effectiveness factor. Alternatively, the substrate concentration that is required to obtain half of the saturation activity was found to be larger than the K_m of the particular enzyme. When the activity of alkaline phosphatase attached to polyacrylamide films of different thicknesses was measured at various substrate concentrations, the shape of the Eadie–Hofstee plots obtained by Van Duijn *et al.* (1967) was very similar to the vertical curves shown in Fig. 24. Nonlinear Burk–Lineweaver plots have also been reported by Bunting and Laidler (1972) with β-galactosidase entrapped in relatively thick polyacrylamide membranes. Kasche *et al.* (1971), on the other hand, found that when α-chymotrypsin was covalently bound to agarose beads, the rate of reaction at high substrate concentrations was proportional to the enzyme concentration in the particles, whereas at low substrate concentration it was proportional to the square root of the enzyme content of agarose. This finding is easily explained in view of the previous theoretical results. At high substrate concentrations diffusional resistances have no effect on the enzyme kinetics; thus, the saturation activity, V_{max}''', is observed. At low concentrations, however, the effectiveness factor is equal to $1/\phi$ when internal diffusional resistances are significant, that is, $\phi > 10$. Then, the observed reaction rate is proportional to $\sqrt{V_{max}'''}$, therefore to the square root of the enzyme concentration.

B. Inhibition of Membrane-Bound Enzymes

The interplay of enzyme inhibition and diffusional resistances can be treated qualitatively in a similar fashion for both external and internal diffusion, provided the appropriate dimensionless numbers are used to represent the results (Moo-Young and Kobayashi, 1972; Thomas and Broun, 1973; Thomas *et al.*, 1974; Engasser and Horvath, 1974b).

When the inhibitor is neither a substrate nor a product, the combined effect of diffusional and chemical inhibition, as expressed by the efficiency factor, η_I, is shown in Fig. 25 as a function of I/K_I for different values of ϕ. Since at a fixed β, η_I is proportional to the rate of reaction, Fig. 25 clearly demonstrates that the effect of increasing inhibitor concentration on the enzyme activity is smaller in the presence than in the absence of diffusional limitations. As in the case of external diffusion, this antienergistic interaction between chemical and diffusional inhibition is readily accounted for by the fact that the inhibition not only reduces the inherent activity of the enzyme by a factor $[1 + (I/K_I)]$, but also diminishes the extent of substrate depletion. Under these conditions the importance of diffusional inhibition is no longer charac-

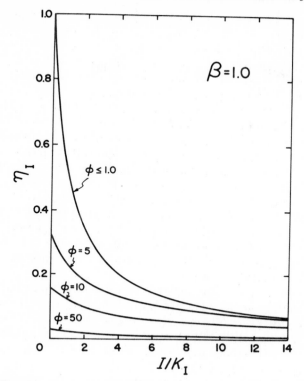

Fig. 25. Effect of noncompetitive inhibitor on the activity of bound enzymes with internal diffusion of the substrate. The efficiency factor, η_I, which expresses the combined effect of chemical and diffusional inhibition, is plotted as a function of the normalized inhibitor concentration I/K_I, at different values of the substrate modulus, ϕ, i.e., at different magnitudes of internal diffusional resistances. (From Engasser and Horvath, 1974b. Reprinted with permission from *Biochemistry* **13**, 3845–3849. Copyright by the American Chemical Society.)

terized by ϕ, but by the effective substrate modulus in the presence of an inhibitor, ϕ_I, given by

$$\phi_I = \phi/[1 + (I/K_I)]^{1/2} \tag{38}$$

Nevertheless, comparison of Figs. 12 and 25 shows that there is a major difference between external and internal diffusional effects. At high values of the external modulus, μ, the rate of the enzymic reaction is solely determined by the rate of external transport; therefore, it is independent of the inherent kinetic parameters of the enzymic reaction and unaffected by the action of a chemical inhibitor. On the other hand, internal diffusion and reaction occur in parallel so that the rate of the enzymic reaction can never be determined by the rate of substrate diffusion alone. Therefore the enzymic activity always depends on the kinetic parameters of the reaction and, unlike the case in external diffusional controlled reactions, the rate of reaction decreases with increasing inhibitor concentrations, even at large values of ϕ.

Experimental evidence of the antienergism was published by Carlsson *et al.* (1972), who used chymotrypsin and trypsin covalently bound to Sephadex, and benzamidine as the inhibitor. Both soluble and Sephadex-bound trypsin, for instance, showed the same inhibition profile when N-benzoyl-L-arginine-*p*-nitroanilide, a substrate with a low turnover number, was used. On the other hand when N-α-benzoyl-L-arginine ethyl ester (BAEE), a more active substrate, was used the bound trypsin was inhibited to a markedly lesser extent than trypsin in free solution, as shown in Fig. 26. This finding is easily explained by the strong diffusion limitations for BAEE which is rapidly hydrolyzed by trypsin. A similar observation has been reported by Thomas *et al.* (1974), who compared the inhibition of lactate dehydrogenase by tartrate with the enzyme in free solution and immobilized in collodion membranes.

Denaturation of an enzyme may be considered as a particular case of inhibition, namely an irreversible inhibition with time, to which the previous theoretical results can also be applied. In this case, the antienergism between diffusional inhibition and denaturation implies that diffusional resistances can greatly enhance the apparent stability of immobilized enzymes (Ollis, 1972). In the limit, when the system is bulk diffusion controlled, the observed activity of the bound enzymes stays constant although the enzyme may be denatured to a large extent. Therefore it is strongly recommended that the possible effect of diffusional phenomena is considered before drawing conclusions about the effect of immobilization on the stability of an enzyme.

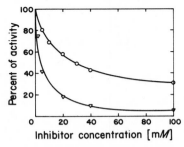

Fig. 26. Inhibitory effect of benzamidine on trypsin in free solution (▽), and on trypsin–Sephadex conjugate (○) as measured with N-α-benzoyl-L-arginine ethyl ester as the substrate at the pH optimum of the enzymic reaction in 0.1 *M* NaCl. Substrate concentration: $2 \times 10^{-3} M$. (From Carlsson *et al.*, 1972.)

When the substrate itself is an inhibitor at high concentrations, multiple steady states can be obtained in a certain concentration range of the substrate, provided internal diffusional resistances are sufficiently large (Moo-Young and Kobayashi, 1972). Product inhibition of membrane-bound enzymes also yields inhibition profiles similar to those shown in Figs. 12 and 13 (Engasser and Horvath, 1974c) except that the appropriate modulus is given by ϕ and the corresponding accumulation factor, α_i, is defined by

$$\alpha_i = D_S^{\text{eff}} K_m / D_P^{\text{eff}} K_P \tag{39}$$

When the reaction rate at changing substrate concentration is plotted according to the Eadie–Hofstee scheme, the graphs obtained are shown in Figs. 27 and 28 for the case of competitive and noncompetitive product inhibition. At low internal diffusional resistances, when $\phi \leqslant 0.1$, the curves for both competitive and noncompetitive inhibition are similar to those obtained previously (Figs. 16 and 18) for $\mu \leqslant 0.1$, except that the effect of product accumulation is much weaker at a given α_i than at the same value of α_e. On the other hand, at $\phi = 10$ the curves for sufficiently low α_i values are sigmoidal. With increasing values of α_i, however, the sigmoidal shape disappears and the resulting curves become similar to those shown in Figs. 16 and 18 for $\mu = 10$.

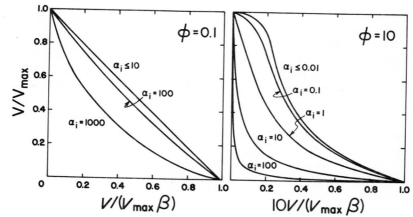

Fig. 27. Effect of competitive product inhibition with internal diffusional resistances by plots of the dimensionless rate of reaction, V/V_{max}, against $V/V_{\text{max}}\beta$, where β is the dimensionless substrate concentration in the macroenvironment, for different values of the internal accumulation factor, α_i, and at two values of the substrate modulus, ϕ. The dimensionless product concentration in the macroenvironment, P/K_P is 0.01. (From Engasser and Horvath, 1974c. Reprinted with permission from *Biochemistry* **13**, 3849–3854. Copyright by the American Chemical Society.)

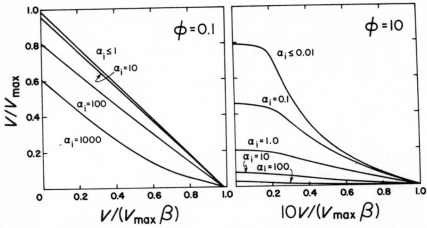

Fig. 28. Effect of noncompetitive product inhibition with internal diffusional resistances by plots of the dimensionless rate of reaction, V/V_{max}, against $V/V_{max}\beta$, where β is the dimensionless substrate concentration in the macroenvironment for different values of the internal accumulation factor, α_i, and at two values of the substrate modulus, ϕ. The dimensionless product concentration in the macroenvironment, P/K_P is 0.01. (From Engasser and Horvath, 1974c. Reprinted with permission from *Biochemistry* **13**, 3849–3854. Copyright by the American Chemical Society.)

Some of the effects of product inhibition in the presence of diffusional limitations have been verified experimentally (Engasser and Horvath, 1974c). Hydrogen ions are the product of the hydrolysis of BAEE and also act as trypsin inhibitors. At pH lower than 8, the effect of H^+ is tantamount to that of a noncompetitive inhibitor with an inhibition constant of $10^{-6}\ M$ as well as to that of a competitive inhibitor with an inhibition constant of $3 \times 10^{-7}\ M$. The rate of BAEE hydrolysis by soluble trypsin and a pellicular polyanionic trypsin conjugate (Horvath, 1974) was measured in the substrate concentration range from 10^{-6} to $10^{-2}\ M$, and the results were plotted according to the Eadie–Hofstee scheme as shown in Fig. 29. The data obtained with trypsin in free solution gave a straight line which indicates that the reaction obeyed Michaelis–Menten kinetics. Assuming that the transport coefficients and internal diffusivities were about the same for both BAEE and H^+, it was estimated that for the competitive inhibition by H^+ the value of both accumulation factors, α_e and α_i, would need to be about 20. Under such conditions, the combined effects of diffusional and competitive product inhibition yield, according to the previous theoretical results, concave plots in a wide range of the substrate moduli μ and ϕ, in accordance with the experimental finding shown in Fig. 29.

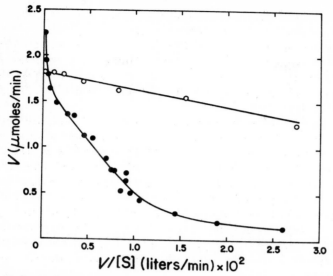

Fig. 29. Eadie–Hofstee plots for the hydrolysis of N-α-benzoyl-L-arginine ethyl ester by trypsin immobilized in a polyanionic gel (●) and by trypsin in free solution (○) at pH 8.0. The reaction rate was measured with the pH stat, and the mixture contained 1.2 M urea, 3.5×10^{-3} M EDTA, 0.3 M NaCl, and 22% (v/v) acetone. (From Engasser and Horvath, 1974c. Reprinted with permission from *Biochemistry* **13**, 3849–3854. Copyright by the American Chemical Society.)

C. Effect of Structural Changes in the Membrane

The activity of an enzyme embedded in a porous medium is affected not only by species that act as chemical effectors of the enzyme, but also by substances that modify the structure of the porous medium and thereby alter the internal diffusivity of the substrate or the product. Many immobilized enzyme systems are gels, which in contact with aqueous solutions undergo structural changes until they reach an equilibrium where the chemical potential of water is the same inside and outside the particles. The water in the interior is subject to a swelling pressure due to the contractive forces of the matrix; at equilibrium, this swelling pressure is equal to the osmotic pressure caused by the difference between the activities of the internal and external water (Helfferich, 1962). Neutral gels, such as agarose, cross-linked dextran, and polyacrylamide, can shrink or swell in response to changes in the composition of the surrounding aqueous solution. Other polymer matrices have fixed ionic charges, and electrostatic interactions between the fixed ionic groups and ions in the external solution can increase the stress on the matrix so that the resin shrinks. It has also been observed that multivalent ions, such as calcium, can cross-link polycarboxylic matrices via complex formation, which results in shrinking of the porous medium. Thus, at sufficiently high concentrations, commonly present ionic species can decrease the effective diffusivity of both the substrate and product inside the gel. Then, as a result of the increased diffusional inhibition, the rate of reaction also decreases.

Such a coarctation of the membrane structure by calcium has been demonstrated with trypsin immobilized in a polycarboxylic membrane (Horvath and Sovak, 1973). As illustrated in Fig. 30, the activity of trypsin in free solution increases with the Ca^{2+} concentration, whereas the activity of bound trypsin goes through a maximum at an intermediate calcium concentration and then decreases. This observation is explained by the binding of Ca^{2+} to the fixed carboxylic groups and the resulting increase in the cross-linking of the polymer network. Thus the effective diffusivity of the substrate, and consequently the rate of reaction in the gel, decreases with increasing Ca^{2+} concentration.

D. Combined Effect of External and Internal Diffusion

The combined effect of external and internal diffusion is now briefly analyzed in order to facilitate an assessment of their relative impor-

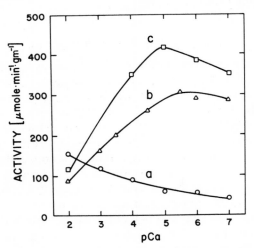

Fig. 30. Effect of Ca^{2+} concentration on the activity of trypsin in free solution (a) and in thin and thick polycarboxylic membranes (b and c, respectively). Substrate: $5 \times 10^{-3} M$ N-α-tosyl-L-arginine methyl ester in $10^{-2} M$ Tris buffer, pH 8.0. (From Horvath and Sovak, 1973.)

tance. The main interest is in the establishment of the conditions under which external diffusion has to be taken into account or can be neglected.

With external diffusional resistances the substrate concentration at the external surface of the porous particles or membrane, S_s, is different from the bulk concentration, S. As a result, Eq. (34) no longer yields the appropriate boundary conditions at the interface. Instead the correct boundary condition is expressed as

$$h_S(S - S_s) = -D_S^{eff}(dS_0/dx) \qquad \text{at} \quad x = 0 \tag{40}$$

which takes into account the continuity of the flux of substrate at the external surface. Equation (40) can be written in dimensionless form as

$$\text{Bi}[\beta - \beta_s] = -(d\beta_0/dz) \qquad \text{at} \quad z = 0 \tag{41}$$

where Bi is the Biot number given by

$$\text{Bi} = h_S l/D_S^{eff} = (D_S/D_S^{eff})(l/\delta) \tag{42}$$

The Biot number was introduced in the chemical engineering literature (Carberry, 1970) to describe the balance of external and internal diffusional resistances. According to Eqs. (41) and (42) the Biot number clearly expresses the relative magnitude of internal and external diffusional resistances for the porous enzymic matrix. The larger is the value of Bi, the smaller is the effect of external diffusion on the overall reaction behavior. The magnitude of external diffusional limitation thus depends on the corresponding value of the bulk concentration, β, the internal modulus, ϕ, and the Biot number, Bi.

From Eqs. (9) and (37) we obtain that the substrate modulus, μ, is related to ϕ and Bi by the following relationship:

$$\mu = \frac{lV_{max}'''}{h_S(K_m \phi/\tanh \phi)} = \frac{\phi \tanh \phi}{\text{Bi}} \tag{43}$$

When μ is smaller than unity, external diffusional resistances can generally be neglected, and the enzyme kinetics are affected only by internal diffusion, as has been discussed previously. At large values of ϕ and μ the effects of both external and internal diffusion have to be taken into account, and they can be assessed from the individual effects shown in the graphs in Figs. 9 and 17. It is seen that, when external or internal diffusional limitations are present alone, the rate of reaction becomes first order and zeroth order at sufficiently low and high substrate concentrations in the macroenvironment. It follows then that the combined effect of internal and external diffusional resistances has to give the same limiting kinetic behavior for membrane-bound enzymes. Thus, when both types of diffusional effects are simultaneously affecting the rate of reaction, the enzyme activity will reach the saturation rate, V_{max}, at sufficiently high substrate concentration and will show first-order kinetics with a rate constant, V_{max}/κ, at sufficiently low substrate concentration. The parameter κ is related to K_m and the first-order overall effectiveness factor, ϵ, by

$$\kappa = K_m/\epsilon \tag{44}$$

and ϵ is given by

$$\epsilon = \frac{\tanh \phi}{\phi\{1 + [(\phi \tanh \phi)/Bi]\}} \tag{45}$$

A more rigorous treatment of the combined effect of external and internal diffusion on Michaelis–Menten kinetics is complicated and involves a large number of dimensionless parameters. Therefore, it requires extensive machine calculations. Such studies have been made in order to analyze the behavior of open tubular heterogeneous enzyme reactors (Horvath *et al.*, 1973a) and of hollow fibers with enzymes immobilized in the porous wall of the fibers (Waterlands *et al.*, 1974).

VI. ANALYSIS OF DIFFUSIONAL EFFECTS AND DETERMINATION OF KINETIC PARAMETERS FROM EXPERIMENTAL DATA

When the kinetics of immobilized enzymes are experimentally investigated, the measured rate and kinetic parameters usually do not reflect the intrinsic kinetic behavior. As the effective rate parameters adequately describe the enzymic behavior only under a given set of experimental conditions, they are of limited value. First, they cannot be directly compared to the intrinsic kinetic parameters of the native enzyme in order to establish the effect of immobilization on the enzyme itself. Only the removal of diffusional and partition effects from the observed results would permit a meaningful comparison between the respective intrinsic parameters. Second, the effective rate parameters determined under certain experimental conditions may not be applicable to another situation when diffusional resistances, which are strongly affected by stirring and particle size, for instance, are different.

Therefore, the extraction of the intrinsic kinetic parameters—the main concern of biochemists—and that of the inherent rate parameters—a main concern of engineers—have received great atten-

tion in the literature. So far, most studies have dealt with the identifi-
cation and removal of diffusional effects. The different methods that
allow the diagnosis, quantitative evaluation and removal of diffusional
limitations are reviewed in this section.

A. Diagnosis of Diffusional Effects

Among the various approaches that are available to demonstrate the
presence of diffusional limitations, the most direct method involves
the change of the transport conditions for the substrate and the product. In view of the previous discussion, external diffusion plays a role if
the bound enzyme activity depends on the efficiency of mixing in the
bulk solution or on the flow rate through the heterogeneous reactor.
Internal diffusional limitations, on the other hand, can be detected by
comparing the effective reaction rates obtained with membranes of
different thicknesses or particles of different diameters. As seen later,
such a comparison can be used to quantitatively determine the inherent kinetic parameters and internal diffusivity.

According to previous results, diffusional limitations also yield
characteristic curves on the classical plots used in enzymology to
analyze Michaelis–Menten kinetics. The Eadie–Hofstee type of plot
has been found particularly appropriate for the diagnosis of external
and internal diffusional resistances. First, the deviations from straight
lines due to diffusional effects are more pronounced on the Eadie–
Hofstee plots than on the commonly used Lineweaver–Burk plots in
agreement with the earlier suggestion of Dowd and Riggs (1965) that
the Eadie–Hofstee plot is more suitable to discern deviations from
straightforward Michaelis–Menten kinetics. Second, Eadie–Hofstee
plots can yield information about the nature of the diffusional effect,
since, at least in the absence of product inhibition, external and internal diffusional limitations manifest themselves in concave and sigmoidal curves, respectively. Third, when the enzymic reaction is inhibited by its product, the shape of the Eadie–Hofstee plot is also
indicative of the relative magnitude of substrate depletion and
product accumulation.

The antienergistic interaction between diffusional and chemical inhibition may serve as an alternative diagnostic tool for diffusional resistances. As observed earlier, when the intrinsic inhibition constants for
the soluble and the bound enzyme are the same, the decrease in the
reaction rate with increasing inhibitor concentration is smaller for the
bound enzyme than for the soluble enzyme in the presence of diffusional limitations.

Finally, the presence of diffusional effects can be inferred from the dependence of the rate of reaction on the temperature and the shape of the Arrhenius plot. At sufficiently low temperature, as long as the reaction is kinetically controlled, the true activation energy of the reaction is observed. At intermediate temperatures, when internal diffusional limitations become significant, the measured activation energy has been shown to be half of the true activation energy (Smith, 1970). When the reaction becomes bulk diffusion controlled, at sufficiently high temperatures, the rate of reaction is practically independent of the temperature so that the apparent activation energy is essentially zero. As a result the examination of the Arrhenius plot obtained by varying the temperature of reaction can be indicative of the influence of diffusional effects at a given temperature, as illustrated in Fig. 31. It is noted, however, that nonlinear Arrhenius plots can arise from temperature-dependent conformational changes or other effects as well. A meaningful inference can be drawn only from the comparison of Arrhenius plots of the soluble and immobilized enzyme. Such studies have been carried out, for instance, by Bernfeld and Bieber (1969) and by Sato *et al*. (1971), who found that the apparent activation energy for the reaction catalyzed by immobilized aminoacylase was half that obtained with soluble enzyme.

B. Characterization and Evaluation of Diffusional Effects

The apparently straight lines on Lineweaver–Burk plots, which can be obtained for diffusion-limited reactions in a relatively narrow range of substrate concentrations, suggest that the interplay of diffusion and

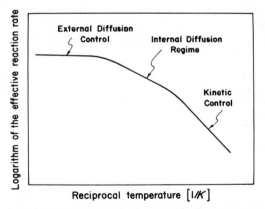

Fig. 31. Schematic illustration of the Arrhenius plot for reactions catalyzed by immobilized enzymes, when diffusional resistances are present in the system. It is assumed that in the temperature range investigated the intrinsic activation energy of the enzyme remains constant.

enzyme kinetics could be conveniently characterized by an apparent K_m value. Indeed, numerous efforts have been made to substantiate such a contrived parameter theoretically on the basis of approximate calculations. According to the graphs shown in previous sections, however, it is obvious that Michaelis–Menten kinetics are generally not obeyed when diffusional resistances affect the bound-enzyme kinetics. Consequently, in the rigorous sense there is no apparent Michaelis constant that would satisfy the kinetic law.

In the literature an "apparent" K_m has been defined in different ways. In some cases an "apparent" K_m was used to express the limiting first-order constant (Hornby *et al.*, 1968; Bunting and Laidler, 1972), which is equivalent to κ defined earlier in this chapter. Its value is readily obtained from the straight line portion of the Lineweaver–Burk plot, which corresponds to the first-order kinetic domain at low concentrations. In other cases, an "apparent" K_m was defined as the concentration at which the rate of the heterogeneous enzymic reaction was half the saturation rate (Sundaram *et al.*, 1972; Sundaram and Pye, 1974; Goldman *et al.*, 1971b). Although both expressions for "apparent" K_m yield the inherent K_m in the absence of diffusional limitations, the above two apparent K_m values significantly differ when diffusional resistances are present.

To avoid any confusion, the term apparent K_m should not be used when the effective reaction rate does not follow the Michaelis–Menten expression. Nevertheless, the kinetics of the immobilized enzyme still can be conveniently characterized either by the concentration that yields half of the saturation rate or by the kinetic parameter κ, that gives the limiting first-order constant, V_{max}/κ. It was shown earlier that the comparison of these two parameters with the inherent K_m of the enzymic reaction provides a quantitative evaluation of the magnitude of diffusional effects.

Undoubtedly, the most appropriate characterizations of diffusional resistances are the internal and external moduli, ϕ and μ. Yet, the precise evaluation of these two parameters requires the knowledge of the kinetic and transport parameters of the system, which are difficult to obtain from experimental data. Broun *et al.* (1972) and Thomas *et al.* (1972), for instance, found ϕ values of 2 and 6 for their glucose oxidase membrane. In Tables III and IV the values of ϕ and μ have been calculated for papain and alkaline phosphatase collodion membranes from the kinetic and transport parameters measured by Goldman *et al.* (1968b, 1971b). It is seen that diffusional limitations are significant with both kinds of enzymic membranes. Internal diffusion plays a predominant role when papain is immobilized in relatively thick

TABLE III

CHARACTERISTIC VALUES OF A COLLODION MEMBRANE WITH IMMOBILIZED PAPAIN
AS CALCULATED FOR THE HYDROLYSIS OF N-BENZOYL-L-ARGININAMIDE
FROM THE DATA[a] OF GOLDMAN *et al.* (1968b)

Thickness of enzyme layer, l (μm)	Internal modulus,[b] ϕ	Biot number,[c] Bi	External modulus,[d] μ	First-order effectiveness factor,[e] ϵ	κ/K_m[f]
470	4.3	47	0.09	0.21	4.8
156	1.45	16	0.09	0.57	1.8
49	0.45	0.5	0.04	0.90	1.1

[a] $K_m = 3.2 \times 10^{-2}$ M, $V'''_{max} = 0.82 \times 10^{-6}$ mole cm^{-3}sec^{-1}; substrate diffusivity in the bathing solution, $D_S = 6 \times 10^{-6}$cm^{2}sec^{-1}; substrate diffusivity in the membrane, $D_S^{eff} = 3 \times 10^{-6}$cm^{2}sec^{-1}; boundary layer thickness, $\delta = 20$ μm.

[b] Measure of the magnitude of internal diffusional resistances. When $\phi \leqslant 0.1$, internal diffusional effects can be neglected.

[c] Expresses the relative magnitude of internal and external diffusional resistances.

[d] Measure of the magnitude of external diffusional resistances. When $\mu \leqslant 0.1$ external diffusional effects are negligible.

[e] When $\epsilon = 1$ substrate transport does not affect the rate of reaction.

[f] Expresses the decrease in the limiting first-order rate constant, V_{max}/κ, due to diffusional limitations.

membranes. With the thin alkaline phosphatase membranes, however, both internal and external diffusional resistances have to be taken into account since the thickness of the membrane is smaller than that of the diffusional boundary layer.

A modified modulus, Φ, has been introduced in the chemical engineering literature, which can be determined from the observed rate of reaction V, so that the knowledge of the inherent kinetic parameters,

TABLE IV

CHARACTERISTIC VALUE OF A COLLODION MEMBRANE WITH IMMOBILIZED ALKALINE
PHOSPHATASE AS CALCULATED FOR THE HYDROLYSIS OF *p*-NITROPHENYL
PHOSPHATE FROM THE DATA[a] OF GOLDMAN *et al.* (1971b)

Thickness of enzyme layer, l (μm)	Internal modulus, ϕ	Biot number, Bi	External modulus, μ	First-order effectiveness, ϵ	κ/K_m
1.6	3.1	0.18	17.4	0.018	54
2.6	5.1	0.29	17.6	0.011	89
8.8	17.2	0.97	17.8	0.003	306

[a] $K_m = 3.4 \times 10^{-5}$ M, $V'''_{max} = 3 \times 10^{-5}$ mole cm^{-3}sec^{-1}; substrate diffusivity in the bathing solution, $D_S = 5 \times 10^{-6}$cm^{2}sec^{-1}; substrate diffusivity in the membrane, $D_S^{eff} = 2.3 \times 10^{-6}$cm^{2}sec^{-1}; boundary layer thickness, $\delta = 20$ μm.

V_{max} and K_m, is not necessary. As suggested by Wagner (1943) and Weisz and Hicks (1962) the modified modulus for a membrane is defined by

$$\Phi = l^2 V / D_S^{eff} S_s \upsilon \tag{46}$$

where υ is the porous catalyst volume. For a sphere, l is replaced by $R/3$. The substrate concentration at the surface, S_s, can be calculated from the bulk concentration, S, the measured activity, V, and the external transport coefficient, h_S, as

$$S_s = S - (V/h_S) \tag{47}$$

Internal diffusional limitations are generally negligible when $\Phi < 1$. For larger values of Φ, the effectiveness factor η, can be graphically determined from charts obtained by Roberts and Satterfield (1965) on the basis of theoretical calculations. If, in addition, the inherent K_m value of the bound enzyme is available, the modulus ϕ can be calculated using the following relationship:

$$\phi^2 = \Phi[(1 + S_s/K_m)/\eta] \tag{48}$$

When K_m is approximately known, ϕ can also be determined from the experimentally measured κ using Eq. (37), provided product inhibition is negligible.

C. Determination of Inherent Kinetic Parameters

The most direct procedure is to carry out the enzymic reaction in the kinetically controlled regime so that diffusional limitations are absent. Under these conditions the inherent V_{max} and K_m are experimentally observed. Often, however, the high activity of the bound enzyme and the low diffusivity of biochemical substances make it impossible to eliminate diffusional interferences. It was already shown that both internal and external diffusional effects can be observed even with membranes a few micrometers thick in a strongly mixed bathing solution. In order to facilitate the evaluation of experimental results, various methods have been therefore developed to allow the determination of inherent parameters in the presence of diffusional resistances.

External diffusional effects can be separated from the observed rate when the kinetic assay is carried out in a system for which the external transport parameters are known. Many such systems have been described in the chemical engineering literature. The procedure is mathematically the simplest when the external surface of the catalyst is, at least approximately, equiaccessible to the substrate. Rotating disks, small membranes and spherical particles in a stirred vessel,

open-tubular and packed-bed reactors at low conversions, are probably the most convenient equiaccessible systems with immobilized enzymes. The corresponding transport coefficients are available for a
wide range of hydrodynamic conditions.

When the transport coefficient of the substrate, h_S, is known, the
dependence of the reaction rate on the external surface concentration,
S_s, is easily determined from the measured rate dependence on the
bulk concentration, S, since Eq. (47) yields the surface concentration,
S_s, for any given activity and bulk concentration. Alternatively, a
graphical method based on Eq. (47) can be used, and this technique is
illustrated in Fig. 32. First the measured activity, V, is plotted against
the bulk concentration, S. Then, at any chosen S a straight line with the
slope equal to $-h_S$ is drawn. This line represents the rate of substrate
transport to the external surface for all S_s values from zero to the given
S. Then a horizontal line is drawn to represent the value of V at this
particular S. The intersection of these two lines yields the actual surface concentration, S_s, at the particular values of V and S. By repeating
the procedure for a number of bulk concentrations, a plot of the enzyme activity against the surface concentration can be constructed.
When the enzyme is attached to an impervious surface, this graphical
method directly yields the inherent kinetic rate, V_{kin}. From this plot,

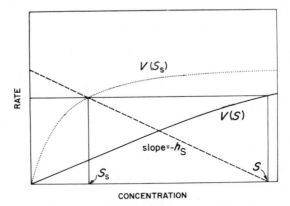

CONCENTRATION

Fig. 32. Graphical determination of the surface concentration, S_s, and the effective
rate, $V(S_s)$, of a heterogeneous enzyme reaction as a function of the surface concentration
from the overall rate of reaction $V(S)$ measured at various substrate concentrations in the
solution, S, when the mass-transfer coefficient for the substrate h_S is known. At a chosen
value of S a straight line of slope $-h_S$ is drawn. The intersection of this line with a
horizontal line drawn at the value of $V(S)$ measured at the same S yields both the
corresponding surface concentration on the abscissa and the rate of the reaction for this
surface concentration on the ordinate. By repeating this procedure for different S values
a plot of $V(S_s)$ against S_s can be constructed. (From Horvath and Engasser, 1974.)

the inherent kinetic parameters, V_{max} and K_m, are obtained by the standard methods of enzymology.

External diffusional effects may also be removed from kinetic data obtained with tubular or packed-bed enzyme reactors at high conversions. In the absence of product inhibition, the saturation activity can be obtained at sufficiently high substrate concentrations, at least theoretically, so that V_{max} is determined in this way. Concomitantly, the first-order constant, V_{max}/κ, is determined at low substrate concentrations using the overall additivity relation for first-order isothermal reactors (Engasser and Horvath, 1974f). This procedure then yields the appropriate V_{max} and K_m values for a surface-bound enzyme directly.

With enzymes entrapped in a porous matrix not only do external diffusional effects have to be removed, but it is also necessary to eliminate the effect of internal diffusion from the kinetic data in order to obtain the inherent kinetic parameters. Different methods have been proposed to determine V_{max} and K_m from the dependence of the reaction rate on the substrate concentration at the external surface of the membrane or particle when product inhibition is negligible. In each case, first the saturation activity is measured at a sufficiently high substrate concentration. Then the first-order constant V_{max}/K_m is evaluated by one of the following two graphical methods, which are based on the measurement of the enzyme activity at low concentrations and on the variation of the characteristic dimension of the enzymic particles. In one case, the triangle method of Weisz and Prater (1954) is used to obtain the modulus ϕ for the membranes or particles of different size. Alternatively, the values of ϕ are determined from the ratios of κ values measured at different membrane thicknesses or particle radii (Engasser and Horvath, 1973). If one of the ϕ values thus obtained is between 1 and 10, both the internal diffusivity, D_S^{eff}, and the first-order rate constant, V_{max}/K_m, can be evaluated by both methods. For larger values of ϕ, the first-order constant can still be obtained from ϕ provided D_S^{eff} is known. A particularly interesting way of varying the membrane thickness is to carry out the experiment first when both sides of the membrane are in contact with the substrate solution, then when only one side is exposed and the other is sealed (Meyer *et al.*, 1970).

Graphical methods have also been proposed for the determination of V_{max}/K_m from the substrate concentration at the external surface that yields the saturation rate (Kobayashi and Laidler, 1973). This procedure, however, necessitates the evaluation of the internal diffusivity from independent measurements. Membranes with both sides exposed to different substrate concentrations provide an alternative evaluation of the first-order rate constant. By measuring the flux of substrate through the membrane in the presence and in the absence of enzyme activity, both D_S^{eff} and V_{max}/K_m have been obtained (Selegny *et al.*, 1971a; DeSimone and Caplan, 1973). Two other methods based on asymptotic solutions and on the nonlinearity of Lineweaver–Burk plots have also been described in the literature (Hamilton *et al.*, 1974).

VII. DYNAMIC EFFECT OF WEAK ACIDS AND BASES ON HETEROGENEOUS ENZYME KINETICS

Weak acids and bases are commonly employed in biochemical experiments as buffers, and their use is essential in the study of enzyme kinetics (Sörensen, 1909). In addition to their coventionally accepted static role as a buffer in a certain pH range, however, they can also play a dynamic role in facilitating the transport of protons, weak acids, and bases (Engasser and Horvath, 1974g). As a result they can significantly affect the course of certain heterogeneous enzyme reactions when diffusional limitations are present in the system.

A. Dynamic Role of Acid–Base Pairs

The reversible binding of a proton by a weak base, B^-, to yield the conjugate acid, BH, accounts for the buffering effect of the acid–base pair in a certain pH range. The same reversible reaction can also provide a mechanism for the facilitated diffusion of protons, weak acids and bases under certain conditions (Gutknecht and Tosteson, 1973; Engasser and Horvath, 1974a).

The concept of facilitated proton transport is schematically illustrated in Fig. 33a. In the heterogeneous system under consideration, hydrogen ions are generated at the source and transported to the sink down their concentration gradient. For instance, protons may be produced at a surface by the enzymic reaction and then diffuse to the bulk solution, where they are neutralized. When an acid–base pair, BH and B^-, is present in the solution the proton transport is augmented by a shuttle mechanism, which involves the movement of the protonated buffer, HB, from the source to the sink and the movement of B^- in the reverse direction.

The shuttle is driven by the same concentration gradient that provides the driving force for diffusive and convective proton transport. Since the concentration of hydrogen ions is higher at the source than at the sink, the concentration of HB is also higher, whereas that of B^- is

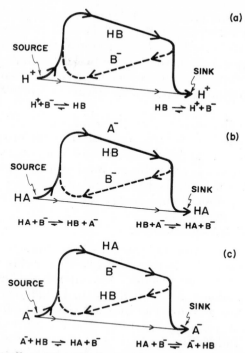

Fig. 33. Schematic illustration of buffer-facilitated transport of protons (a), weak acids (b), and weak bases (c). The concentration gradients of the species are illustrated by straight lines. (From Engasser and Horvath, 1975b.)

lower at the source than at the sink. The fluxes of HB and B^- driven by the corresponding concentration gradients create a shuttle that opens an alternative route for proton transport. In essence, the proton is bound to the carrier B^- at the source and the resulting HB moves down its gradient to the sink, where it dissociates. Then B^- returns to the source, where it is protonated and the cycle repeats itself.

The concept of the buffer shuttle can also be applied to the transport of weak acids and bases in heterogeneous systems. The reversible interaction between the buffer and the weak acid, HA, is expressed by the following reaction:

$$HA + B^- \rightleftarrows A^- + HB$$

As illustrated in Fig. 33b, HA yields A^- and the protonated buffer, HB, at the source. Both species move down their gradients to the sink, where A^- is reprotonated and removed from the system as HA. Then

B^- returns to the source, where it deprotonates HA so that a transport shuttle is established. Thus, the buffer-facilitated transport of weak acids is essentially a facilitated proton transport with B^- as the carrier. Figure 33c shows the scheme for the buffer-facilitated transport of a weak base, A^-. Here the transport of A^- from the source to the sink is facilitated by proton transport with B^- as the carrier in the reverse direction.

In many enzymic reactions the substrates or the products are protons, acids, or bases. Their transport to or from the enzyme microenvironment can thus be facilitated by the buffer or other substances such as EDTA and cysteine, which can bind protons reversibly. In the following section the effect of buffers is quantitatively illustrated on the pH profile of some bound enzymes.

B. pH Profile of Bound Enzymes

Among the many enzymic reactions that produce hydrogen ions, ester hydrolysis is probably the most significant in kinetic studies. The pH dependence of such reactions has often been investigated with immobilized enzymes also and found to be affected by diffusional limitations for the protons. The unusual shape of certain pH-activity profiles, which have been reported in the literature (Goldman *et al.*, 1968a; Silman and Karlin, 1967), can be explained by taking into account the facilitation of the proton transport by buffers or other proton acceptors. The effect of buffer-facilitated proton transport on the pH profile is briefly discussed as follows.

Let us assume that hydrogen ions are produced in a bound enzyme at the saturation rate, and the intrinsic pH dependence of the reaction is bell shaped, so that it is described by the following mathematical expression:

$$V_{max} = \frac{V^*}{1 + (H_0^+/K_1) + (K_2/H_0^+)} \tag{49}$$

where H_0^+ is the hydrogen ion concentration at the surface, V^*, K_1 and K_2 are the three kinetic parameters. At the saturation rate, the diffusion of the substrate has no effect on the rate, thus only the diffusion of protons formed in the reaction has to be considered. In the absence of any acid–base pair, the transport of protons takes place solely by molecular and convective diffusion. Therefore, at steady state the following relationship holds:

$$\frac{V^*}{1 + (H_0^+/K_1) + (K_2/H_0^+)} = h_H(H_0^+ - H^+) \tag{50}$$

where h_H is the proton transport coefficient and H^+ the macroenvironmental concentration of the hydrogen ions in the bulk solution. The magnitude of the diffusional resistances for protons is expressed by the dimensionless proton modulus, μ, as

$$\mu = V^*/h_H K_1 \tag{51}$$

The pH profiles obtained by solving Eqs. (49) and (50) are shown in Fig. 34, where the normalized enzyme activity is plotted against the bulk pH for different values of μ. The magnitude of K_1 and K_2 is arbitrarily chosen as 10^{-4} and 10^{-8} M, respectively. It is seen that only at $\mu \leq 10^{-5}$ is the pH profile of the bound enzyme unaffected by diffusional limitations. When $\mu > 10^{-4}$, protons accumulate at the surface, and as a result the enzyme activity plateaus at a sufficiently high pH indicating that the surface pH becomes independent of the bulk pH.

In the presence of a conjugate acid–base pair, B^- and BH, the hydrogen ions generated at the surface either diffuse freely or are carried by the buffer. At steady state, the rate of H^+ generation must be equal to the total rate of proton transport, i.e., to the sum of H^+ and BH transport. Thus

$$\frac{V^*}{1 + (H_0^+/K_1) + (K_2/H_0^+)} = h_H(H_0^+ - H^+) + h_{BH}(BH_0 - BH) \tag{52}$$

where h_{BH} is the transport coefficient for the protonated base; BH_0 and BH are the concentration at the surface and in the bulk, respectively.

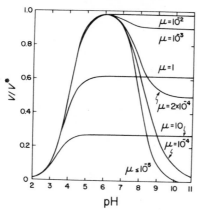

Fig. 34. pH profiles of a surface bound enzyme as a function of the pH of the bulk solution at different values of the proton modulus, μ. The curve for $\mu \leq 10^{-5}$ represents the pH profile of the enzyme without diffusional limitations. (From Engasser and Horvath, 1974a.)

In the case of a buffer having a concentration of $10^{-2} M$ and a pK_A value of 8, the pH profile of the bound enzyme is shown in Fig. 35 for different values of μ. As seen, the activity always decreases after reaching the maximum value and vanishes at a high enough bulk pH. On the other hand, at sufficiently high μ, the pH profile is S-shaped before reaching the maximum. This dependence of the enzymic activity on the bulk pH is readily accounted for by the dynamic role of the acid–base pair. At low pH, the concentration of B^- is too low to affect proton transport. As a result, H^+ accumulates at the surface and the activity begins to plateau as it does in the absence of buffer. With increasing pH, however, the generation of B^- results in a sharp decrease in H_0^+; therefore, the enzymic activity sharply increases as shown in Fig. 35. The activity reaches its maximum value when H_0^+ equals 10^{-6} and then decreases with the further increase in pH.

The shape of the pH activity profiles is strongly influenced by the concentration and the pK_A of the buffer (Engasser and Horvath, 1974a). At high enough buffer concentrations, the shuttle mechanism is efficient enough to prevent any significant accumulation of protons in the enzyme microenvironment and the intrinsic bell-shaped profile is obtained. Hydroxyl ions, which are always present in aqueous solution, can also act as proton carriers and facilitate proton transport at alkaline pH values.

Experimentally, pH profiles similar to those shown in Fig. 35 have been obtained by Goldman *et al.* (1968a), who studied the hydrolysis

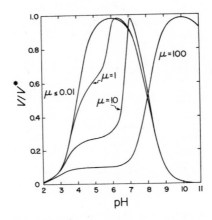

Fig. 35. pH profiles of a surface-bound enzyme as a function of the pH of the bulk solution at different values of μ in the presence of 10^{-2} *M* acid–base pair having a pK_A value of 8.0. The curve for $\mu \leq 10^{-2}$ represents the pH profile of the enzyme without diffusional limitations. (From Engasser and Horvath, 1974a.)

of N-α-benzoyl-L-arginine ethyl ester (BAEE) and N-benzoylglycyl ethyl ester (BGEE) by papain in free solution and immobilized in collodion membranes. Figure 36 shows some of the resulting pH profiles. In the presence of 0.1 M phosphate and 0.4 M Tris the pH profiles for BAEE and BGEE hydrolysis with immobilized papain were found to be very similar to the corresponding bell-shaped profiles obtained with soluble papain. When the solution contains only 5×10^{-3} M cysteine and 2×10^{-3} M EDTA, however, the pH profile for BAEE hydrolysis by the immobilized enzyme is S-shaped, flattening out in the neutral pH region and then rising at alkaline values up to pH 9.6. Under the same conditions the rate of BGEE hydrolysis reaches a constant value between pH 4 and 6, increases sharply at pH values up to 8.5, then decreases rapidly with the further increase of pH.

In view of the previous theoretical results, these experimentally obtained S-shaped pH profiles can be attributed to the facilitated transport of protons by cysteine and EDTA. In spite of the relatively low concentration of these species, they appear to facilitate the transport of hydrogen ions from the interior of the membrane to the bulk solution. At relatively high buffer concentrations, on the other hand, proton accumulation in the membrane is considerably reduced by the buffer shuttles. Under these conditions, the bell-shaped pH profiles characteristic of the absence of diffusional limitations are observed.

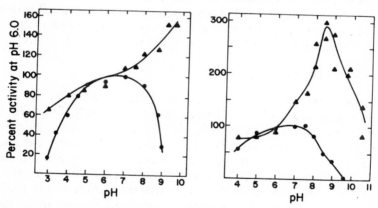

Fig. 36. pH activity profiles obtained with a papain membrane (▲) and with papain in free solution (●). The data on the left-hand graph were obtained by the pH-stat method with a reaction mixture of the following composition: 0.05 M N-α-benzoyl-L-arginine ethyl ester, 0.005 M cysteine, and 0.002 M ethylenediaminetetraacetic acid. The data on the right-hand graph were obtained in the same way with the following reaction mixture: 0.015 M N-benzoyl-L-glycine ethyl ester, 0.024 M 2.3-dimercaptopropanol, and 0.33 M KCl. (From Goldman *et al.*, 1968a. Reprinted with permission from *Biochemistry* **7**, 486–500. Copyright by the American Chemical Society.)

VIII. ELECTROSTATIC EFFECTS ON BOUND-ENZYME KINETICS

Enzymes are frequently immobilized on charged membranes or embedded in porous media containing fixed charges. At the same time many substrates and effectors are ionogenic or ionized in solution. Therefore, as discussed in Section I of this chapter, partitioning effects due to electrostatic interactions between the fixed charges and the mobile ions can have a significant influence on the kinetic behavior of bound enzymes.

The interaction between the charged support and mobile charged species often produces a nonuniform distribution of the charged substrate or effector between the micro- and macroenvironment of the bound enzyme. This partitioning effect between the enzyme medium and the bulk solution is discussed for the simplest case when diffusional limitations are negligible, i.e., when the reaction is kinetically controlled. The relationship between the intrinsic and inherent kinetic behavior under such conditions has been well established both theoretically and experimentally. On the other hand, electrical effects are likely to be significant also when the transport of substrate and product is driven, in addition to the concentration gradients, by an electrical potential gradient. So far, owing to its complexity, the influence of electrical potential on the magnitude of diffusional resistances has received only limited theoretical consideration (Shuler *et al.*, 1972, 1973; Hamilton *et al.*, 1973) and will not be discussed in this chapter.

A. Donnan Distribution

The distribution of charged species between a porous matrix with fixed ionic groups and the external solution is usually described by the Donnan equilibrium (Helfferich, 1962). Figure 37 schematically illustrates the concentration profiles for a positively charged substrate, S^+, a cation A^+, and an anion B^- in a membrane containing fixed negatively charged groups and in the bathing solution. In the following, the partition of the diffusible charged species will be quantitatively analyzed.

When the difference between the activity coefficients in the exterior and interior as well as the swelling pressure of the matrix are neglected, the potential difference at the membrane interface, E_D, is expressed by the ionic concentrations as

$$E_D = \frac{RT}{F} \ln \frac{S^+}{S_0^+} = \frac{RT}{F} \ln \frac{A^+}{A_0^+} = \frac{RT}{F} \ln \frac{B_0^-}{B^-} \tag{53}$$

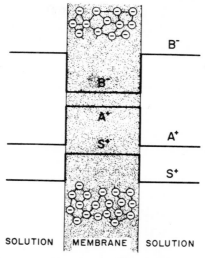

Fig. 37. Schematic illustration of the concentration profiles caused by the partition of a positively charged substrate, S^+, an anion, B^-, and a cation, A^+, between a membrane containing fixed negative charges and the bathing solution as a result of the Donnan distribution.

where F is the Faraday constant and the subscript 0 refers to the concentrations inside the membrane.

We define an electrostatic partition coefficient, Λ, by

$$\Lambda = S_0^+/S^+ = A_0^+/A^+ = B^-/B_0^- \tag{54}$$

which is then related to E_D, the Donnan potential, by

$$\Lambda = \exp(E_D/RT) \tag{55}$$

Electroneutrality in both the solution and the membrane requires that

$$S^+ + A^+ = B^- \tag{56}$$

and

$$S_0^+ + A_0^+ \pm X^\pm = B_0^- \tag{57}$$

where $X^\pm$ is the concentration of the fixed univalent ions in the matrix which can be positively or negatively charged.

From Eqs. (54), (56), and (57) it follows that for a given set of ionic concentrations Λ can be calculated by the second-order equation

$$B^-\Lambda^2 \pm X^\pm\Lambda - B^- = 0 \tag{58}$$

so that

$$\Lambda = \frac{\pm X^{\pm} + (X^{\pm 2} + 4B^{-2})^{1/2}}{2B^{-}} \tag{59}$$

As seen for the positively charged substrate, Λ is larger and smaller than unity when X is a negative or positive ion, respectively.

B. Effect on Michaelis–Menten Kinetics

The partition coefficient introduced previously can be used to determine the effect of electrostatic interactions on the Michaelis–Menten kinetics when the substrate is positively charged. The effective rate of the enzymic reaction, V, is given by

$$V = V_{\max}S_0^+/(K_m + S_0^+) \tag{60}$$

where K_m is the intrinsic Michaelis constant of the bound enzyme. S_0^+, the substrate concentration in the membrane, is calculated for any given external concentration S^+ from Eq. (54).

Since the partition coefficient Λ is a function of the concentrations of both the fixed charges, $X^{\pm}$, and the anion, B^-, in the solution, the dependence of the enzyme activity on the substrate concentration, S^+, is not the same when the concentration B^- is constant or varies with the substrate concentration.

1. Kinetics at Constant Ionic Strength

When the concentration of the positively charged substrate is varied in the exterior, the concentration B^-, and thus the ionic strength, can be kept constant either by adjusting the value of A^+ or by maintaining it at a sufficiently high level so that $A^+ \gg K_m$, that is A^+ is always much higher than S^+. Under these conditions Λ is constant and, as seen from Eq. (60), the dependence of the reaction rate on the external substrate concentration follows the Michaelis–Menten law with the same saturation rate but with the inherent Michaelis constant, K_m^{el}, given by

$$K_m^{el} = K_m/\Lambda \tag{61}$$

According to Eqs. (59) and (61), electrostatic effects can yield an inherent Michaelis constant that is smaller or larger than the intrinsic K_m depending on whether the substrate and the matrix carry like or unlike charges. At high ionic strength when $B^- \gg X^{\pm}$, however, electrostatic partition effects become negligible since Λ and K_m^{el} approach unity and K_m, respectively.

In previously published experiments the ionic strength was generally maintained constant either by adjusting the salt concentration in the reaction mixture (Goldstein *et al.*, 1964; Wharton *et al.*, 1968) or by working at salt concentrations much higher than the substrate concentration (Hornby *et al.*, 1966). Therefore, the observed reaction rates obeyed the Michaelis–Menten kinetic law. Table V shows that with uncharged supports both the intrinsic and inherent Michaelis constants have been found to be the same. In contradistinction, unlike charges on the substrate and the support enhance the substrate concentration in the microenvironment relative to that in the macroenvironment, and therefore cause the inherent K_m to be smaller than the intrinsic K_m. Conversely, like charges on the substrate and support result in an increase in the inherent Michaelis constant, since the substrate concentration in the vicinity of the enzyme is lower than in the bulk solution.

In many studies the magnitude of electrostatic effects is also strongly dependent on the ionic strength of the medium, i.e., on the total ionic concentration. As seen in Fig. 38, the intrinsic K_m for the hydrolysis of BAEE by bromelain in free solution is essentially constant over a wide range of ionic strength. The inherent K_m with bromelain attached to CM-cellulose, on the contrary, strongly increases with the ionic strength, then plateaus at high ionic strength. This experimental finding fully agrees with the previous theoretical results. Since BAEE is positively charged and CM-cellulose is negatively charged at pH 7, the partitioning of the substrate between the enzyme micro- and macroenvironment results in an inherent K_m smaller than the intrinsic K_m. Electrostatic effects, however, diminish with increasing ionic strength until the intrinsic Michaelis constant of the bound enzyme is obtained. Figure 38 also shows that soluble bromelain and insoluble CM-cellulose–bromelain complex yield different intrinsic Michaelis constants, and this observation suggests that the intrinsic properties of the enzyme have also been changed upon immobilization.

2. Kinetics at Changing Ionic Strength

When K_m and A^+ are of commensurable magnitude, the ionic strength, and as a result the electrostatic partition coefficient, varies with the substrate concentration at a fixed value of A^+. Since under such conditions Λ is a function of S^+, the reaction no longer obeys the Michaelis–Menten kinetic law (Engasser and Horvath, 1975a), as illustrated in Figs. 39 and 40 for $K_m = 5 \times 10^{-2}$ M and different values of A^+. As seen, the effect of electrostatic partitioning of the substrate between the membrane and the bulk solution cannot be simply characterized by an inherent Michaelis constant, K_m^{el}. Figure 39 shows that the dependence of the enzymic activity on the macroenvironmen-

TABLE V

EFFECT OF THE INTERACTION BETWEEN CHARGED SUBSTRATE AND CHARGED SUPPORT ON THE K_m OBSERVED
EXPERIMENTALLY WITH VARIOUS ENZYMES IN FREE SOLUTION AND IN IMMOBILIZED FORM[a]

Enzyme	Support	Charge	Substrate	Charge	Observed $K_m[M]$	References
ATP creatine phosphotransferase	None	0	ATP	−	6.5×10^{-4}	Hornby et al. (1968)
	p-Aminobenzyl cellulose	0	ATP	−	8.0×10^{-4}	
	CM-cellulose-90	−	ATP	−	7.0×10^{-3}	
Trypsin	None		BAA	+	6.8×10^{-3}	Goldstein et al. (1964)
	Maleic acid-ethylene	−	BAA	+	2.0×10^{-4}	
Chymotrypsin	None		ATEE	0	2.7×10^{-4}	C. Money and E. M. Crook (in Hornby et al., 1968)
	CM-cellulose-70	−	ATEE	0	5.6×10^{-4}	
Papain	None		BAEE	+	1.9×10^{-2}	Silman et al. (1966)
	p-Aminophenylalanine-L-leucine copolymer	0	BAEE	+	1.9×10^{-2}	
Ficin	None		BAEE	+	2.0×10^{-2}	Hornby et al. (1966)
	CM-cellulose-70	−	BAEE	+	2.0×10^{-3}	

[a] Abbreviations: ATP, adenosine triphosphate; CM, carboxymethyl; BAA, N-α-benzoyl-L-arginine amide; BAEE, N-α-benzoyl-L-arginine ethyl ester; ATEE, acetyl-L-tyrosine-ethyl ester.

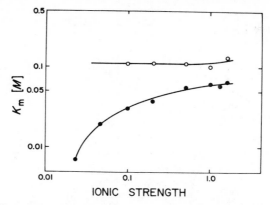

Fig. 38. Plot illustrating the effect of the ionic strength on the K_m value measured with bromelain in free solution (○) and with bromelain immobilized on carboxymethyl cellulose (●) for the hydrolysis of N-α-benzoyl-L-arginine ethyl ester at pH 7.0. The ionic strength was adjusted with KCl. (Adapted from data by Wharton *et al.*, 1968.)

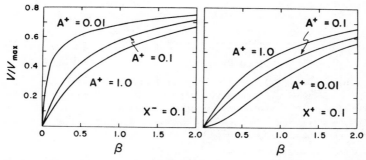

Fig. 39. Activity of a bound enzyme against the dimensionless concentration of a cationic substrate at different ionic concentrations of A^+ in the external solution. The enzymic microenvironment contains either negatively charged X^- or positively charged X^+ fixed groups. The intrinsic K_m of the enzymic reaction is $5 \times 10^{-2} M$. (From Engasser and Horvath, 1975a.)

tal substrate concentration is much more complex and may even become sigmoidal when both the substrate and membrane carry like charges and A^+ is sufficiently low. The comparison of the Lineweaver–Burk and Eadie–Hofstee-type plots in Fig. 40 demonstrates, in agreement with previous results obtained with diffusional effects, that the Eadie–Hofstee-type plots are more useful to diagnose deviations from the Michaelis–Menten kinetics and can yield characteristic curves, such as that for the remarkable sigmoidal kinetics.

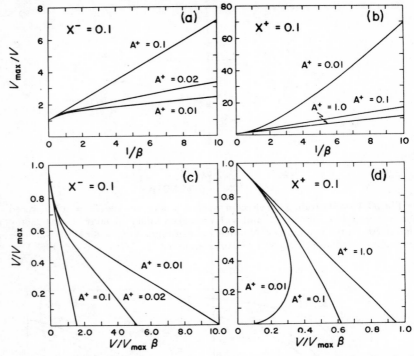

Fig. 40. Lineweaver–Burk, (a) and (b), and Eadie–Hofstee, (c) and (d), plots for the enzymic reaction with a cationic substrate in a charged membrane at different ionic concentrations of A^+ in the external solution. The intrinsic K_m value is $5 \times 10^{-2} M$. The fixed charges are negative X^- in (a) and (c) and positive X^+ in (b) and (d), respectively. (From Engasser and Horvath, 1975a.)

C. Effect on Inhibition and pH Profiles

Electrostatic partition can also affect the inhibition of bound enzymes. For instance, when the solution contains a positively charged noncompetitive inhibitor, I^+, together with other ions A^+ and B^-, the inhibitor concentration in the membrane, I_0^+, and in the solution, I^+, are related by

$$I_0^+ = \Lambda I^+ \tag{62}$$

where the partition coefficient Λ is given by Eq. (59). At high enough substrate concentration the reaction rate is expressed by

$$V = V_{max}/[1 + (I_0^+/K_I)] \tag{63}$$

where K_I is the intrinsic inhibition constant of the bound enzyme.

As in the case of the positively charged substrate, Λ is practically independent of I^+ when the ionic strength is constant. Then the kinetic effect of the fixed charged can be characterized by an inherent inhibition constant, K_I^{el}, which is given by

$$K_I^{el} = K_I/\Lambda \tag{64}$$

The kinetic behavior of bound enzymes, however, is more complex when K_I and A^+ are the same order of magnitude. This is shown in Fig. 41, which illustrates the dependence of the observed rate of reaction on the microenvironmental inhibitor concentration when the enzyme is immobilized in a neutral, a positively or a negatively charged membrane. As seen, the effect of the charged inhibitor is greatly modified by the fixed charges, and the sensitivity of the enzyme to the inhibitor concentration depends on the sign of the respective charges. In agreement with the theoretical results, the inhibition of immobilized trypsin by soybean trypsin inhibitor was found to be strongly dependent on the nature of the fixed charges (Levin *et al.*, 1964); similar effects have been observed on the inhibition of immobilized cholinesterase (Axén *et al.*, 1969).

Electrostatic partition effects can also account for the different redox dependence of papain activity in solution and on charged kaolinite

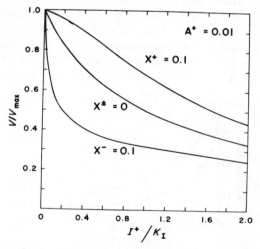

Fig. 41. Activity of a bound enzyme against the dimensionless concentration of cationic inhibitor, I^+/K_I, in the macroenvironment. The value of the inhibition constant, K_I, is $5 \times 10^{-2} M$. The enzymic microenvironment is either neutral, $X^{\pm} = 0$, or contains fixed positive charges X^+ or negative charges X^-. $A^+ = 0.01$. (Engasser and Horvath, 1975a.)

surface (Benesi and McLaren, 1975). As shown in Fig. 42, the relative activity of papain, a SH-dependent enzyme, decreases with the increasing ratio of oxidizing disulfide to reducing thiol. When papain is adsorbed on clay particles, however, the effect is more pronounced than in free solution. This finding is easily explained by the increased oxidizing potential in the microenvironment of the adsorbed enzyme due to the preferential attraction of disulfide, which has a double positive charge, to the negatively charged clay surface.

The pH-activity profile of bound enzymes has been thoroughly investigated both from the experimental and theoretical point of view. As early as in 1957, McLaren and Estermann compared the pH-activity profiles of chymotrypsin obtained in free solution and with the enzyme on negatively charged kaolinite particles. They found that the pH of half-maximum activity of the immobilized chymotrypsin was shifted about two units toward higher pH values (McLaren and Estermann, 1957). Goldstein *et al.* (1964, 1970), on the other hand, have shown that the pH-activity profiles of the polyanionic derivatives of several proteolytic enzymes are displaced toward more alkaline pH values at low ionic strength. Conversely, with polycationic derivatives of these enzymes the pH-activity profiles were shifted toward more acidic pH values.

This effect is illustrated by the data of Goldstein and Katchalski (1968) in Fig. 43, where the pH-activity profiles of soluble chymotrypsin are compared with those obtained with chymotrypsin immobilized on a negatively charged ethylene–maleic acid copolymer and a positively charged polyornithyl support. Similar shifts in pH profile have been reported with many other immobilized enzyme systems (Gold-

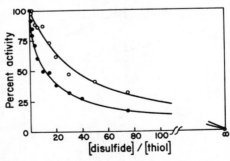

Fig. 42. Activity of papain in free solution at pH 6.1 (O) and of a papain–kaolinite complex at pH 6.9 in the bulk solution (●) against the concentration ratio of dithiodiglycol (oxidizing disulfide) and β-aminoethylmercaptan (reducing thiol) in the reaction mixture as measured by the rate of hydrolysis of N-α-benzoyl arginine ethyl ester. Ionic strength = 0.07. (Reprinted with permission from Benesi and McLaren, *Soil Biology and Biochemistry.* © 1975, Pergamon Press.)

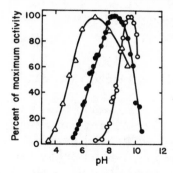

Fig. 43. pH-activity profiles obtained at low ionic strength (0.008) for chymotrypsin (●), for a polyanionic derivative of chymotrypsin (○) and for a polycationic derivative of chymotrypsin (△) using acetyl-L-tyrosine ethyl ester as the substrate. The polyanionic conjugate was prepared by immobilizing chymotrypsin with an ethylene–maleic anhydride copolymer. The polycationic derivative is polyornithyl chymotrypsin. (From Goldstein and Katchalski, 1968.)

stein, 1970; Hornby *et al.*, 1966; Patel *et al.*, 1969; McLaren and Babcock, 1959).

These results are easily explained by the unequal partition of the hydrogen ions between the enzyme micro- and macroenvironment. The usual bell-shaped pH profile of the enzymic activity is the result of the activation and the inhibition of the enzyme by H^+ at concentrations lower and higher than that at the pH optimum, respectively. Therefore, it can be characterized by the pertinent activation and inhibition constants, whose pK values are approximately at the pH values corresponding to the two half-maximum activities of the activity profile. Since the two constants are usually much smaller than the total ionic concentration of the solution, the ionic strength remains practically constant when only the H^+ concentration is varied in the pH domain of the activation and inhibition constants. Under such conditions both intrinsic constants are multiplied by the same factor to yield the corresponding inherent activation and inhibition constants according to Eq. (64). As the multiplication of the two constants by the same factor represents the same incremental change in their pK values, that is, in the pH of the half-maximum activities, the shape of the pH profile of the bound enzyme is not affected but is shifted toward higher or lower pH values depending on whether the matrix is negatively or positively charged. Displacements of the pH activity profile have been found to be most pronounced at low ionic strength in agreement with the theoretical results. Similarly at high ionic strength, when proton concentration differences between the solution and the membrane are negligible, the intrinsic pH activity profile of the bound enzyme has been obtained.

IX. COIMMOBILIZED MULTIENZYME SYSTEMS

The previous sections dealt exclusively with the kinetic behavior of heterogeneous systems containing a single enzyme, thus reflecting the fact that so far the overwhelming majority of experimental and theoretical studies have been focused on individual immobilized enzymes.

It is believed, however, that more complex multienzyme systems containing several coimmobilized enzymes that catalyze a series of consecutive reactions will receive increasing attention. First, many analytical and industrial processes that are likely to be carried out with immobilized enzymes are expected to involve a sequence of enzymic reactions. Therefore, multienzyme reactors and other systems, such as multienzyme electrodes, can facilitate the development of convenient and efficient processes for such applications. Although the use of a reactor comprising a combination of several individual immobilized enzymes in different particles or membranes can be of advantage under certain conditions (Mosbach *et al.*, 1974b), our discussion is restricted to heterogeneous enzyme systems that contain all the enzymes coimmobilized on the same support.

The second reason for our interest in multienzyme systems is their physiological significance. Most metabolic pathways consist of a series of reactions in which the product of an enzymic reaction is the substrate of another. The participating enzymes are usually bound to cellular membranes and organelles in a way that the consecutive reaction steps take place in proximity. Thus, coimmobilized multienzyme systems can serve as more refined models for the compartmentation of intracellular enzymes than can single immobilized enzymes, and their kinetic behavior is expected to yield valuable information about phenomena involved in the regulation and control of metabolic processes. So far only a few studies have been addressed to the behavior of coimmobilized enzymes; the results are briefly reviewed in this section.

Mosbach and Mattiasson (1970) bound hexokinase and glucose-6-phosphate dehydrogenase together onto polymer particles and compared the kinetics of this immobilized two-enzyme system with that of the same two enzymes in free solution. The reaction investigated was the conversion of glucose to 6-phosphogluconolactone according to the following scheme:

$$\text{glucose} \xrightarrow{\text{hexokinase}} \text{glucose -6-phosphate} \xrightarrow[\text{dehydrogenase}]{\text{glucose-6-phosphate}} \text{6-phosphogluconolactone}$$

As illustrated in Fig. 44, the steady-state activity is essentially the same in both cases, but the rate of reaction with the bound enzymes reaches the steady state much faster. This phenomenon is readily explained by the more rapid buildup of the steady-state substrate concentration in the microenvironment of the second coimmobilized enzyme. An even more significant reduction of the transient period was observed when β-galactosidase, hexokinase, and glucose-6-phosphate dehydrogenase, which catalyze three consecutive irreversible reactions, were bound to the same matrix (Mattiasson and Mosbach, 1971).

When the first of two consecutive enzymic reactions is reversible, on the other hand, the binding of the two enzymes on the same support may also result in an increase in the steady-state rate. This effect was observed by Srere *et al.* (1973), who coimmobilized malate dehydrogenase and citrate synthase. These enzymes catalyze the formation of oxaloacetate and citrate from malate according to the following reaction scheme:

$$\text{malate} \xrightarrow{\text{malate dehydrogenase}} \text{oxaloacetate} \xrightarrow{\text{citrate synthase}} \text{citrate}$$

Because in this system the first reaction is thermodynamically unfavorable in the direction of oxaloacetate formation, the coimmobilized enzymes showed a 2-fold increase in the steady-state rate of citrate formation as compared to the rate catalyzed by the same enzymes in free solution.

Both the observed decrease in response time and increase in steady-state activity upon coimmobilization of several enzymes in the same matrix have been attributed to the presence of diffusional resistances between the microenvironment of the bound enzymes and the bulk solution. Under such conditions, the intermediates of the reaction

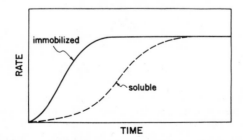

Fig. 44. Comparison of the overall transient rates of consecutive enzymic reactions catalyzed by the coimmobilized enzymes and the enzymes in free solution.

sequence have higher concentrations in the microenvironment of the immobilized enzymes than in the vicinity of the same enzymes in free solution. The transient behavior of two consecutive reactions catalyzed by two enzymes both in solution and in coimmobilized form has been analyzed by Goldman and Katchalski (1971).

Diffusional resistances for the intermediates in the enzymic environment can also explain the observed displacement of pH optima of immobilized multienzyme systems with respect to that of the same enzymes in free solution (Gestrelius *et al.*, 1972, 1973). Figure 45a shows the separate pH activity profiles of amyloglucosidase (pH optimum 4.8) and glucose oxidase (pH optimum 6.4). These profiles are not affected by immobilization of the enzymes on Sepharose particles per se. However, in the following reaction, catalyzed by the two coimmobilized enzymes,

$$\text{maltose} \xrightarrow{\text{amyloglucosidase}} \text{glucose} \xrightarrow{\text{glucose oxidase}} \text{gluconolactone}$$

the pH optimum is shifted to a higher value than that obtained with the corresponding enzymes in free solution, as shown in Fig. 45b. The magnitude of the displacement in the pH optimum is strongly dependent on the relative activity of the two enzymes present. With a large excess of glucose oxidase, the first reaction catalyzed by amyloglucosidase is the rate-limiting step, and consequently the pH optimum for the combined reaction catalyzed by both the soluble and bound enzymes is at pH 4.8, the pH optimum for amyloglucosidase. By decreasing the relative amount of glucose oxidase, however, the pH optima of both the matrix-bound and soluble enzymes increase as

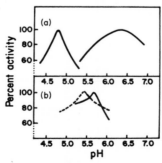

Fig. 45. Profiles of pH activity of two enzymes, amylo-α-1,4-α-1,6-glucosidase and glucose oxidase, (a) individually and (b) together. The individually determined pH profiles of each enzyme, shown on the upper graph, are the same in free solution and when the enzymes are bound to Sepharose. The pH profiles obtained with the two enzymes together, however, are different in free solution (dashed line) and in coimmobilized form (solid line) as shown on the lower graph. (From Gestrelius *et al.*, 1972.)

the second reaction becomes also rate determining. As seen in Fig. 45b, the pH optimum of the bound enzyme is shifted toward higher pH values because of the increased glucose concentration in the microenvironment of the bound glucose oxidase due to diffusional resistances.

In the coimmobilized multienzyme systems discussed above, each enzyme was randomly distributed throughout the polymer matrix just as in the free solution. Another kinetic behavior, as well as some specific spatial effects, may be expected when the different enzymes are confined to particular loci in the matrix. Thus, Selegny *et al.* (1970) could demonstrate the active transport of glucose with a structured double-layer bienzymic membrane. As shown in Fig. 46, the membrane consists of two adjacent catalytically active layers, one with immobilized hexokinase that catalyzes the phosphorylation of glucose with ATP, the other with immobilized phosphatase that catalyzes the hydrolysis of glucose 6-phosphate to glucose. Both external sides of the bienzymic membrane are covered with a selective carrier layer, which is permeable to glucose but impermeable to glucose 6-phosphate. Since glucose is consumed in the first layer and then reproduced in the second, the interplay of reaction and diffusion yields a sinusoidal profile of glucose concentration across the membrane. The concentration gradient thus established at the two surfaces of the

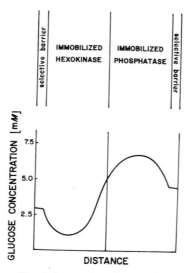

Fig. 46. Concentration profile of glucose in a bienzymic membrane system that demonstrates the active transport of glucose. (From Broun *et al.*, 1972.)

membrane causes the glucose to enter the membrane on the hexokinase side and leave it on the phosphatase side. The overall effect of the two enzymic reactions between the two selective barriers is such that glucose is transported across the membrane against its concentration difference between the contacting solutions at the external surfaces. As shown, the active transport of glucose requires not only that the two enzymes be immobilized in a membrane according to a precise geometrical distribution, but also that energy, in the form of ATP, be supplied to the system.

X. PERSPECTIVES

A. Immobilized Enzyme Kinetics

Although the complexity of heterogeneous enzyme systems is widely recognized, most theoretical results have been obtained with relatively simple models, namely, a single enzyme as well as a single substrate and product pair and at steady state. The effect of electrostatic interactions on transport has generally been neglected. These simplifications notwithstanding, theoretical models of these heterogeneous enzyme systems have been very useful to gain insight into the underlying physicochemical phenomena and to interpret the kinetic data obtained with many artificially or naturally bound enzymes. Nevertheless, further progress requires that more complex enzyme systems be studied both theoretically and experimentally. It is expected, therefore, that the scope of the investigations of bound-enzyme kinetics will be extended to include reaction schemes with multiple substrates and products, the transport of charged substrates and products in electrical fields, as well as the analysis of the transient behavior of multienzyme systems.

1. More Than One Substrate or Product; Cofactors

The majority of enzymic reactions involve at least two substrates and two products. Many reactions require a cofactor, which can also be considered as an additional substrate for our purpose. The previous theoretical results obtained with a single substrate are valid when the other reactant is in such a great excess that the rate of reaction is essentially independent of its concentration. This is the case in many enzymic reactions, especially with hydrolytic enzymes, because the concentration of water in aqueous solution is essentially constant, and its effect on the reaction kinetics can be neglected.

So far only a few experimental and theoretical studies have dealt with heterogeneous enzymic reactions whose rate is affected by two substrates. Cho and Swaisgood (1974) investigated the properties of rabbit muscle lactate dehydrogenase bound to porous glass. They found that the apparent Michaelis constant for NADH was greater with the immobilized than with the soluble enzyme, probably owing to diffusional limitations. On the other hand, the apparent K_m for pyruvate was lower, and this was attributed to conformational changes induced by the matrix. Atkinson and Lester (1974) theoretically investigated the kinetics of glucose oxidase immobilized in gel particles under conditions where both glucose and oxygen concentrations were limiting the rate of reaction. Their findings corroborate the expectations that such heterogeneous systems, in which two substrates are limiting the rate of reaction, are strongly influenced by the concerted effects of the two diffusional inhibitions. Similar effects are anticipated when the two products inhibit or activate the bound enzyme.

2. Effect of Electrical Field

In Section VIII, B and C, electrostatic effects were examined for enzymes entrapped in a charged membrane when the reaction rate was kinetically controlled, and therefore was not affected by the diffusion of the substrate. Under these conditions, the behavior of the bound enzyme was shown to be significantly changed by the partition of a charged substrate or effector between the bulk solution and the membrane.

Electrostatic phenomena, however, can also affect the transport of charged species and, consequently, may have an even greater influence on the overall kinetics when diffusional resistances are present in the heterogeneous enzyme system. So far only a few theoretical studies have considered the transport of substrate under the influence of both a concentration and an electrical potential gradient. The models employed were restricted to an enzyme bound to an impervious surface and the transport of a single charged species (Shuler *et al.*, 1972, 1973; Hamilton *et al.*, 1973; Kobayashi and Laidler, 1974). Without any doubt a broader theory that takes into account both the behavior of enzymes in charged matrices and the internal diffusion of charged substrate and products would be of great interest because most porous media employed with immobilized enzymes carry fixed charges and most biological substances are ionogenic. The decrease in the activity coefficient of charged species in charged membranes also deserves more attention (Laurent, 1971) in order to reflect the effect of electrostatic phenomena on the kinetics of the reaction.

3. Transient Analysis

In the past, only a few transient analyses of the kinetics in heterogeneous enzyme systems have been reported (Sundaram *et al.*, 1970; Selegny *et al.*, 1971b; Lin, 1972; Choi and Fan, 1973; Bruns *et al.*, 1973), and most theoretical and experimental studies have been based on the assumption of steady state. It is expected, however, that transient kinetics will gain more practical importance in the future as the theoretical approaches to the control of immobilized enzyme reactors, the optimization of enzyme electrodes, and the design of chromatographic enzyme reactors necessitate a departure from time invariant models. Our understanding of the regulation of metabolic processes in the heterogeneous cellular milieu would also be greatly enhanced by the analysis of transient models. Of particular biological significance are oscillations that can occur when an enzymic reaction is coupled with a membrane transport process. An enzyme-membrane oscillator has already been obtained (Naparstek *et al.*, 1973) with a glass electrode coated with a coreticulated papain–albumin membrane. The electrode was connected to a pH meter and immersed in a solution of *N*-benzoyl-L-arginine ethyl ester at relatively high pH. It was observed that the pH at the membrane–glass interface of the electrode oscillated at a certain hydrogen ion and substrate concentration in the bulk solution as well as at given stirring rate. Theoretical modeling of such systems (Zabusky and Hardin, 1973; Caplan *et al.*, 1973) showed that the oscillations were caused by the transient interaction of the autocatalytic enzyme reaction and diffusional inhibition.

4. Multienzyme Systems

It was shown in Section IX that when several different enzymes, which catalyze consecutive reactions, are immobilized in the same microenvironment, the resulting multienzyme system may be more efficient in catalyzing the overall reaction then the same enzymes in free solution. Decrease in the response time, increase in the steady-state activity, displacement of pH optima are some of the most interesting effects that have been observed with multienzyme systems and attributed to diffusional limitations. It is believed that coimmobilized multienzyme systems will find increasing application in enzymic reactors as well as in analytical and therapeutic devices and will also serve as valuable models of membrane bound enzymes *in vivo*. In order to fully understand the behavior of these systems, however, a significant amount of experimental and theoretical research is still necessary. For instance, a detailed quantitative analysis of the effect of diffusional resistances and the spatial distribution of the different enzymes would facilitate the interpretation of experimental results, the design of heterogeneous reactors and the elucidation of the mechanisms of certain metabolic processes.

B. Kinetics of Cellular Processes

Many metabolic pathways have been deciphered in terms of the elementary chemical reactions involved. The enzymes that catalyze consecutive steps in a metabolic pathway have been isolated, and their kinetic properties in free solution have been established. The results of such studies, together with those obtained from the measurement of average metabolite concentrations in cells, have then been used to establish the various steps and the concomitant control mechanisms in a particular pathway.

The regulation of metabolic processes, which involve one or more rate-limiting steps, has been inferred from the kinetics of the individual processes with the assumption that the rate of transport of the metabolites from one enzyme to the other in the pathway is relatively fast with respect to the rate of enzymic reactions. With diffusional resistances in the system, however, a transport step can also be rate limiting, and as a result concentration gradients can be present. Therefore, the rate and regulation of the pathway can be significantly different from those inferred from the behavior of the enzymes in free solution and the average cellular concentrations of the various metabolites. In view of this, the study of immobilized enzyme systems and the effect of transport phenomena on the overall kinetic behavior can shed light on certain so far neglected aspects in cellular physiology.

1. Diffusional Resistances in Metabolic Processes

Most studies of metabolic processes have been focused on enzymic reactions mainly because the investigation of transport processes has been beset with great experimental difficulties. Thus, at the present time, precise data on diffusional resistances inside the cells are not available. The diffusional resistances encountered by certain metabolites, however, are likely to be of physiological importance in view of the great heterogeneity of the cellular structure. It is now well established that cells contain a large number of organelles and that many enzymes are bound to membranous structures or subcellular particles. As a result, the interior of cells is characterized by an elaborate compartmentation of the various enzymes and metabolites either in the cytoplasm or in the organelles (Brown and Chattopadhyay, 1972; Coleman, 1973; Greville, 1969).

Only a few experimental bits of evidence for diffusional limitations *in vivo* have been reported. For instance, the transport of sugars and amino acids form the extracellular fluid to the intracellular milieu has been found to be the rate-determining step in many muscle cells (Morgan *et al.*, 1961; Odessey and Goldberg, 1972). The attenuating effect of "unstirred layers" on the rate of uptake of different species across biological membranes has also been demonstrated (Naftalin, 1971; Wilson *et al.*, 1971; Hersey and High, 1972; Zander and Schmid-Schoenbein, 1972). For intracellular processes, the evaluation of the possible values of the substrate modulus was used to estimate the magnitude of diffusional resistances (Engasser and Horvath, 1974b). With average values of the kinetic parameters of enzymic reaction and metabolite diffusivity in aqueous solution, the estimated modulus was smaller than unity. In agreement with the calculations by Shuler *et al.* (1973), this result showed that diffusional resistances are likely to be negligible for reactions that take place in the cytoplasm. When a metabolite has to diffuse across a membrane, however, high values of the modulus can be expected since the diffusivity through biological membranes can be several orders of magnitude smaller than in aqueous solution (Davson and Danielli, 1952). Therefore, in the heterogeneous structure of cells, diffusional resistances can indeed be a factor to be considered in cellular physiology.

2. Regulation of Cellular Processes

Because diffusional resistances are assumed to be localized in the cellular membranes, the concentration of metabolites may be quite different in various cell compartments. As a result, the rate and regulation of a given metabolic process is determined not by the average concentration of the metabolite in the cell, but by its local concentration in the microenvironment of the regulating enzyme. This chapter provides ample illustrations that the response of the enzymic reaction to changes in substrate, product, or effector concentrations in the macroenvironment is greatly affected by diffusional resistances. Under these conditions, the interplay of transport phenomena with enzymic reactions can result in either an attenuation or an amplification of the regulatory effect that could be inferred from changes in the average cellular concentrations. For instance, owing to the antienergism between diffusional and chemical inhibition, an enzyme may effectively become insensitive to an inhibitor. On the other hand, the regulation with substrate inhibition can be greatly magnified by diffusional resistances that result in multiple steady states.

When the transport of a metabolite is one of the rate-determining steps, the pathway can be regulated not only by modulation of the activity of the appropriate enzymes, but also by changes in the transport rate of the metabolite. It is well known that the permeability of cellular membranes may be altered by many species, such as hormones, calcium, and cholesterol. The measurement of changes in permeability, however, yields direct information only on the regulation of the transport step, not on the control of the pathway when an enzymic reaction also affects the overall rate of the metabolic process. Therefore, the precise regulatory role of species that modulate the transport of metabolites across membranes can be assessed only by taking into account the interaction of the rate-determining transport and reaction steps.

One of the pecularities of living systems is the fact that transport across membranes often occurs not by molecular or convective diffusion, but by a carrier or some other transport system. The theoretical analysis of this behavior leads then to nonlinear transport equations (Stein, 1967). Active transport, which requires an additional energy source, is also involved in many metabolic processes. Unlike molecular and convective diffusion, however, the interplay of such mediated transport phenomena with enzyme kinetics has received limited attention with regard to quantitative analysis (Post *et al.*, 1961; Cirillo, 1970; Engasser and Horvath, 1974e). Yet, such studies are expected to shed light on the detailed mechanism of metabolic regulation when transport is involved, particularly with respect to the action of hormones, which have the ability of modulating the carrier affinity or capacity in mediated transport (Randle *et al.*, 1966; Manchester, 1970; Haynes, 1972; Newsholme and Start, 1973).

Some of the theoretical and experimental results obtained with heterogeneous enzyme systems that are affected by transport phenomena have been used to elucidate certain aspects of the control of cellular processes. For example, the interplay of transport and reaction has been theoretically analyzed in order to gain insight into the uptake of glucose by muscle cells (Post *et al.*, 1961; Engasser and Horvath, 1974e), the regulation of the rate of oxidation in the Krebs cycle (Srere *et al.*, 1973), and the regulation of glycolysis by phosphofructokinase (Engasser and Horvath, 1974d). Further exploration of the effect of cell heterogeneity and compartmentation on the regulation of metabolism, however, would require an extension of the present theory to more complex systems, which involve multiple reactions and transport steps, multiple substrates and effectors, as well as mediated and active transport. The test of the theoretical results would require precise experimental data on the geometrical distribution of enzymes, the local concentration of metabolites in the different parts of the cell, the kinetics of intracellular transport processes, and the metabolic rates in various pathways *in vivo*.

LIST OF SYMBOLS[2]

A^+	Cation concentration in the macroenvironment (mole/cm³)
A_0^+	Cation concentration in the microenvironment (mole/cm³)
B^-	Anion concentration in the macroenvironment (mole/cm³)
B_0^-	Anion concentration in the microenvironment (mole/cm³)
BH	Concentration of undissociated buffer in the macroenvironment (mole/cm³)
BH_0	Concentration of undissociated buffer in the microenvironment (mole/cm³)
Bi	Biot number
D	Diffusivity (cm²/sec)
D_S	Substrate diffusivity in bulk solution (cm²/sec)
D_S^{eff}	Effective substrate diffusivity in porous medium (cm²/sec)
D_P^{eff}	Effective product diffusivity in porous medium (cm²/sec)
d	Characteristic length (cm)
E_D	Donnan potential (V)
F	Faraday constant (cal/V)
H^+	Hydrogen ion concentration in the macroenvironment (mole/cm³)
H_0^+	Hydrogen ion concentration in the microenvironment (mole/cm³)
h	Transport coefficient (cm/sec)
h_S	Transport coefficient for substrate (cm³/sec)
h_P	Transport coefficient for product (cm³/sec)
h_H	Transport coefficient for hydrogen ion (cm³/sec)
h_{BH}	Transport coefficient for undissociated buffer (cm³/sec)
I	Inhibitor concentration (mole/cm³)
I^+	Concentration of a positively charged inhibitor in the macroenvironment (mole/cm³)
I_0^+	Concentration of a positively charged inhibitor in the microenvironment (mole/cm³)
J_S	Flow rate of substrate to the surface (mole/sec)
J_P	Flow rate of product to the surface (mole/sec)
K	Ratio of the kinetic parameters for substrate inhibition, K_1/K_2
K_1	Kinetic parameter (mole/cm³)
K_2	Kinetic parameter (mole/cm³)
K_I	Inhibition constant (mole/cm³)
K_I^{el}	Inherent inhibition constant with electrostatic effect (mole/cm³)
K_m	Michaelis constant (mole/cm³)
K_m^*	Effective Michaelis constant, Eq. (22) (mole/cm³)

---

[2] Most results in this chapter are expressed in terms of dimensionless numbers. This derivation requires the use of a consistent system of units. In order to avoid misunderstanding, the dimensions of the various quantities are shown in CGS units. It is noted that these units are not necessarily the same as those commonly used in biochemistry. For instance, all concentrations are expressed in terms of mole/cm³, whereas the usual unit of concentration is mole/liter.

K_m^{el} Inherent Michaelis constant with electrostatic effect (mole/cm³)
K_P Product inhibition constant (mole/cm³)
l Membrane thickness (cm³)
Nu   Nusselt number
P Product concentration in the macroenvironment (mole/cm³)
P_0 Product concentration in the microenvironment (mole/cm³)
R Radius of a spherical particle (cm)
R Gas constant (cal/°K)
S Substrate concentration in the macroenvironment (mole/cm³)
S_0 Substrate concentration in the microenvironment (mole/cm³)
S^+ Concentration of positively charged substrate in the macroenvironment (mole/cm³)
S_0^+ Concentration of positively charged substrate in the microenvironment (mole/cm³)
S_s Surface concentration of substrate (mole/cm³)
T Temperature (°K)
V Effective rate of reaction (mole/sec)
V_{max} Saturation rate of enzymic reaction (mole/sec)
V_{max}''' Saturation rate of enzymic reaction per unit volume (mole/sec cm³)
V^* Kinetic parameter related to bell-shaped pH activity profile (mole/sec)
V_{kin} Inherent rate of reaction (mole/sec)
V_{diff} Inherent rate of substrate transport to enzymic surface (mole/sec)
$\mho$ Volume of porous medium (cm³)
x Distance in the membrane (cm)
$X^\pm$ Concentration of fixed positive or negative charges in the porous enzymic medium (mole/cm³)
z Dimensionless distance in the membrane

Greek Symbols

α_e Accumulation factor with external diffusion
α_i Accumulation factor with internal diffusion
β Dimensionless substrate concentration in the bulk solution
β_0 Dimensionless substrate concentration in the porous enzymic medium
β_s Dimensionless substrate concentration at the membrane interface
δ Thickness of diffusion boundary layer
ϵ Effectiveness factor for first-order reaction
ϵ Void fraction
η Effectiveness factor
η_I Efficiency factor for inhibition
κ Kinetic parameter for first-order reaction with diffusion limitation
Λ Partition coefficient
μ Substrate modulus for external diffusion
μ_I Substrate modulus with inhibition for external diffusion
Φ Modified modulus for internal diffusion
ϕ Substrate modulus for internal diffusion
Φ_I Substrate modulus with inhibition for internal diffusion
τ Tortuosity factor

216   *Jean-Marc Engasser and Csaba Horvath*

REFERENCES

Aris, R. (1957). *Chem. Eng. Sci.* **6**, 262.
Aris, R. (1972). *Math. Biosci.* **13**, 1–8.
Atkinson, B., and Lester, D. E. (1974). *Biotechnol. Bioeng.* **16**, 1321–1343.
Axén, R., and Ernback, S. (1971). *Eur. J. Biochem.* **18**, 351–360.
Axén, R., Heilbronn, E., and Winter, A. (1969). *Biochim. Biophys. Acta* **191**, 478–481.
Axén, R., Myrin, P. A., and Jansson, J. C. (1970). *Biopolymers* **9**, 401–413.
Bass, L., and McIlroy, D. K. (1968). *Biophys. J.* **8**, 99–108.
Benesi, A., and McLaren, A. D. (1975). *Soil Biol. Biochem.* **7**, 379–381.
Bernfeld, P., and Bieber, R. E. (1969). *Arch. Biochem. Biophys.* **131**, 587–595.
Bird, R. B., Stewart, W. E., and Lightfoot, E. D. (1960). "Transport Phenomena," pp. 396–405. Wiley, New York.
Blaedel, W. J., Kissel, T. R., and Boguslaski, R. C. (1972). *Anal. Chem.* **44**, 2030–2307.
Blum, J. J. (1956). *Biochim. Biophys. Acta* **21**, 155–166.
Blum, J. J., and Jenden, D. J. (1957). *Arch. Biochem. Biophys.* **66**, 316–332.
Brams, W. H., and McLaren, A. D. (1974). *Soil Biol. Biochem.* **6**, 183–189.
Broun, G., Thomas, D., and Selegny, E. (1972). *J. Membr. Biol.* **8**, 313–332.
Brown, H. D., and Chattopadhyay, S. K. (1972). *In* "Chemistry of the Cell Interface" (H. D. Brown, ed.), Part A, pp. 73–203. Academic Press, New York.
Bruns, D. D., Bailey, J. E., and Luss, D. (1973). *Biotechnol. Bioeng.* **15**, 1131–1145.
Bunting, P. S., and Laidler, K. J. (1972). *Biochemistry* **11**, 4477–4483.
Caplan, S. R., Naparstek, A., and Zabusky, N. J. (1973). *Nature (London)* **245**, 364–366.
Carberry, J. J. (1970). *Catal. Review* **3**, 61–91.
Carbonell, R. G., and Kostin, M. D. (1972). *AIChE J.* **18**, 1–12.
Carlsson, J., Gabel, D., and Axén, R. (1972). *Hoppe-Seyler's Z. Physiol. Chem.* **353**, 1850–1854.
Cha, S. (1970). *J. Biol. Chem.* **245**, 4814.
Chang, T. M. S. (1964). *Science* **146**, 524–625.
Cho, I. C., and Swaisgood, H. (1974). *Biochim. Biophys. Acta* **334**, 243–256.
Choi, P. S. K., and Fan, L. T. (1973). *J. Appl. Chem. Biotechnol.* **23**, 531–548.
Cirillo, V. C. (1970). *J. Protozool.* **17**, 178–181.
Coleman, R. (1973). *Biochim. Biophys. Acta* **300**, 1–30.
Cresswell, P., and Sanderson, A. R. (1970). *Biochem. J.* **119**, 447–451.
Damköhler, G. (1937). *Deut. Chem. Ing.* **3**, 430.
Davson, H., and Danielli, J. F. (1952). "The Permeability of Natural Membranes." Cambridge Univ. Press, London and New York.
Degani, J., and Miron, T. (1970). *Biochim. Biophys. Acta* **212**, 362–364.
DeSimone, J. A., and Caplan, S. R. (1973). *Biochemistry* **12**, 3032–3039.
Doscher, M. S., and Richards, F. M. (1963). *J. Biol. Chem.* **238**, 2399–2406.
Dowd, J. E., and Riggs, D. S. (1965). *J. Biol. Chem.* **240**, 863–869.
Engasser, J.-M., and Horvath, C. (1973). *J. Theor. Biol.* **42**, 137–155.
Engasser, J.-M., and Horvath, C. (1974a). *Biochim. Biophys. Acta* **358**, 178–192.
Engasser, J.-M., and Horvath, C. (1974b). *Biochemistry* **13**, 3845–3849.
Engasser, J.-M., and Horvath, C. (1974c). *Biochemistry* **13**, 3849–3854.
Engasser, J.-M., and Horvath, C. (1974d). *Biochemistry* **13**, 3855–3859.
Engasser, J.-M., and Horvath, C. (1974e). *Arch. Biochem. Biophys.* **164**, 37–42.
Engasser, J.-M., and Horvath, C. (1974f). *Chem. Eng. Sci.* **29**, 2259–2262.
Engasser, J.-M., and Horvath, C. (1974g). *Physiol. Chem. Physics* **6**, 541–543.
Engasser, J.-M., and Horvath, C. (1975a). *Biochem. J.* **145**, 431–435.
Engasser, J.-M., and Horvath, C. (1975b). *Physiol. Chem. Phys.* **6**, 541–543.
Filippusson, H., and Hornby, W. E. (1970). *Biochem. J.* **120**, 215–219.

Fink, D. J., Na, T., and Schultz, J. S. (1973). *Biotechnol. Bioeng.* **15**, 879–888.
Frank-Kamenetskii, D. A. (1969). "Diffusion and Heat Transfer in Chemical Kinetics," pp. 53–57. Plenum, New York.
Gabel, D., and Hofsten, B. V. (1971). *Eur. J. Biochem.* **15**, 410.
Gatfield, I. L., and Stute, R. (1972). *FEBS Lett.* **28**, 29–31.
Gestrelius, S., Mattiasson, B., and Mosbach, K. (1972). *Biochim. Biophys. Acta* **276**, 339–343.
Gestrelius, S., Mattiasson, B., and Mosbach, K. (1973). *Eur. J. Biochem.* **36**, 89–96.
Glassmeyer, C. K., and Ogle, J. D. (1971). *Biochemistry* **10**, 786–792.
Goldman, R., and Katchalski, E. (1971). *J. Theor. Biol.* **32**, 243–257.
Goldman, R., Kedem, O., Silmon, I. H., Caplan, S. R., and Katchalski, E. (1968a). *Biochemistry* **7**, 486–500.
Goldman, R., Kedem, O., and Katchalski, E. (1968b). *Biochemistry* **7**, 4518–4531.
Goldman, R., Goldstein, L., and Katchalski, E. (1971a). *In* "Biochemical Aspects of Reactions on Solid Supports" (G. R. Stark, ed.), p. 36. Academic Press, New York.
Goldman, R., Kedem, O., and Katchalski, E. (1971b). *Biochemistry* **10**, 165–172.
Goldstein, L. (1970). *In* "Proteolytic Enzymes" (G. Perlmann and L. Lorand, eds.), Methods in Enzymology, Vol. 19, pp. 935–962. Academic Press, New York.
Goldstein, L. (1972). *Biochemistry* **11**, 4072–4084.
Goldstein, L., and Katchalski, E. (1968). *Z. Anal. Chem.* **243**, 375.
Goldstein, L., Levin, Y., and Katchalski, E. (1964). *Biochemistry* **3**, 1913–1919.
Goldstein, L., Pecht, M., Blumberg, S., Atlas, D., and Levin, Y. (1970). *Biochemistry* **9**, 2322.
Greville, G. D. (1969). *In* "Citric Acid Cycle. Control and Compartmentation" (J. M. Lowenstein, ed.), pp. 1–105. Dekker, New York.
Gutknecht, J., and Tosteson, D. C. (1973). *Science* **182**, 1258–1261.
Hamilton, B. K., Stockmeyer, L. J., and Colton, C. K. (1973). *J. Theor. Biol.* **41**, 547–560.
Hamilton, B. K., Gardner, C. R., and Colton, C. K. (1974). *AIChE J.* **20**, 503–510.
Haynes, R., and Walsh, K. A. (1969). *Biochem. Biophys. Res. Commun.* **36**, 235.
Haynes, R. C. (1972). *In* "Energy Metabolism and the Regulation of Metabolic Processes in Mitochondria" (M. A. Mehlman and R. W. Hanson, eds.), pp. 239–252. Academic Press, New York.
Helfferich, F. (1962). "Ion Exchange," pp. 96–100. McGraw-Hill, New York.
Hersey, S. J., and High, W. L. (1972). *Am. J. Physiol.* **223**, 903.
Hicks, G. P., and Updike, S. J. (1966). *Anal. Chem.* **38**, 726.
Hill, A. V. (1928). *Proc. R. Soc. London, Ser. B* **104**, 39–95.
Hornby, W. E., Lilly, M. D., and Crook, E. M. (1966). *Biochem. J.* **98**, 420–425.
Hornby, W. E., Lilly, M. D., and Crook, E. M. (1968). *Biochem. J.* **107**, 669–674.
Horvath, C. (1974). *Biochim. Biophys. Acta* **358**, 164–177.
Horvath, C., and Engasser, J.-M. (1973). *Ind. Eng. Chem., Fundam.* **12**, 229–235.
Horvath, C., and Engasser, J.-M. (1974). *Biotechnol. Bioeng.* **16**, 909–923.
Horvath, C., and Solomon, B. A. (1972). *Biotechnol. Bioeng.* **14**, 885–914.
Horvath, C., and Sovak, M. (1973). *Biochim. Biophys. Acta* **298**, 850–860.
Horvath, C., Shendalman, L. H., and Light, R. T. (1973a). *Chem. Eng. Sci.* **28**, 375–388.
Horvath, C., Solomon, B. A., and Engasser, J.-M. (1973b). *Ind. Eng. Chem., Fundam.* **12**, 431–439.
Jacobs, M. H. (1935). *Ergeb. Biol.* **12**, 1–155.
Kasche, H., Lundquist, H., Bergmann, R., and Axén, R. (1971). *Biochem. Biophys. Res. Commun.* **41**, 615–621.
Katchalski, E., Silman, I., and Goldman, R. (1971). *Adv. Enzymol. Relat. Areas Mol. Biol.* **34**, 445–536.
Kobayashi, T., and Laidler, K. J. (1973). *Biochim. Biophys. Acta* **302**, 1–12.

Kobayashi, T., and Laidler, K. J. (1974). *Biotechnol. Bioeng.* **16**, 77–97.

Kobayashi, T., and Moo-Young, M. (1973). *Biotechnol. Bioeng.* **15**, 47–67.

Kobayashi, T., Van Dedem, G., and Moo-Young, M. (1973). *Biotechnol. Bioeng.* **15**, 27–45.

Koch, A. L., and Coffman, R. (1970). *Biotechnol. Bioeng.* **12**, 651–677.

Krogh, A. (1919). *J. Physiol. (London)* **52**, 409–415.

Lasch, J. (1973). *Mol. Cell. Biochem.* **2**, 79–86.

Laurent, T. C. (1971). *Eur. J. Biochem.* **21**, 498–506.

Levenspiel, O. (1972). "Chemical Reaction Engineering," p. 464. Wiley, New York.

Levich, V. G. (1962). "Physicochemical Hydrodynamics," pp. 39–136. Prentice-Hall, Englewood Cliffs, New Jersey.

Levin, Y., Pecht, M., Goldstein, L., and Katchalski, E. (1964). *Biochemistry* **3**, 1905–1913.

Lilly, M. D., Hornby, W. E., and Crook, E. M. (1966). *Biochem. J.* **100**, 718–723.

Lin, S. H. (1972). *Biophysik* **8**, 264–270.

McLaren, A. D., and Babcock, K. L. (1959). In "Subcellular Particles" (T. Hayashi, ed.), pp. 23–36. Ronald Press, New York.

McLaren, A. D., and Estermann, E. F. (1957). *Arch. Biochem. Biophys.* **68**, 157–160.

McLaren, A. D., and Packer, L. (1970). *Adv. Enzymol. Relat. Areas Mol. Biol.* **33**, 245–308.

McLaren, A. D., and Peterson, G. H. (1967). "Soil Biochemistry." Dekker, New York.

Manchester, K. L. (1970). In "Biochemical Actions of Hormones" (G. Litwack, ed.), Vol. 1, pp. 267–320. Academic Press, New York.

Marsh, D. R., Lee, Y. Y., and Tsao, G. T. (1973). *Biotechnol. Bioeng.* **15**, 483–492.

Mattiasson, B., and Mosbach, K. (1971). *Biochim. Biophys. Acta* **235**, 253–257.

Melrose, G. J. H. (1971). *Rev. Pure Appl. Chem.* **21**, 83.

Meyer, J., Sauer, F., and Woermann, D. (1970). *Ber. Bunsenges. Phys. Chem.* **74**, 245–250.

Mogensen, A. O., and Vieth, W. R. (1973). *Biotechnol. Bioeng.* **15**, 467–481.

Moo-Young, M., and Kobayashi, T. (1972). *Can. J. Chem. Eng.* **50**, 162–167.

Morgan, H. E., Cadenas, E., Regen, D. M., and Park, C. R. (1961). *J. Biol. Chem.* **236**, 262–268.

Mosbach, K., and Mattiasson, B. (1970). *Acta Chem. Scand.* **24**, 2093–2100.

Mosbach, K., and Mosbach, R. (1966). *Acta Chem. Scand.* **20**, 2807–2810.

Mosbach, K., Larsson, P. O., Brodelius, P., Guilford, H., and Lindberg, M. (1974a). In "Enzyme Engineering" (E. K. Pye and L. B. Wingard, eds.), Vol. 2, pp. 237–242. Plenum, New York.

Mosbach, K., Mattiasson, B., Gestrelius, S., and Srere, P. A. (1974b). In "Insolubilized Enzymes" (M. Salmona, C. Saronio, and S. Garattini, eds.), pp. 123–133. Raven, New York.

Naftalin, R. J. (1971). *Biochim. Biophys. Acta* **233**, 635.

Naparstek, A., Thomas, D., and Caplan, S. R. (1973). *Biochim. Biophys. Acta* **323**, 643–646.

Newsholme, E. A., and Start, C. (1973). "Regulation in Metabolism," pp. 329–337. Wiley, New York.

Odessey, R., and Goldberg, A. L. (1972). *Am. J. Physiol.* **223**, 1376–1383.

Ollis, D. F. (1972). *Biotechnol. Bioeng.* **14**, 871–884.

O'Neill, S. P. (1972). *Biotechnol. Bioeng.* **14**, 675–678.

Patel, A. B., Pennington, S. N., and Brown, H. D. (1969). *Biochim. Biophys. Acta* **178**, 626.

Porath, J., Axen, R., and Ernback, S. (1967). *Nature (London)* **215**, 1491–1492.

Post, R. L., Morgan, H. E., and Park, C. R. (1961). *J. Biol. Chem.* **236**, 269–272.

Racker, E. (1967). *Fed. Proc., Fed. Am. Soc. Exp. Biol.* **26**, 1335–1340.

Randle, P. J., Garland, P. B., Hales, C. N., Newsholme, E. A., Denton, R. M., and Pogson, C. I. (1966). *Recent Prog. Horm. Res.* **22**, 1–48.

Rashevsky, N. (1948). "Mathematical Biophysics." Univ. of Chicago Press, Chicago, Illinois.

Reiner, J. M. (1969). "Behavior of Enzyme Systems," p. 82. Van Nostrand-Reinhold, New York.

Roberts, G. W., and Satterfield, C. N. (1965). *Ind. Eng. Chem., Fundam.* **4**, 288–293.

Rony, P. R. (1971). *Biotechnol. Bioeng.* **13**, 431–447.

Roughton, F. J. W. (1932). *Proc. R. Soc. London, Ser. B* **111**, 1.

Rovito, B. J., and Kittrell, J. R. (1973). *Biotechnol. Bioeng.* **15**, 143–161.

Satterfield, C. N. (1970). "Mass Transfer in Heterogeneous Catalysis," p. 33. MIT Press, Cambridge, Massachusetts.

Sato, T., Mori, T., Tosa, T., and Chibata, I. (1971). *Arch. Biochem. Biophys.* **147**, 788–796.

Selegny, E., Broun, G., and Thomas, D. (1970). *C. R. Acad. Sci.* **271**, 1423–1426.

Selegny, E., Broun, G., and Thomas, D. (1971a). *Physiol. Veg.* **9**, 25–50.

Selegny, E., Kernevez, J. P., Broun, G., and Thomas, D. (1971b). *Physiol. Veg.* **9**, 51–63.

Sharp, A. K., Kay, G., and Lilly, M. D. (1969). *Biotechnol. Bioeng.* **11**, 363–380.

Shuler, M. L., Aris, R., and Tsuchiya, H. M. (1972). *J. Theor. Biol.* **35**, 67–76.

Shuler, M. L., Tsuchiya, M., and Aris, R. (1973). *J. Theor. Biol.* **41**, 347–356.

Silman, H. I., and Karlin, A. (1967). *Proc. Natl. Acad. Sci. U.S.A.* **58**, 1664–1668.

Silman, H. I., Albu-Weissenberg, M., and Katchalski, E. (1966). *Biopolymers* **4**, 441–448.

Sluyterman, L. A. A., and De Graaf, M. J. M. (1969). *Biochim. Biophys. Acta* **171**, 277–288.

Smith, J. M. (1970). "Chemical Engineering Kinetics," p. 444. McGraw-Hill, New York.

Sörensen, S. P. L. (1909). *Biochem. Z.* **21**, 131–304.

Srere, P. A., Mattiasson, B., and Mosbach, K. (1973). *Proc. Natl. Acad. Sci. U.S.A.* **70**, 2534–2538.

Steers, E., Cuatrecasas, P., and Pollard, H. (1971). *J. Biol. Chem.* **246**, 196.

Stein, W. D. (1967). "The Movement of Molecules Across Cell Membranes," pp. 126–176. Academic Press, New York.

Sundaram, P. V. (1973). *Biochim. Biophys. Acta* **321**, 319–328.

Sundaram, P. V., and Laidler, K. J. (1972). *In* "Chemistry of the Cell Interface" (M. D. Brown, ed.), pp. 255–296. Academic Press, New York.

Sundaram, P. V., and Pye, E. K. (1974). *In* "Enzyme Engineering" (E. K. Pye and L. B. Wingard, eds.), Vol. 2, pp. 449–452. Plenum, New York.

Sundaram, P. V., Tweedale, H., and Laidler, K. J. (1970). *Can. J. Chem.* **48**, 1498–1504.

Sundaram, P. V., Pye, E. K., Chang, T. M. S., Edwards, V. H., Humphrey, A. E., Mosbach, K., Patchornik, A., Porath, J., Weetall, H. H., and Wingard, L. B. (1972). *In* "Enzyme Engineering" (L. B. Wingard, ed.), Vol. 1, pp. 15–18. Wiley (Interscience), New York.

Suzuki, H., Ozawa, U., and Maeda, H. (1966). *Agric. Biol. Chem.* **30**, 807.

Swaisgood, H. E., and Horton, H. R. (1974). *In* "Enzyme Engineering" (E. K. Pye and L. B. Wingard, eds.), Vol. 2, pp. 169–177. Plenum, New York.

Taylor, J. B., and Swaisgood, H. E. (1972). *Biochim. Biophys. Acta* **284**, 268–277.

Thiele, E. W. (1939). *Ind. Eng. Chem.* **31**, 916–920.

Thomas, D., and Broun, G. (1973). *Biochimie* **55**, 975–984.

Thomas, D., Broun, G., and Selegny, E. (1972). *Biochimie* **54**, 229–244.

Thomas, D., Bourdillon, D., Broun, G., and Kernevez, J. P. (1974). *Biochemistry* **13**, 2995–3000.

Valenzuela, P., and Bender, M. L. (1971). *Biochim. Biophys. Acta* **250**, 538–548.

Van Duijn, P., Pascoe, E., and VanderPloeg, M. (1967). *J. Histochem. Cytochem.* **15,** 631–645.

Vieth, W. R., and Venkatasubramanian, K. (1974). *Chem. Technol.* 303–320.

Vieth, W. R., Mendiratta, A. K., Mogensen, A. O., Saini, R., and Venkatasubramanian, K. (1973). *Chem. Eng. Sci.* **28,** 1013–1020.

Wagner, C. (1943). *Z. Phys. Chem., Abt. A* **193,** 1.

Warburg, O. (1923). *Biochem. Z.* **142,** 317–335.

Waterlands, L. R., Michaels, A. S., and Robertson, C. R. (1974). *AIChE J.* **20,** 50–59.

Webb, J. L. (1963). "Enzyme and Metabolic Inhibitors." Academic Press, New York.

Weetall, H. H., and Hersh, L. S. (1970). *Biochim. Biophys. Acta* **206,** 54.

Weisz, P. B., and Hicks, J. S. (1962). *Chem. Eng. Sci.* **17,** 265.

Weisz, P. B., and Prater, C. D. (1954). *Adv. Catal.* **6,** 143–170.

Wharton, C. W., Crook, E. M., and Brocklehurst, K. (1968). *Eur. J. Biochem.* **6,** 572–578.

Wheeler, A. (1951). *Adv. Catal.* **3,** 249–327.

Wilchek, M., and Rotman, M. (1970). *Isr. J. Chem.* **8,** 172.

Wilson, F. A., Sallee, V. L., and Dietschy, J. M. (1971). *Science* **174,** 1031.

Wilson, J. H., Kay, G., and Lilly, M. D. (1968). *Biochem. J.* **108,** 845–853.

Wingard, L. B. (1972). *Adv. Biochem. Eng.* **2,** 1–48.

Wong, J. T. F. (1965). *J. Am. Chem. Soc.* **87,** 1788.

Zaborsky, O. (1973). "Immobilized Enzymes." CRC Press, Cleveland, Ohio.

Zabusky, N. J., and Hardin, R. H. (1973). *Phys. Rev. Lett.* **31,** 812–815.

Zander, R., and Schmid-Schoenbein, H. (1972). *Pfluegers Arch.* **335,** 58.

Design and Analysis of Immobilized-Enzyme Flow Reactors

W. R. Vieth, K. Venkatasubramanian,
A. Constantinides, and B. Davidson
Department of Chemical and Biochemical Engineering,
Rutgers University, New Brunswick, New Jersey

I. INTRODUCTION

Despite a rather prodigious amount of research effort expended in the area of immobilized enzymes, relatively little attention has been paid to the process-engineering aspects of supported enzyme systems. For their efficient practical utilization, it is necessary to characterize enzyme–carrier conjugates in an appropriate reactor environment. The activity of the fixed enzyme and its relation to (a) the reactor configuration, (b) the mass transfer effects adjacent to and within the carrier phase, (c) the electrostatic partitioning of substrate between the reactor fluid and carrier phases, and (d) the residence time distribution within the reactor are some of the factors that need to be quantified in order to place the prediction of reactor performance on a reliable basis. Only in recent years, have some attempts been made in this direction.

Although there are several reviews on methods of enzyme immobilization, activities, stabilities, and applications of supported enzymes (Goldstein, 1969; Goldman *et al.*, 1971; Melrose, 1971; Katchalski *et al.*, 1971; Smiley and Strandberg, 1972; Weetall and Messing, 1972; Sundaram and Laidler, 1971; Vieth and Venkatasubramanian, 1973, 1974a,b), discussions on reactor-engineering aspects are conspicuously minimal in all but a few of them. A recent book on immobilized enzymes (Zaborsky, 1973) devotes less than 6% of its content to enzyme reactors. Wingard (1972), O'Neill (1972a), and Hultin *et al.* (1972) initiated the efforts to review some aspects of enzyme reactor analysis. In an earlier review article (Vieth and Venkatasubramanian, 1974c), we also made a similar attempt. However, the scope of these reviews did not permit an in-depth consideration of the subject. In this chapter, we have attempted to knit several sporadic pieces of information to provide an integrated approach to enzyme-reactor analysis and design. It is hoped that this would be a propitious effort in helping to pave the way to an overall solution to the problems of the design of immobilized-enzyme reactors.

We address the enzyme reactor engineering problems in some detail, beginning with a consideration of different reactor types, criteria necessary for the choice of a reactor type, operational parameters of a reactor, and elementary design equations for idealized cases. Inter- and intracarrier mass transport, together with enzymic catalysis in the reactor, are discussed next. Since there is another chapter in this volume which analyzes extensively the diffusional effects in immobilized enzymes, our effort here is restricted to the mass-transfer effects as they relate to reactor design and analysis. Modeling of both single and multienzyme systems with complete and approximate kinetic expressions has been presented, and reactor design procedures developed. Process optimization formats are included where applicable. Both steady-state and transient operations are considered. Within the latter, the effects of simultaneous enzyme denaturation and desorption, as well as the buildup of reaction intermediates are indicated. The efficiency of several reactor designs are compared. In many instances, work carried out at our laboratory is featured in somewhat greater detail mainly to facilitate development of a coherent theme. In developing such a theme, we have emphasized concepts and have presented most generally useful algorithms rather than cataloguing a compendium of different procedures.

II. ENZYME REACTOR TYPES AND THEIR CHOICE

A. Salient Features of Different Reactor Types

1. Batch Enzyme Reactors

Different types of reactors can be used for process-scale operations with enzymes in their free or immobilized forms. Some of these reactors are schematically shown in Fig. 1. Based upon the mode of charging/discharging, enzyme reactors may be broadly classified as (1) batch and (2) continuous-flow reactors. The batch reactor is simple, needs little supporting equipment, and is therefore very suitable for small-scale experimental studies. The use of free enzymes is generally restricted to a batch, stirred-tank operation. In these cases, the free enzyme is charged into the reactor along with the substrate and the reaction is carried out to the desired degree of conversion. Usually, no attempt is made to recover the enzyme from the reaction product since the cost of enzyme recovery is generally prohibitive. However, the enzyme is often inactivated by thermal or other means. Many enzymic reactions employing free enzymes or crude enzyme preparations are carried out in this manner in the food and beverage industries.

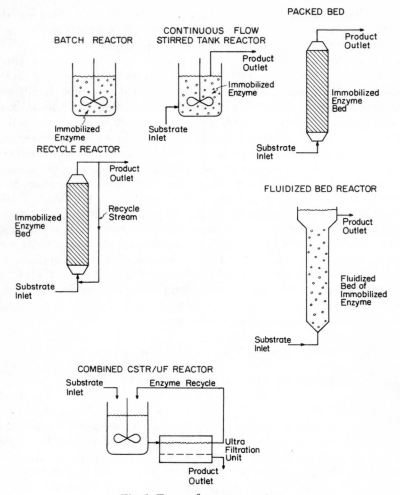

Fig. 1. Types of enzyme reactors.

When immobilized enzymes are used in a batch reactor, the immobilized preparation should be separated from the product stream by a subsequent step. Recovery procedures, whether by filtration or by ultracentrifugation, are likely to cause appreciable loss of expensive immobilized enzyme. Furthermore, the enzyme might be inactivated when subjected to such repeated recovery cycles (O'Neill, 1972a). Therefore, batch reactors have limited potential in industrial immobilized-enzyme catalysis. There are but a few reports on the use of batch immobilized-enzyme reactors in the literature; e.g., Lilly *et al*. (1972) and Warburton *et al*. (1973) have carried out batch reactor studies with immobilized penicillin amidase.

2. Packed-Bed Reactors

The two main types of continuous reaction equipment are the packed-bed reactor and the continuous-flow stirred-tank reactor (CSTR). A hybrid of these two types is the fluidized-bed reactor. When the immobilized enzyme is in the form of spheres, chips, disks, sheets, beads, or pellets, it can be packed readily into a column. Most of the published supported-enzyme reactor studies are on packed-bed reactors. Some representative examples of such reactor systems include enzymes attached to (a) porous glass beads (Weetall, 1971, 1973; Weetall and Messing, 1972; Weetall *et al*., 1974a; Marsh *et al*., 1973; Weibel and Bright, 1971), (b) beads of ion-exchange resins such as DEAE-cellulose and DEAE-Sephadex (Chibata *et al*., 1972; Tosa *et al*., 1973; Bachler *et al*., 1970), (c) polyacrylamide gel disks (Bunting and Laidler, 1972), (d) sheets or plates of enzyme-containing polymers (Sampson *et al*., 1972), and (e) chips of collagen membranes (Goldberg, 1974; Saini and Vieth, 1975). Spherical enzyme-containing microcapsules (Chang, 1972; Mogensen and Vieth, 1973; Giniger, 1973) have also been used in packed columns. In our laboratory, we have also developed novel tubular reactor configurations in which collagen-enzyme membranes are wound to form spiral multipore reactor modules (Wang and Vieth, 1973; Vieth and Venkatasubramanian, 1973, 1974c; Vieth *et al*., 1972a, 1974).

In a packed-bed reactor, there is a steady movement of the substrate across a bed of immobilized enzyme in a chosen spatial direction. If the fluid velocity profile is perfectly flat over the cross section, the reactor is said to operate as a plug-flow reactor (PFR) under this ideal condition. In other words, the fluid elements are visualized to move through the reactor in a pluglike fashion. Of course, in actual reactor systems, the fluid-flow pattern tends to be different from this idealized condition. The deviations from plug-flow may occur owing to (a) the existence of velocity gradients normal to the direction of flow, (b) diffusion of substrate in the axial direction, and (c) the existence of temperature gradients normal to the flow direction (Denbigh, 1965). The effects of these nonideal flow characteristics on the reactor design will be considered in a later section.

Several other enzyme reactor configurations, which may be approximated as packed-bed reactors, have been proposed. Tubular reactors packed with filter paper (Kay *et al*., 1968), porous sheets (Self *et al*.,

1969; Sharp *et al.*, 1969; Reynolds, 1972), and porous blocks, such as open-pored polyether or polyurethane foams (Lilly and Dunnill, 1972; Maldonado, 1974), have been used as a possible means of improving fluid–solid contact efficiency. However, these reactors are likely to suffer from severe pressure-drop problems.

Reactors packed with semipermeable hollow fibers, which permit the passage of only the reactant and product but not the enzyme, have recently made their appearance on the scene. They provide a high catalyst surface area in a given reactor volume; reactant selectivity can be achieved by using fibers of dissimilar permeability characteristics. Hollow fibers can be arranged in a reaction system in different ways. The enzyme solution can be entrapped within the hollow fiber and a fiber bundle housed in a reactor shell. The substrate is passed over the fiber surface while it diffuses simultaneously into the fiber lumen where the enzyme catalysis takes place. The products of the reaction migrate back to the bulk fluid stream (Dinelli, 1972; Rony, 1972; Corno *et al.*, 1972; Marconi *et al.*, 1973, 1974). Another possibility is to retain the enzyme solution outside the fiber surface while passing the substrate through its core (Lambert and Chambers, 1973). Shortcomings of these systems include (a) practical difficulty in fabricating hollow fiber membranes of adequate permeability to substrate and product and (b) the lack of proper control of packing to assure uniform flow distribution within the fiber bundles, resulting in significant mass transfer resistance to substrate and product transport. These difficulties can be alleviated at least in part by employing anisotropic hollow fiber systems. The enzyme solution is constrained within an annular porous support structure of the fiber, which is separated from a substrate flowing through the fiber lumen by a dense membrane permeable only to substrate and product (Robertson and Waterland, 1973).

3. Continuous-Flow Stirred-Tank Reactors

In an ideal CSTR, the contents of the reactor are perfectly mixed. Consequently, all elements of the reactor have essentially the same composition, and this is the same as the composition of the outflow. Therefore, the reaction rate is determined by the composition of the exit stream from the reactor. While in a PFR the substrate concentration is maximized with respect to final conversion at every point in the reactor, it is minimized at every point in a CSTR. Thus in a CSTR, the average reaction rate is lower than it would be in a tubular reactor. On the other hand, the open construction of the CSTR permits ready replacement of immobilized enzyme catalyst. It also facilitates easy control of temperature and pH. Denbigh and Page (1954) have discussed the advantageous features of CSTRs and their applicability in simulating the behavior of biological systems.

Continuous-flow stirred-tank reactors have been used for particulate immobilized enzyme systems (Smiley, 1971; O'Neill *et al.*, 1971; Weetall and Havewala, 1972; Ryu *et al.*, 1972; Dutta *et al.*, 1973). CSTR systems require a means of retaining the supported enzyme particle within the reactor. This may be achieved simply by providing a filter at the reactor outlet. Enzyme-support particles can also be separated by means of ultrafiltration (Ryu *et al.*, 1972) or by a separate settling stage. The possibility exists of magnetic separation or retention with enzymes attached to magnetically active supports (Heden, 1972; Robinson *et al.*, 1973). Enzymically active polymer disks, mounted on an agitator shaft, provide another alternative approach to retaining the supported enzyme in the reactor (Worthington, 1972).

4. Continuous Flow Stirred-Tank/Ultrafiltration Membrane Reactors

Continuous processing with free enzymes in a CSTR can be accomplished by using an ultrafiltration membrane in the process loop. The ultrafiltration membrane provides a semipermeable barrier allowing the passage of product and unreacted substrate while retaining the high-molecular-weight enzyme. The technical feasibility of such a reactor system has been demonstrated for continuous enzymic saccharification of cellulose (Ghose and Kostick, 1970), for starch hydrolysis by α-amylase and glucoamylase (Butterworth *et al.*, 1970), for sucrose hydrolysis by invertase (Bowski *et al.*, 1972), for enzymic solubilization of fish protein concentrate (Cheftel *et al.*, 1971), and for conversion of benzylpenicillin to 6-aminopenicillanic acid by penicillin amidase (Ryu *et al.*, 1972). A combined system of CSTR and ultrafiltration membrane can also be used with a disperse soluble immobilized enzyme (i.e., by attaching the enzyme to a soluble high-molecular-weight polymer such as dextran). Hydrolysis of casein by chymotrypsin (O'Neill *et al.*, 1971a) and starch hydrolysis by α-amylase (Wykes *et al.*, 1971) have been investigated in this manner.

CSTR/UF membrane reactors also provide the possibility of separating a low-molecular-weight product from a high-molecular-weight substrate. For instance, glucose and maltose can be separated in this manner from starch or cellulose. The use of tubular ultrafiltration membrane reactors in which the membrane constitutes a permeable reactor wall have been explored for such combined reaction/separation schemes (Closset *et al.*, 1974; Tachauer *et al.*, 1974). Enzymes coupled directly to plane or hollow-fiber ultrafiltration membranes offer a different approach (Gregor and Rauf, 1974). Uses of other dual-functional membrane systems, such as combined reaction and separation by reverse osmosis, are also being investigated (Davidson *et al.*, 1974). Microencapsulated enzymes (Chang and Malave, 1970; Chang *et al.*, 1971) and enzyme-containing liquid membranes (Li, 1972; Mohan and Li, 1974) have also been proposed for combined reactor–separator devices.

5. *Fluidized-Bed Reactors*

In a fluidized-bed reactor, the substrate is passed upward through the immobilized enzyme bed at a velocity high enough to lift the particles; however, the velocity is not so high as to sweep away the particles from the reactor itself. The fluid flow pattern provides a degree of mixing that falls somewhere between complete backmixing, as in a CSTR, and no backmixing, as in a PFR. In chemical engineering practice, a fluidized bed reactor is traditionally of importance when excellent heat and mass transfer characteristics are required. In particular, it eliminates local hot spots within the reactor in highly exothermic reactions. But most enzymic reactions are essentially isothermal processes.

Fluidized-bed reactors are likely to find some application in supported-enzyme catalysis where viscous, particulate substrates are to be handled. Fluidized-bed reactor systems with immobilized enzyme particles or chips have been reported for the hydrolysis of lactose present in cheese whey (Coughlin *et al.*, 1973), starch hydrolysis (Barker *et al.*, 1971; Kent and Emery, 1974), and glucose isomerization (Goldberg, 1974). At least in some of these cases, the reactor operation was as an expanded bed rather than as a completely fluidized bed. Expanded-bed operation has also been reported by O'Neill *et al.* (1971d). Enzymes covalently coupled to magnetic iron oxide particles can apparently be easily maintained in the fluidized state magnetically (Anonymous, 1974).

6. *Other Reactor Types*

a. Recycle Reactors: Many variations and combinations of the basic types of enzyme reactors discussed here are possible. For example, recycle reactors may find application when an insoluble substrate is to be processed. In this type of reactor, a portion of the outflow is recycled and mixed with the inlet stream to the reactor. This permits operation of the reactor at high fluid velocities, which minimizes bulk mass transfer resistance to the transport of substrate to the immobilized enzyme surface. Even though high flow rates reduce the contact time of the substrate in the reactor (per pass), the recycling process effectively provides high enough contact time to achieve desired conversions. Recycle reactor systems for starch hydrolysis (Kingma, 1966) and for treating body fluids (Horvath *et al.*, 1973a) have been reported. They are also excellent tools for kinetic studies (Ford *et al.*, 1972; Venkatasubramanian *et al.*, 1972; Constantinides *et al.*, 1973).

b. Tubular Reactors with Enzymically Active Walls: Enzymes can be bound directly to the inner wall of a tubular reactor (Sundaram and Hornby, 1970; Hornby and Filippusson, 1970). They can also be attached to a suitable matrix to form an enzymically active porous annulus at the tube wall (Horvath and Solomon, 1972; Horvath *et al.*, 1972; 1973a). Such reactors are very useful in biomedical applications as extracorporeal shunts to effect specific enzyme-catalyzed biochemical changes in body fluids. The advantages of these reactors are the relatively low pressure drop and favorable flow pattern, i.e., unhindered fluid flow through the reactor tube without local turbulence effects, thus lowering the possibilities of body fluid coagulation. These reactors, however, are not useful in industrial systems because of their very low catalytic surface per unit reactor volume.

c. Other Reactor Configurations: Preliminary investigations on the use of slurry reactors, trickle-bed, and slug-flow reactors for immobilized enzyme catalysis have been attempted. In a slurry reactor, the supported enzyme particles are maintained as a slurry in the flow field of the substrate; thus, it offers easier temperature and pH controls (Laurence *et al.*, 1973). Witt *et al.* (1970) employed covalently cross-linked papain precipitate in a system similar to a slurry reactor. Desugaring of egg products with immobilized glucose oxidase and catalase in a trickle-bed reactor (Hultin *et al.*, 1974) has been reported. A tubular reactor can be operated with slug flow by introducing an inert gas stream (along with the substrate) into the reactor. Slug flow has been found to augment radial mass transfer in reactors with catalytically active walls resulting in greater reaction rates (Horvath *et al.*, 1972). The enhancement of radial mass transport in small-diameter tubes under laminar flow conditions can be explained by the secondary flow pattern within the liquid slugs, which increases the transport of substrate to the catalytic surface (Horvath *et al.*, 1973b). The salient features of different reactor types are summarized in Table I.

B. Choice of Reactor Type

A number of factors influence the choice of a particular type of reactor as listed in Table II. Reactors that approach plug-flow characteris-

230   **W. R. Vieth et al.**

TABLE I
SALIENT FEATURES OF DIFFERENT REACTOR TYPES

1. Combined continuous-flow stirred-tank reactor (CSTR) ultrafiltration system for free or soluble immobilized enzymes
   Colloidal or insoluble substrates can be processed
   Poor stability of the enzyme for long-term operation
   Enzyme loss by adsorption onto the membrane
   Concentration polarization of the enzyme at the membrane surface
2. Continuous-flow stirred-tank reactors
   Easy to control pH when necessary
   Colloidal or insoluble substrates can be processed
   Easy replacement of catalyst
   Higher yields for substrates undergoing substrate inhibition
3. Plug-flow reactors
   Higher conversion efficiency (compared to CSTR)
   Higher yields for substrates undergoing product inhibition
4. Fluidized-bed reactors
   Better heat and mass transfer characteristics
   Freedom from plugging
   Insoluble substrates can be used
   Low pressure drop
   Large power requirement for fluidizing the bed
   Uncertainties in reactor scale-up

tics may be expected to dominate immobilized enzyme applications because of their innate kinetic advantage over CSTRs. The influence of reaction kinetics on reactor performance will be discussed in detail in the next section. Suffice it to say here that a packed bed reactor has intrinsic kinetic advantages over a CSTR for most reaction types. However, in the case of substrate-inhibited enzyme systems, CSTR performance is superior to that of a packed column. When the immobilized enzyme is particulate in nature, it can readily be packed

TABLE II
FACTORS INFLUENCING THE CHOICE OF REACTOR TYPE

1. Form of the immobilized enzyme: particulate, membranous, fibrous
2. Nature of the substrate: soluble, particulate, colloidal
3. Operational requirements of the reaction, e.g., pH control
4. Reaction kinetics
5. Carrier loading capacity
6. Catalytic surface-to-reactor volume ratio
7. Mass-transfer characteristics: microdiffusional and macrodiffusional efficiencies
8. Ease of catalyst replacement, regeneration
9. Ease of fabrication
10. Reactor costs

into a column or used in a CSTR. Membranes or fibrous forms are more readily suitable for packed-bed operation. However, membranes can be chipped and used in a CSTR also.

Plug-flow reactor columns packed with immobilized enzyme in the form of relatively deformable and/or aggregative particles suffer from high pressure drops and inability to achieve sufficient flow rates for large-scale operation. Increasing the pressure differential across the packed bed in an attempt to augment flow rates is often fruitless because of the compaction and further reduction in hydraulic permeability of the resulting bed. Fluidized-bed reactors can handle smaller particles with a concomitant increase in available catalytic surface area.

The nature of the substrate—whether it is soluble, particulate, or colloidal—is another factor to be considered. Particulate and colloidal substrates are likely to plug-up packed columns. This can be alleviated by using a recycle reactor or a fluidized-bed reactor; in these reactors, the high fluid velocity through the bed of immobilized enzyme reduces the possibility of segregation, settling, and plugging of substrate particles. A CSTR can also be used for processing particulate substrates, provided the substrate and the immobilized enzyme are kept suspended at fairly high agitation rates. But high agitation rates may also cause shearing of immobilized enzyme from its carrier matrix. Restricted-flow problems in the case of starch hydrolysis have been avoided in a CSTR (Smiley, 1971). An experimental comparison of a CSTR and a PFR has been reported by O'Neill and co-workers (1971b). These authors found higher conversion efficiency of maltose to glucose in a CSTR than in a PFR. Another experimental comparison of these two reactor types has also been reported (Woychik and Wondolowski, 1973).

Operational requirements of the reaction might dictate the choice of reactor type. For example, a CSTR is more suitable for reaction systems that require the addition of acid or alkali to control pH. Similarly, oxygen transfer can be more readily accomplished in a CSTR. The amount of catalytic surface that can be housed in a given reactor volume is another important factor. Membranes and hollow fibers afford reactor configurations of high surface-to-volume ratios. Apropos this point, mass-transfer characteristics of an immobilized-enzyme reactor can be significantly altered depending on the reactor type.

Columns packed with small immobilized-enzyme particles necessitate column operation at low fluid velocities. This results in channeling and dispersion of the substrate, which lower the conversion. These difficulties are overcome in a recycle or fluidized-bed reactor. On the other hand, fluidized-bed reactors have a large power requirement for fluidizing the bed. They are also very difficult to model, and therefore reactor design must be based on empirical correlations. In scaling-up of fluidized bed reactors, one often encounters exceptional difficulties (Denbigh, 1965).

In a CSTR, the immobilized enzyme can be replaced without interrupting the process. For many other reactor systems the reactor must be shut down. However, in some cases, the catalyst can be regenerated *in situ*. Chibata and co-workers (1972) have described a catalyst regeneration scheme for a packed-bed reactor utilizing an enzyme (aminoacylase) ionically bound to an ion-exchange resin carrier. Experience in our laboratory indicates that prefabricated cartridges of collagen–enzyme membranes can be regenerated or easily replaced in a reactor whenever necessary.

Considering reactor costs, stirred vessels are the cheapest because of their relatively simpler construction. In addition, they have excellent operational flexibility. Other types of reactors have to be designed and built for a specific process. While discussing reactor costs, one should take into account the cost of the immobilized enzyme itself, since the reactor type may exert some influence on enzyme stability. For instance, a CSTR-ultrafiltration membrane reactor system using free enzyme is likely to contribute to enzyme loss either through adsorption onto the ultrafiltration membrane or by concentration polarization of the enzyme at the membrane surface. Enzyme inactivation at the membrane surface in the shear field of the flowing liquid has been observed (Charm and Lai, 1971). High-flux ultrafiltration units that utilize a liquid shear effect to minimize concentration polarization suffer from loss of enzyme activity caused by high shear stress (Charm and Wong, 1970; Charm and Matteo, 1971).

The extent of shearing to which a conjugated enzyme is subjected would depend on the reactor type. A CSTR offers far greater chance of enzyme loss through shearing than other reactor types. Attrition of the immobilized-enzyme particle would cause a gradual reduction in the mean particle size; it has been observed that such finer particles could be lost by carry-over from the reactor (Regan *et al.*, 1974). Modifications in reactor design might reduce the intensity of shearing. For instance, the enzyme can be bound directly to the agitator shaft. Weetall and Havewala (1972) employed glass beads in rigid screen baskets attached to a stirrer. The rotating baskets, while providing agitation for the tank, protected the beads from shearing. Enzymes attached to metallic screens (Weetall and Hersh, 1970) or to rotating drums (Smiley and Strandberg, 1972) or to disks mounted on the agitator (Worthington, 1972) are other possible variations. The effect of shearing on the stability of a covalently coupled enzyme in a packed-bed reactor has recently been examined (Weetall *et al.*, 1974c). These authors conclude that the shear energy density required to rupture covalent bonds is much greater than that generated in a packed-bed immobilized-enzyme column.

From the foregoing discussion, it is clear that there are no simple rules for choosing a particular reactor type. For an efficient reactor system, one should endeavor to combine the advantageous features of the different types. An example of such a reactor design developed in our laboratory is briefly described below. The supported membranous form of collagen–enzyme complexes mentioned above can be arranged to form a multipore, spiral, biocatalytic reactor as shown in Fig. 2. The spiral reactor configuration is formed by coiling alternate layers of the enzyme–membrane and a backing material around the central spacer element. An inert polymeric netting, Vexar, is used as the backing material. It separates the successive layers of enzyme–collagen membrane, thus preventing overlapping of the membrane layers. The coiled cartridge contains a plurality of flow compartments, which provide a large surface area for efficient bulk mass interchange with the substrate. The cartridge is fitted into an outer shell, which is affixed to two end plates provided with an inlet and outlet for the flow of the substrate over the membrane surface. A uniform axial distribution of

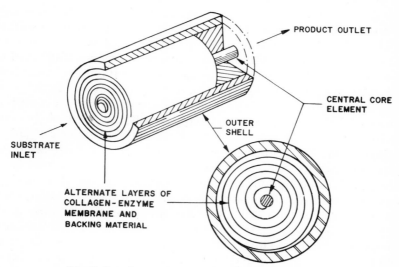

Fig. 2. Spiral-wound multipore biocatalytic module.

the substrate is achieved by metering the flow through an appropriate distributor plate.

Such a reactor configuration offers a number of advantages. The spiral module segregates the flow units into a number of individual flow compartments. By routing the substrate through such a large number of flow compartments of essentially equal hydraulic resistance, each fluid element is made to spend the same residence time in contact with the membrane unit as any other element. This leads to good contact efficiency by eliminating channeling and dispersion of the substrate. The membranous mesh spacer augments bulk mass transfer by promoting radial mixing in each of the flow compartments, thus acting as a local turbulence promoter. The small fluid channels also provide a large membrane area per unit reactor volume and sustain high throughputs. This reactor design can be readily modified to permit forced permeation of the substrate under hydraulic pressure (Venkatasubramanian and Vieth, 1973). This offers as well a means of reducing internal diffusional resistances within the carrier matrix.

III. IDEALIZED ENZYME REACTOR SYSTEMS

A. Background

In developing mathematical models for enzyme reactors, certain general assumptions can be made. These assumptions, which are valid in many cases of practical interest, simplify the mathematical analysis considerably. They are as follows: (a) Since most enzyme-catalyzed reactions take place in the physiological temperature range and exhibit low enthalpies of reaction, isothermal conditions are maintained in an enzyme reactor. (b) The immobilized enzyme particles are packed in a column or suspended in a stirred vessel in a uniform manner so that there are no statistically significant variations between two different parts of the reactor. (c) From the arguments set forth by Denbigh (1965) for isothermal, packed-bed reactors, the relevant assumptions can be made concerning plug-flow conditions and a negligibly small contribution of longitudinal turbulent dispersion in comparison with the transport due to the bulk flow. Further, for an isothermal reactor with a flat velocity profile and with uniform distribution of catalyst particles over the cross section, there is no concentration gradient in the radial direction. Hence, radial dispersion in the reactor cannot be a factor in the analysis under the conditions of radial symmetry. (d) The CSTR system may be assumed to be perfectly

mixed and the reactor residence time distribution may be characterized by a single mean residence time.

Deviations from these ideal flow patterns resulting in a residence
time distribution of different fluid elements may be in the form of
channeling or backmixing of fluid, recycling of fluid or the existence of
stagnant regions in the reactor (Levenspiel, 1972). The problems of
nonideal flow affect the reactor scaling-up procedures profoundly because it is very difficult to assess the magnitude of the nonideality of
fluid flow.

By the term "ideal enzyme reactor," it is herein meant that (a) the
reactor meets the above-mentioned assumptions regarding isothermal
conditions, uniform enzyme distribution, plug-flow fluid dynamics
with no axial dispersion or radial gradients, and perfect mixing for a
CSTR unit, (b) the immobilized enzyme experiences no external or
internal mass transport limitations, and (c) no significant partitioning of
substrate occurs between the solid (immobilized enzyme particle) and
the fluid phases. Although in practice very few reactor systems qualify
to be designated as ideal, it is useful to examine these idealized cases
to illustrate the principles of biochemical reactor theory. For instance,
the effect of reaction kinetics on the performance of different reactor
types can readily be scrutinized with ideal reactor models. Thus, the
concentration history of the ideal batch and plug-flow reactors can be
shown to be identical in form. While the concentration of substrate and
product change continuously with respect to reaction time in the ideal
stirred batch reactor, the concentration changes occur along the length
of the ideal plug-flow reactor.

1. Reactor Design Parameters

The important design parameters which govern the performance of
an idealized immobilized enzyme reactor are (a) reactor space time,
(b) inlet concentration of substrate, (c) the amount of immobilized
enzyme used in the reactor, and (d) the temperature and pH of the
reaction. The reactor space time τ is defined as

$$\tau = V_R/Q \tag{1}$$

where V_R is the volume of the reactor and Q is the flow rate through the
reactor. Space time is the time required to process one reactor void
volume of substrate measured at specified conditions. [Note that in an
idealized reactor, all the fluid elements would spend the same time in
the reactor. But in most real reactor systems, there would be a distribution of residence times of different fluid elements due to nonideal flow
characteristics. Thus, space time is a parameter which describes only

the macroscopic flow behavior.] Three other parameters defined below are useful in evaluating reactor performance. The fractional conversion X is given by

$$X = (S_0 - S)/S_0 \tag{2}$$

where S_0 and S are the inlet and outlet substrate concentrations, respectively. Reactor productivity Pr is defined as

$$\text{Pr} = XS_0/\tau \tag{3}$$

Pr gives an indication of the amount of product produced per unit time per unit reactor volume. Reaction capacity C_R of a reactor is given by

$$C_R = V_m \epsilon V_R \tag{4}$$

where V_m is the maximum reaction rate $= k_2E_0$, k_2 is the rate constant, E_0 is the enzyme concentration in the reactor, and ϵ is the porosity of the reactor. ϵ is the ratio of the fluid volume to the total volume of the reactor (V_R). ϵ may vary from 0.5 to 0.7 in the case of a plug-flow reactor and is near 1.0 for a CSTR.

2. Enzyme Kinetics

The simplest kinetic expression describing free enzyme catalysis is given by the Briggs–Haldane monoenzyme, monosubstrate, steady-state model:

$$r = -(dS/dt) = k_2E_0S/(K_m + S) \tag{5}$$

where r is the reaction rate, S is the substrate concentration, and K_m is the Briggs–Haldane (popularly known as the Michaelis–Menten) constant of the enzyme.

The quantity k_2E_0 is the maximum reaction rate (V_m) obtainable for a given system. Equation (5) is derived based on the assumption (known as the steady-state hypothesis) that the enzyme and the substrate form a complex whose concentration reaches a steady state within a short time interval. The conditions under which the above assumption is valid have been investigated thoroughly. Wingard (1972) has presented an excellent discussion of these investigations with reference to their application to immobilized-enzyme systems. Aris (1972) has examined theoretically the implications of the steady-state hypothesis when it is applied to a single catalyst particle. For most practical purposes, Eq. (5) is a reasonably satisfactory rate expression. Detailed discussion on the kinetics of other complex enzymic reactions can be found in several texts (e.g., Laidler and Bunting, 1973; Dixon and Webb, 1964) and in a recent review as it relates to immobilized-enzyme catalysis (Carbonell and Kostin, 1972).

Reactions catalyzed by supported-enzyme systems can also be described by kinetic expressions of the form of Eq. (5). However, when an enzyme is attached to a solid support, the kinetic pattern of reaction changes considerably. This leads to changes in the values of the kinetic parameters K_m and V_m. The kinetics of such reactions are obscured by several factors, such as (a) change in enzyme conformation, (b) steric hindrances, (c) microenvironmental effects, and (d) bulk and internal diffusion effects, so that the observed kinetic parameters are only "apparent" or "effective" parameters lumped into modified kinetic constants K'_m and V'_m for the immobilized enzyme. Influences of these factors on the reaction kinetics are discussed in the chapter by Engasser and Horvath.

B. Idealized Reactor Performance Equations

1. Michaelis–Menten Kinetics

Using Eq. (5), simple steady-state reactor performance equations may be obtained readily. A steady-state reactor material balance yields the following equations for a CSTR and PFR.
CSTR:

$$\tau = S_0 X / r \tag{6}$$

PFR:

$$\tau = S_0 \int_0^X \frac{dX}{r} \tag{7}$$

X here refers to the degree of conversion measured in the reactor outlet. By substituting Eq. (5) in Eqs. (6) and (7), the following performance equations are derived with K_m changed to K'_m to indicate apparent or effective parameter values.
CSTR:

$$S_0 X + K'_m [X/(1 - X)] = k_2 E_0 \tau \tag{8}$$

PFR:

$$S_0 X - K'_m \ln (1 - X) = k_2 E_0 \epsilon \tau \tag{9}$$

The equation for a batch reactor is essentially the same as Eq. (9) when τ is replaced by the actual reaction time and ϵ is set to 1.0. Equation (9), originally derived by Bar-Eli and Katchalski (1963), was found to be in reasonable agreement with experimental data for a few small, laboratory packed-bed columns (Hornby *et al.*, 1966). However, this equation is inadequate for most other systems because of diffusional barriers to the transport of substrate and product. Experimental data obtained from a CSTR with amyloglucosidase conjugated to DEAE-cellulose were found to agree with Eq. (9) in one case (O'Neill *et al.*, 1971b). Based on Eqs. (8) and (9), theoretical plots have been developed that describe the variations of reaction capacities of a CSTR and a PFR with identical input concentration of substrate and the value of K'_m (Lilly and Sharp, 1968). Plots of this type are useful in getting an overview of the reaction process as it is affected by these parameters.

When the feed substrate concentration is much higher than K'_m, the reaction rate becomes zero order with respect to substrate concentration and Eqs. (8) and (9) become identical, as shown by Eq. (10), when ϵ is near unity:

$$\tau = XS_0/k_2E_0 \tag{10}$$

In this case, the required volume of the reactor or the amount of enzyme required for a given space time is the same irrespective of the reactor type. The applicability of Eq. (10) has been demonstrated for DEAE-Sephadex–aminoacylase (Tosa *et al.*, 1969) and for ethylene–maleic anhydride–trypsin (Bar-Eli and Katchalski, 1963) columns.

When K'_m is much higher than S_0, the reaction rate is portrayed by first-order kinetics with respect to substrate concentration. Now Eqs. (8) and (9) reduce to Eqs. (11) and (12), respectively:

$$K'_m[X/(1 - X)] = k_2E_0\tau \tag{11}$$

$$-K'_m \ln (1 - X) = k_2E_0\epsilon\tau \tag{12}$$

In the above equations, the quantity $(k_2E_0)/K'_m$ may be considered as a pseudo first-order rate constant k_f. In our laboratory, we found that, under certain conditions, the kinetic behavior of some collagen-immobilized enzymes could be approximated by a pseudo first-order equation (Vieth *et al.*, 1972a,b; Venkatasubramanian *et al.*, 1972). The form of Eq. (12) was therefore applicable in those cases. The ratio of the amounts of immobilized enzyme required in a CSTR and a PFR can be deduced from the above equations for first-order kinetics as follows:

$$\frac{E_{\text{CSTR}}}{E_{\text{PFR}}} = -\frac{X\epsilon}{(1 - X) \ln (1 + X)} \tag{13}$$

From Eq. (13) it is clear that the higher the required conversion level, the higher is the relative amount of enzyme required in a CSTR. Assuming $\epsilon = 0.5$, to achieve 90% and 99% conversion levels, approximately 4.5 and 25 times more enzyme would be necessary in a CSTR, at the same reactor space time for systems obeying first-order kinetics. The ratio of the relative amounts of enzyme required are shown in Fig. 3 as a function of fractional conversion when Michaelis–Menten kinetics is applicable. Provided the reaction kinetics obey the rate expression considered here, i.e., Michaelis–Menten kinetics, Fig. 3 would be useful in selecting a reactor type based on the amount of enzyme required to effect a desired conversion, or, alternatively, to calculate the enzyme requirement for a desired conversion level.

2. Substrate and Product Inhibition Kinetics

When the enzyme is subjected to inhibition by the substrate and/or the product, the reactor performance is altered significantly. The amount of enzyme required to achieve the same degree of conversion in a plug-flow reactor at a given space time may be 2–500 times larger in the presence of (noncompetitive) product inhibition than in the absence of inhibition effects (Lilly and Dunnill, 1972). The kinetic expressions for substrate and product inhibitions and the reactor performance equations based on them are shown in Table III. Quantitative comparisons of the relative amounts of enzyme required in a CSTR and a PFR for enzyme catalysis with substrate and product inhibition have been presented elsewhere (O'Neill, 1972a). For substrate inhibition kinetics, the fractional conversion is drastically reduced in a plug-flow reactor compared to a CSTR, especially at high reactant concentrations. This is because the enzyme in a CSTR always

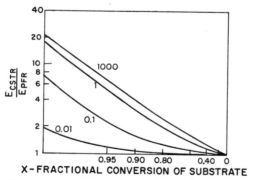

Fig. 3. Variation of the amount of enzyme required (for Michaelis–Menten kinetics) in a continuous-flow stirred-tank reactor (CSTR) to that in a plug-flow reactor (PFR), E_{CSTR}/E_{PFR}, with fractional conversion, for various values of inlet substrate concentration, at the same K'_m and space time. Numbers on the curves represent the dimensionless substrate concentration, β, which is the ratio of the Michaelis constant K'_m to the feed concentration S_0. (Adapted from Lilly and Sharp, 1968.)

TABLE III

KINETIC EXPRESSIONS FOR SUBSTRATE AND PRODUCT INHIBITIONS AND THE REACTOR PERFORMANCE
EQUATIONS, EQS. (14)–(21a), BASED ON THEM

Kinetic form	Kinetic expression	Reactor performance equation[a]	
		CSTR	PFR
Substrate inhibition	$r = \dfrac{k_2 E_0}{1 + (K'_m/S) + (S/K_S)}$	(14) $\; S_0 X + K'_m\left(\dfrac{X}{1-X}\right)$ $+ \dfrac{S_0^2}{K_S}(X - X^2)$ $= k_2 E_0 \tau$ (15)	$S_0 X - K'_m \ln(1 - X)$ $+ \dfrac{S_0^2}{2K_S}(2X - X^2)$ $= k_2 E_0 \tau$ (16)
Product inhibition (competitive)	$r = \dfrac{k_2 E_0}{1 + (K'_m/S)[1 + (P/K'_P)]}$	(17) $\; S_0 X + K'_m\left(\dfrac{X}{1-X}\right)$ $+ \dfrac{K'_m}{K'_P}\dfrac{S_0 X^2}{(1-X)}$ $= k_2 E_0 \tau$ (18)	$S_0 X\left(1 - \dfrac{K'_m}{K'_P}\right)$ $- K'_m \ln(1-X)$ (19)
Product inhibition (noncompetitive)	$r = \dfrac{k_2 E_0}{[1 + (K'_m/S)][1 + (P/K'_P)]}$	(20) $\; S_0 X + K'_m\left(\dfrac{X}{1-X}\right)$ $+ \dfrac{K'_m}{K'_P}\dfrac{S_0 X^2}{(1-X)}$ $+ \dfrac{S_0^2 X^2}{K'_P} = k_2 E_0 \tau$ (21)	$S_0 X\left(1 - \dfrac{K'_m}{K'_P}\right)$ $- K'_m \ln(1-X)\left[1 + \dfrac{S_0}{K'_P}\right]$ $+ \dfrac{S_0^2 X^2}{2K'_P}$ $= k_2 E_0 \tau$ (21a)

[a] CSTSR = continuous-flow stirred tank reactor; PFR = plug-flow reactor.

operates at a substrate concentration equal to that in the product stream. Similarly, the adverse effect of product inhibition is more pronounced in a CSTR.

Some enzyme systems undergo complex inhibition effects, in which case the reaction rate expressions become complicated. Therefore, it often requires numerical solutions of reactor equations to predict reactor performance. The hydrolysis of penicillin by immobilized penicillin amidase is such a system, in which the enzyme is subjected to complex inhibition patterns by both the products of the hydrolysis reaction. Ryu *et al.* (1972) presented performance curves of a CSTR and a PFR for this system that were in good agreement with experimental data. Mathematical simulation of other complex reaction schemes in CSTRs have been reported (Bowski *et al.*, 1972; Lilly *et al.*, 1972). Quantitative comparisons of a CSTR and a PFR with complex reaction kinetics have also been attempted (Warburton *et al.*, 1973). Experimental data on the hydrolysis of acid whey by immobilized β-galactosidase obey the reactor performance equation [Eq. (19)] for Michaelis–Menten kinetics with competitive inhibition (Weetall *et al.*, 1974b).

C. Reactor Type and Operational Stability

A crucial design parameter in the design of immobilized-enzyme reactors is the operational stability of the supported enzyme. Therefore the reactor-design equation should incorporate this parameter so that prediction of variations of conversion level could be attempted. The loss of catalytic activity in a supported-enzyme reactor may be due to (a) the inactivation of the enzyme, (b) enzyme desorption from the carrier matrix, and (c) disintegration or solubilization of support. Loss of activity by enzyme elution from the carrier matrix would particularly be true of enzymes that are immobilized by adsorption onto a carrier; in these cases, the enzyme might continuously desorb from the matrix (Goldman *et al.*, 1965; Weetall and Hersh, 1970; Lilly *et al.*, 1973). With enzymes attached to collagen membranes by complexation, there is sometimes encountered an initial decline in activity due to the desorption of (overloaded) loosely bound enzyme prior to reaching a stable, matrix-saturated limit of activity (Vieth *et al.*, 1972a; Wang and Vieth, 1973; Venkatasubramanian *et al.*, 1972, 1974a; Constantinides *et al.*, 1973).

In general, elution of enzyme from the carrier matrix often occurs when a high-molecular-weight substrate is processed. When the substrate and/or carrier is of the charged polyelectrolyte type, this effect is even further compounded. Even though the extent to detachment of the enzyme would be expected to be a function of the strength of the immobilizing bonds, this phenomenon has been observed even with covalently bound enzymes (Epton and Thomas, 1972; Sampson et al., 1972). The incorporation of this important, but often neglected, mechanism into the reactor model will be considered in detail in a later section.

Enzyme inactivation during reactor operation may occur owing to one or more of several phenomena, such as thermal or pH shock, slow thermal denaturation, continuous exposure to inhibitory reaction product and/or substrate, formation of irreversible inactive complexes with substrate, microbial attack, and foreign materials, such as trace metals present in reaction solution. Further discussions on these mechanisms can be found in other reviews (Hultin et al., 1972; Vieth and Venkatasubramanian, 1974b).

The time scale of enzyme inactivation is generally much longer than the reactor operation time scale so that the reaction system itself may be assumed to be in a quasisteady condition. Enzyme stability, among various other factors, depends on substrate concentration. Most enzymes are known to be more stable in the presence of their substrate (Laidler and Bunting, 1973). Since the pattern of fluid flow through the enzyme reactor determines the substrate concentration profile in the reactor, the extent of protection of enzyme by substrate would be governed by the choice of reactor operating conditions. Recently, O'Neill (1972b,c) has considered quantitatively the effects of enzyme inactivation in ideal CSTR and PFR systems.

For the case of substrate-independent thermal denaturation of enzyme, the rate of catalyst decay under isothermal reactor operation may be approximated by first-order kinetics:[1]

$$-(dE_0/dt) = k_d E_0 \tag{22}$$

where E_0 is the effective enzyme concentration in the reactor, k_d is the enzyme decay constant, and t is the reactor operating time. Equation (22), in conjunction with Eqs. (8) and (9), yields the following equations, which describe reactor performance in the presence of enzyme decay.

CSTR:

$$\ln\left[\frac{S_0 X + K'_m[X_0/1 - X_0]}{S_0 X_t + K'_m[X_t/(1 - X_t)]}\right] = k_d t \tag{23}$$

[1] Equations (14) through (21) are to be found in Table III.

PFR:

$$\ln \left[\frac{S_0 X_0 - K'_m \ln (1 - X_0)}{S_0 X_t - K'_m \ln (1 - X_t)} \right] = k_d t \tag{24}$$

where X_0 and X_t are the steady-state conversions when $t = 0$ and when the reactor has been operating for a time t, respectively. Similar equations have been developed by Lasch and Koelsch (1973) also.

From Eqs. (23) and (24), it may be deduced that in the limiting case of zero-order regime, there is no difference between the CSTR and the PFR in the decay of immobilized enzyme activity. As the reaction order increases from zero to one, the difference in conversion decay rates between PFR and CSTR is magnified. The output of the PFR decreases more rapidly than a CSTR; thus, it is more sensitive to enzyme inactivation than the CSTR. However, it is worth reiterating here that, under the same operating conditions, the CSTR requires a far greater amount of immobilized enzyme than is required by the PFR to effect the same degree of conversion in the quasisteady condition.

When enzyme inactivation is substrate concentration dependent, the rate of loss of enzyme activity can be described as (O'Neill, 1972d)

$$-(dE_0/dt) = (k'_d/S)E_0 \tag{25}$$

where S is the substrate concentration and k'_d is a lumped deactivation rate constant. From Eqs. (25) and (8), the following equation can be derived for a CSTR:

$$S_0(X_t - X_0) + S_0 \ln \left(\frac{X_0}{X_t} \right) + K'_m \ln \left[\frac{K'_m + S_0(1 - X)}{K'_m + S_0(1 - X_t)} \right] = k'_d t \tag{26}$$

In the CSTR, substrate concentration is uniform at any operating time so that all the enzyme would decay at the same rate. But in the PFR, the substrate concentration varies along the length of the reactor; thus, the enzyme is inactivated at different rates in different sections of the reactor, which introduces some complexity in modeling the process. However, by considering the PFR to be a series of CSTRs with equal amounts of immobilized enzyme in each tank, this difficulty can be overcome. Such an analysis by O'Neill (1972c) indicates that in both the limiting cases of zero- and first-order reaction kinetics the PFR is more efficient, as it has a lower decay rate of enzyme. It must be pointed out here that the above analysis is useful only in illustrating broadly the implications of enzyme inactivation for the fluid flow pattern through the immobilized enzyme reactor. Equations (22) and (25) both are extremely simplified descriptions of the enzyme inactivation process. The inadequacy of Eq. (22) has recently been established experimentally (Weetall and Havewala, 1972). More realistic approaches to modeling enzyme denaturation and elution from the carrier matrix will be considered in a later section.

IV. STEADY-STATE ANALYSIS OF MASS TRANSFER AND BIOCHEMICAL REACTION IN ENZYME REACTORS

When an enzyme is attached to a carrier matrix, its properties are altered significantly. The role of diffusional barriers, electrostatic fields, enzyme conformational changes and steric hindrances in causing these modifications of enzyme activity have been discussed in detail elsewhere in this volume. Our primary focus in this section is to connect these effects to the reactor environment. Perhaps the most important of all these effects is mass transport of substrate and product(s); therefore the enzyme reactor design as affected by the mass transfer problem is discussed in detail here.

Unlike a soluble enzyme, the solid-supported enzyme reacts in a heterogeneous phase. The heterogeneous aspects of enzyme-catalyzed reactions have been summarized in a recent review (McLaren and Packer, 1970). Many principles and procedures of conventional heterogeneous catalysis are therefore applicable to immobilized-enzyme reactor systems. There are at least five distinct steps in the overall enzymic reaction: (a) transport of substrate from the bulk solution to the immobilized enzyme carrier surface; (b) diffusion of substrate to the enzyme site; (c) enzyme-catalyzed reaction; (d) transport of product from the domain of the enzyme to the outer surface of the carrier; (e) transport of product to the bulk phase.

At steady state, the rates of the individual steps will be identical. This equality is used to develop a global rate expression in terms of concentrations of the bulk fluid. If one of these reaction steps is significantly slower than the rest, then this step offers the greatest resistance, and hence it is considered to be the step controlling the overall reaction rate. Where all the steps are of similar relative importance, detailed models are necessary to describe the simultaneous mass transfer and biochemical reaction in the enzyme reactor system.

Data obtained from laboratory enzyme reactors usually reveal the presence of mass-transfer resistances to the transport of substrate adjacent to—and within—the immobilized enzyme system (Hornby *et al.*, 1966; Sharp *et al.*, 1969; O'Neill *et al.*, 1971a; Vieth *et al.*, 1972c,

1973; Kobayashi and Moo-Young, 1973; Constantinides *et al.*, 1973; Venkatasubramanian *et al.*, 1974b). The apparent reaction kinetics have been observed to be dependent on the flow rate (linear velocity) of substrate over the immobilized enzyme bed. The apparent Michaelis constant K'_m varies with the flow rate, approaching the K_m value of the free enzyme at very high flow rates (Wilson *et al.*, 1968; Hornby *et al.*, 1968). Similarly, the dependence of reaction rate on agitation speed in a CSTR has been demonstrated (Lilly and Sharp, 1968). These studies are indicative of the significance of external or bulk diffusional resistance to substrate transport. Changes in the particle sizes of the immobilized derivatives also profoundly influence reaction rates (Quiocho and Richards, 1966; Tosa *et al.*, 1967; Kay and Lilly, 1970; Bunting and Laidler, 1972; Rovito and Kittrell, 1973; Marsh *et al.*, 1973), thus establishing the role of pore diffusional effects. A detailed experimental study has clearly shown the inadequacy of the idealized reactor performance equations based on simple Michaelis–Menten kinetics alone to explain the experimental observations (Weetall and Havewala, 1972). In this rather extensive study, the effects of flow rate, pore size of immobilized enzyme particle, and enzyme loading were investigated in both packed-bed and continuous-flow stirred-tank reactors. It is therefore mandatory to develop comprehensive models that account quantitatively for the diffusional effects.

A. External Film Diffusion

The local rate of diffusion of the substrate from the fluid bulk to the surface of the immobilized enzyme particle may be considered to be proportional to the area for mass transfer and the driving force for mass transfer, i.e., the concentration difference between the bulk and the catalyst external surface:

$$r_m = k_L a_m (S_F - S_S) \tag{27}$$

where r_m is the local mass transfer rate of substrate, k_L is the mass transfer coefficient, a_m is the surface area for mass transfer, S_F is the substrate concentration in the bulk, and S_S is the substrate concentration at the surface of immobilized enzyme.

A very simple approach to include the bulk diffusional effect in the reactor equation is to assimilate it into the lumped parameter K'_m. Assuming a linear gradient of substrate concentration across the boundary layer, the apparent Michaelis–Menten constant can be expressed as (Hornby *et al.*, 1968)

$$K'_m = K_m + (k_2 E_0 \delta/D) \tag{28}$$

where K_m is the Michaelis–Menten constant of native enzyme, δ is the thickness of the Nernst boundary layer, and D is the molecular diffusivity of substrate in the reaction solution.

With this lumped parameter, K'_m, the same reactor performance equations discussed in the last section may be employed, provided D and δ can be estimated accurately. However, experimental determination and unambiguous correlation of δ has not been feasible. Therefore, Eq. (28) has little practical value.

1. Mass-Transfer Correlations

Several correlations for mass transfer from liquid to solid are available in the chemical engineering literature. Karabelos et al. (1971) have presented an extensive survey of these correlations. A convenient, generalized formulation for bulk mass-transfer correlations is of the form

$$j_D = C(N_{Re})^{-n} \tag{29}$$

C and n are characteristic constants of a particular correlation. N_{Re} is the Reynolds number defined as

$$N_{Re} = d_P G/\mu \tag{30}$$

where d_P is the particle diameter (characteristic dimension of the immobilized enzyme particle), G is the superficial mass velocity $(= u\rho)$, u is the fluid velocity, μ is the substrate viscosity, and ρ is the density of substrate.

j_D is a factor that incorporates the mass-transfer coefficient k_L and reactor variables, such as flow rate, particle diameter, fluid density, and viscosity, as follows. It simplifies further, as shown, for constant Schmidt number values:

$$j_D = (k_L\rho/G)(\mu/\rho D)^{2/3} = C(N_{Re})^{-n} \tag{31}$$

The dimensionless group in parentheses is called the Schmidt number. D is the diffusivity of substrate. Different mass transfer correlations have different values of C and n.

Rovito and Kittrell (1973) have shown that the McCune and Wilhem (1949) correlation of mass-transfer coefficient is applicable to a packed bed of porous glass beads:

$$j_D = 1.625(N_{Re})^{-0.507} \tag{32}$$

For a packed bed of chips of collagen–whole microbial cell membrane (containing glucose isomerase activity), the correlation of Ergun (1952) has been successfully applied (Saini and Vieth, 1975). The correlation proposed by Chu *et al.* (1953) shown below was successfully employed in estimating bulk mass-transfer coefficients for a packed bed of collagen–enzyme chips (Davidson *et al.*, 1974; Fernandes *et al.*, 1974):

$$j_D = 5.7(N_{Re})^{-0.78} \tag{33}$$

Using the correlation of Satterfield (1970), the following equation for k_L has been obtained for a fixed-bed of spherical, microencapsulated enzymes (Mogensen and Vieth, 1973):

$$k_L = \frac{0.000464G^{1/3}}{\epsilon d_P^{2/3}} \tag{34}$$

Wilson and Geankoplis (1966) have proposed a correlation for external transport in packed beds at low particle Reynolds numbers (as is the usual case for immobilized enzyme reactors) which has been used by Ford *et al.* (1972):

$$k_L = \frac{6.54D^{2/3}G^{1/3}}{\epsilon d_P^{5/3}\rho^{1/3}} \tag{35}$$

The following theoretical relationships reported by Kunii and Suzuki (1967) have been applied to a packed-bed reactor of immobilized enzyme (Kobayashi and Moo-Young, 1971, 1973) using a channeling model and are applicable primarily for Peclet numbers of 10 or less:

$$N_{Sh} = (1/ad_P\xi)N_{Pe} \tag{36}$$

where N_{Sh} is the Sherwood number $(= k_L d_P/D)$, N_{Pe} is the Peclet number $(= ud_P/D)$, a is the external surface area of supported enzyme per unit reactor volume, and ξ is the ratio of average channeling length to particle diameter. For their experimental system, the authors found that ξ can be expressed as

$$\xi = 0.02d_P^{-1.5} \tag{37}$$

Mogensen and Vieth (1973) have suggested mass-transfer correlations for spherical microcapsules in a batch reactor. Simple correlations for suspended catalyst particles in stirred vessels have also been proposed (O'Neill, 1972d).

2. Combined Bulk Mass-Transfer and Biochemical Reaction

By using mass-transfer correlations such as those cited above, the relative importance of (external) diffusional resistance and reaction kinetics can be ascertained. This analysis would be applicable to systems with negligible pore diffusional barriers, e.g., nonporous supports. Combining Eqs. (27) and (31), an expression for the drop in substrate concentration between the bulk and the carrier surface, i.e., $(S_F - S_S)$ can be obtained:

$$(S_F - S_S) = \frac{r_{obs}(N_{Sc})^{2/3}\rho}{a_m j_D G} \tag{38}$$

r_{obs} is the observed reaction rate, i.e., the rate experimentally determined from bulk concentration measurements. When $(S_F - S_S)$ approaches zero, there is negligible bulk-transport resistance. High values of this difference implys that the overall reaction is diffusion controlled and Eq. (27) would be applicable to evaluate reaction rates. Since in this case, $S_F \gg S_S$, Eq. (27) may be simplified as

$$r_{obs} = k_L a_m S_F \tag{39}$$

For intermediate values of $(S_F - S_S)$, both mass-transfer and chemical reaction kinetics must be considered to predict reactor performance. For simplicity, one might consider first-order kinetics.

For many engineering calculations, first-order kinetics is an acceptable approximation. The Michaelis–Menten reaction scheme reduces to first-order kinetics at the exit concentrations obtained in reactors of practical interest. The integrated form of the Michaelis–Menten equation for a packed reactor is

$$\tau' = (K'_m/V'_m)[- \ln (1 - X) + (S_0 X/K'_m)] \tag{40}$$

where τ' is the reactor space time based on reactor fluid volume:

$$\tau' = (V_R \epsilon)/Q$$

The pseudo first-order assumption yields

$$\tau' = (K'_m/V'_m)[- \ln (1 - X)] \tag{41}$$

where the ratio (V'_m/K'_m) is the pseudo first-order rate constant, k_f. Consider a required conversion level of 90%. With $X = 0.9$ and taking $S_0/K'_m = 0.3$, the error in underestimating the reactor space time required by assuming first-order kinetics would be only $[S_0 X/K'_m]/[S_0 X/K'_m - \ln(1 - X)]$ or $0.27/2.573$, i.e., 10.5%. Thus, mass transfer-kinetic models incorporating first-order kinetic equations will provide good approximations for immobilized enzyme reactor systems of interest.

In the presence of significant external film diffusional resistance, the rates of mass-transfer and chemical reactions would be equal at the steady state. Under these conditions, Eq. (41) can be used for design purposes. But the pseudo first-order constant is now modified to include k_L:

$$k_f' = \frac{k_f k_L a_m}{k_f + k_L} \tag{42}$$

where k_f' is the combined bulk mass-transfer kinetic coefficient and k_f is the first-order rate constant ($= K_m'/V_m'$).

By using this approach, Rovito and Kittrell (1973) obtained an experimental correlation between k_f' and fluid velocity in the reactor. From Eqs. (30) and (31), k_L can be expressed as

$$k_L = C'G^N \tag{43}$$

where C' is a collection of terms tantamount to

$$C' = (C/\rho)(N_{Sc})^{-2/3}(d_P/\mu)^{-n}; \qquad N = 1 - n \tag{44}$$

Equation (42) may then be written in combination with Eq. (43) as

$$1/k_f' = (1/C'a_m)(1/G^N) + (1/k_f a_m) \tag{45}$$

An experimental correlation of $1/k_f'$ against $1/G^N$ is shown in Fig. 4. As predicted by Eq. (45), data for several reactor systems fall in a straight line. The best correlation is obtained for a value of $N = 0.5$. This correlation was found to be applicable over the particle Reynolds number range of 0.2 to 25. Similar correlations have also been developed for the porous glass–catalase system (Rovito and Kittrell, 1974).

B. Diffusive and Electrostatic Effects

The problem of boundary layer diffusional resistance can often be compounded by partitioning of substrate due to electrostatic forces. The apparent activity of an immobilized enzyme would be affected by the nature and the extent of charge on (a) the carrier surface and (b) the substrate. Electrostatic phenomena in conjugated enzyme systems were first identified by Goldstein *et al.* (1964) in polyanionic and polycationic supports and have since been reported by several investigators. The origin of these effects and their influence on the fixed enzyme activity have been summarized in different review articles (Goldstein, 1969; Goldman *et al.*, 1971; Melrose, 1971; Katchalski *et al.*, 1971). The primary concern here is to quantitate the combined diffusive and electrostatic effects and to outline how they affect reactor design.

The partitioning of substrate between the bulk solution phase and the carrier phase can be described by a partition coefficient, H. If S_S is

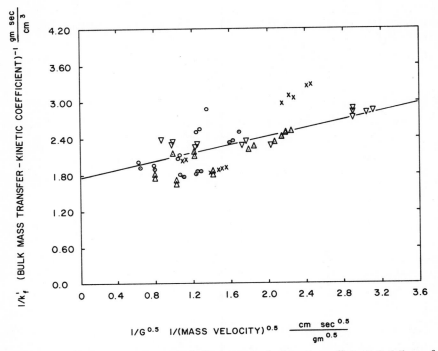

Fig. 4. Experimental correlation for bulk mass transfer-kinetic coefficient. $\bigcirc$, 1.5 gm of catalyst; $\odot$,1.0 gm of catalyst; $\triangle$, 1.0 gm of catalyst; $\times$, 0.5 gm of catalyst (reactor size: 5 mm diameter in all these cases); $\triangledown$, 0.5 gm of catalyst; $\triangle$, 1.0 gm of catalyst (reactor size: 7 mm diameter). (From Rovito and Kittrell, 1973.)

the substrate concentration at the carrier surface, then HS_s would be the effective concentration experienced by the immobilized enzyme surface. Goldstein and Katchalski (1968) have expressed the partition coefficient H as,

$$H = \exp\left(\bar{Z}\epsilon_d\psi/\bar{k}T\right) \tag{46}$$

where $\bar{Z}$ is the sign of charge (± 1), ϵ_d is the dielectric constant, $\bar{k}$ is the Boltzmann constant, T is the absolute temperature, and Ψ is the electrostatic potential prevailing in the vicinity of the matrix-bound enzyme. Thus the electrostatic effects can be lumped into a coefficient H. Wilson *et al.* (1968), Sundaram *et al.* (1970), and Kobayashi and Moo-Young (1971, 1973) have utilized such an approach in developing equations to describe the diffusive and electrostatic phenomena. However, it is difficult to determine H accurately.

Another way to account for substrate partition resulting from electrostatic effects is to incorporate it into the apparent K_m value (Hornby *et al.*, 1968). This simplified approach leads to a modified Michaelis constant that takes into account both bulk diffusional and electrostatic effects thus:

$$K'_{\mathrm{m}} = \left[K_{\mathrm{m}} + \frac{k_2 E_0 \delta}{D} \right] \left[\frac{\bar{R}T}{\bar{R}T - e\delta\mathrm{F} \text{ grad } \psi} \right] \tag{47}$$

where δ is the thickness of the diffusion layer, $\bar{R}$ is the gas constant, T is the absolute temperature, e is the electric charge, F is the Faraday constant, Ψ is the electrostatic potential, and K_{m} is the Michaelis–Menten constant for free enzyme. It may be noted that the above equation results from Eq. (28) to which the electrostatic effects term has been added. With this redefinition of K_{m}, the same reactor performance equations discussed earlier are applicable.

The diffusive and electrostatic effects have been analyzed in some detail by Shuler *et al.* (1972) for a nonporous (i.e., no pore diffusional limitation) immobilized-enzyme system. The mass-transport fluxes of substrate and product in the fluid phase aré represented by the Nernst–Planck equation:

$$J_{\mathrm{s}} = D \frac{dS}{dx} + S \frac{eDF}{\bar{R}T} \frac{d\psi}{dx} \tag{48}$$

$$J_{\mathrm{P}} = D \frac{dP}{dx} + P \frac{eDF}{\bar{R}T} \frac{d\psi}{dx} \tag{49}$$

where J_{S} and J_{P} are the mass fluxes of substrate and product, respectively, and S and P are the substrate and product concentrations, respectively.

The diffusivities of both substrate and product are assumed to be D. The first term in the above expressions is the Fickian mass transport under a concentration gradient. The second term represents the migration under an electrostatic potential gradient. At steady state,

$$J_{\mathrm{s}} + J_{\mathrm{P}} = 0 \tag{50}$$

Equations (48) and (49) can be added and the result integrated with the appropriate boundary conditions from $x = 0$ to $x = \delta$, the thickness of the Nernst boundary layer being δ (Kobayashi and Laidler, 1974a). Shuler *et al.* (1972) obtained the following approximate solution for J_{S}:

$$J_{\mathrm{s}} = M k_{\mathrm{L}}[S_{\mathrm{F}} - S_{\mathrm{s}} \exp \gamma] \tag{51}$$

where k_{L} is the mass transfer coefficient ($\approx D/\delta$),

$$\gamma = eF \Psi(0)/RT \tag{52}$$

and M is the electrostatic potential modifier defined as

$$M = \delta \Big/ \int_0^\delta \exp\left\{\gamma[\psi(x)/\psi(0)]\right\} dx \tag{53}$$

At the immobilized enzyme surface, the diffusive transfer rate is equal to the reaction rate under steady-state conditions, i.e.,

$$J_{\mathrm{S}} = Mk_{\mathrm{L}}[S_{\mathrm{F}} - S_{\mathrm{S}} \exp \gamma] = k_2 E_0 S_{\mathrm{S}}/(K_{\mathrm{m}}' + S_{\mathrm{S}}) \tag{54}$$

From the solution of the above equation, an effectiveness factor, η_{DE}—similar to effectiveness factors for internal diffusive effects—can be defined as the ratio of the actual reaction rate to that which would be obtained in the absence of diffusive and electrostatic effects. The effectiveness factor η_{DE} is a function of three parameters: (a) the electrostatic parameter γ, (b) the ratio of maximum reaction rate to the maximum mass transfer rate Ξ, and (c) the Michaelis–Menten constant. For small values, the effectiveness factor can be approximated as

$$\eta_{\mathrm{DE}} = \frac{(1 + \beta)}{1 + \beta \exp \gamma\{1 + [\Xi/(1 + \beta \exp \gamma)]\}} \tag{55}$$

where β is the dimensionless Michaelis constant ($= K_{\mathrm{m}}'/S_{\mathrm{F}}$).

$$\Xi = k_2 E_0 \, \delta/MDS_{\mathrm{F}} \tag{56}$$

where M is the electrostatic potential modifier defined as before.

Once η_{DE} is evaluated in this manner, the rest of the modeling is identical to that discussed in Section IV,C. However, to calculate η_{DE}, the potential distribution function must be known. Using the Gouy diffuse double-layer potential distribution given by

$$\psi(x) = \psi(0) \exp(-\kappa x) \tag{57}$$

where $1/\kappa =$ thickness parameter of the diffuse double-layer, effectiveness factor values have been calculated (Shuler *et al.*, 1972). However, it must be noted that Eq. (57) is valid only for systems of small surface potentials [$\Psi(0) \leqslant 25$ mV].

Recently, Hamilton *et al.* (1973) have extended the above analysis to higher surface potentials by employing the complete Gouy–Chapman potential distribution given below:

$$\psi(x) = \frac{2\bar{R}T}{eF} \ln \left[\frac{C_1 + C_2 \exp(-\kappa x)}{C_1 - C_2 \exp(-\kappa x)} \right] \tag{58}$$

where

$$C_1 = \exp(\gamma/2) + 1 \tag{59}$$
$$C_2 = \exp(\gamma/2) - 1 \tag{60}$$

Equation (58) reduces to Eq. (57) when $|\gamma| \ll 1$. These authors show theoretically that the effectiveness factor can attain magnitudes greater than unity at high surface potential values as shown in Fig. 5.

It can be noted from this figure that η_{DE} levels off after a certain surface charge density; i.e., the surface potential becomes essentially infinite. The dependence of effectiveness factor values on bulk substrate concentration has also been examined; it is shown that η_{DE} goes through a maximum. These results have significance for efficient design of immobilized enzyme systems for processing charged substrates. The reaction rate may be augmented by selection of a carrier that carries a charge sign opposite to that of substrate, so that there would be a favorable charge distribution. The type of result shown in Fig. 5 provides a criterion for the choice of a maximally effective surface charge density. Diffusive and electrostatic effects with substrate-inhibition kinetics have been presented recently (Shuler *et al.*, 1973).

Similar to the effectiveness factor η_{DE}, a utilization factor was defined by Kobayashi and Laidler (1974a) as the ratio of the mass transfer rate to the rate of substrate consumption in the bulk phase, i.e., the reaction rate based on bulk concentration. Equations to evaluate the utilization factor are given for different kinetic situations, such as Michaelis–Menten kinetics, substrate inhibition, and product inhibition. Expres-

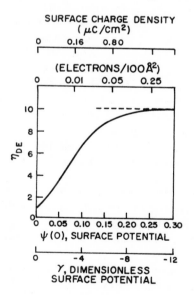

Fig. 5. Effectiveness factor, η_{DE}, as a function of surface potential or surface charge density. (From Hamilton *et al.*, 1973.) For parameter values used in simulating this profile, refer to the original article.

sions for the lumped apparent Michaelis–Menten constant and criteria for insignificant diffusion effects have also been presented.

At this point, it is worth reiterating that most of these results are purely theoretical. They are yet to be verified directly by experimental observations. The theoretical calculations are based on parameter values which may be representative of only a very limited number of immobilized enzyme systems. It must also be recognized that the idealized Nernst–Planck expression used here may not be valid under some circumstances (Lightfoot and Scattergood, 1965).

C. Internal (Pore) Diffusion

The internal diffusional effects can be quantitated through the use of an effectiveness factor, η. It is defined as the ratio of actual reaction rate in the matrix to the (maximum) rate obtainable if the substrate concentration were uniformly equal to its value at the carrier surface; i.e., without pore diffusional resistances. The observed reaction rate may therefore be expressed as

$$r = \eta \frac{k_2 E_0 S_8}{K'_m + S_8} \tag{61}$$

Different approaches to the estimation of the effectiveness factor have been the subject of several papers in the literature. An earlier chapter in this volume covers these approaches in more depth.

For many laboratory enzyme reactor systems, the rate-controlling step has been found to be the internal diffusion step (Bunting and Laidler, 1972; Marsh et al., 1973; Miyamoto et al., 1973). For a given reaction system, the effectiveness factor depends inversely on the particle size. It has been estimated that extremely fine particles of 30 μm or less in diameter and average pore size less than 2000 Å must be used to avoid pore diffusional barriers in an immobilized porous glass–glucose oxidase system (Rovito and Kittrell, 1973). These particle sizes would cause severe pressure-drop problems in industrial-scale packed-bed reactors. The problem is even more exacerbated when substrates of low diffusivities are to be processed. Therefore, suitable means must be devised to alleviate the microdiffusional problems. In order to do this, it is essential to develop models to describe the diffusion–reaction problem.

The simultaneous mass transfer and reaction of the substrate in the (internal) matrix of a single catalyst element can be analyzed by a second-order differential equation:

$$D_e(\partial^2 S/\partial x^2) - r = 0 \tag{62}$$

The solution of this equation yields the effectiveness factor in terms of the Thiele modulus ϕ, which is defined as

$$\phi = l\left[\frac{k_{\text{true}}S_S^{m-1}}{D_e}\right] \tag{63}$$

where x is the distance from the center of the catalyst particle, l is the characteristic dimension of catalyst particle, D_e is the effective diffusivity, m is the reaction order, and k_{true} is the true kinetic constant without mass-transfer disguises.

D_e is the effective diffusivity in the pores of the supported enzyme particle. It may be quite different from the bulk phase diffusivity of the substrate due to the tortuous diffusion paths within the pores. Empirical correlations for describing such effects—while available for a large number of inorganic systems (for review, see Satterfield, 1970; Satterfield and Sherwood, 1963)—have not been extended to immobilized enzymes.

It may be noted from Eq. (63) that the Thiele modulus expresses the ratio of the true reaction rate to the diffusion rate. For zero and first-order rate equations, simple expressions are obtained for ϕ. Therefore, analytical solutions for the combined intraparticle diffusion-reaction equations are readily derived. Several authors have chosen this approach (Goldman *et al.*, 1968; Sundaram *et al.*, 1970; Rony, 1971; Ollis, 1972; Lasch, 1973; Rovito and Kittrell, 1973; Vieth *et al.*, 1973; DeSimone and Caplan, 1973a). Mogensen and Vieth (1973) have derived an effectiveness factor expression for microencapsulated enzymes. In addition to the diffusional resistance of the solid membrane, the transport impedance through the internal liquid is also considered in this case.

For first-order reactions, the modulus ϕ is independent of substrate concentration, as is evident from Eq. (63). The reactor design equations are therefore readily derived. For an intraparticle diffusion-controlled first-order reaction in a packed-bed reactor containing spherical particles,

$$\tau = -\frac{\ln(1-X)}{k_f}\left[\frac{1}{\phi}\left(\frac{1}{\tanh 3\phi} - \frac{1}{3\phi}\right)\right] \tag{64}$$

The term in parentheses is the effectiveness factor with $\phi = R(k_f/D_e)^{0.5}$. For a packed-bed of chips, Eq. (64) becomes,

$$\tau = -\frac{\ln(1-X)}{k_f}\left[\frac{\tanh\phi}{\phi}\right] \tag{65}$$

From the above analysis, Rovito and Kittrell (1973) and Bunting and Laidler (1974) obtained experimental effectiveness factors for different particle sizes. These values are compared to the theoretical predictions of Eqs. (64) and (65) in Fig. 6. The data for the pore diffusion studies seem to fit reasonably well the pseudo first-order model.

Analytical solutions cannot be readily obtained with Michaelis–Menten kinetics due to the nonlinearity of the kinetic expression. Following the treatment of Langmuir–Hinshelwood kinetics (which is formally identical to Michaelis–Menten kinetics) by Roberts and Satterfield (1965), many attempts have been made to evaluate effectiveness factors for Michaelis–Menten kinetics. In this treatment, the modulus is changed in such a way that concentration- and nonconcentration-dependent parameters are separated. The concentration term is included in a dimensionless Michaelis constant $\beta = K'_m/S_0$. The modified Thiele modulus is defined by

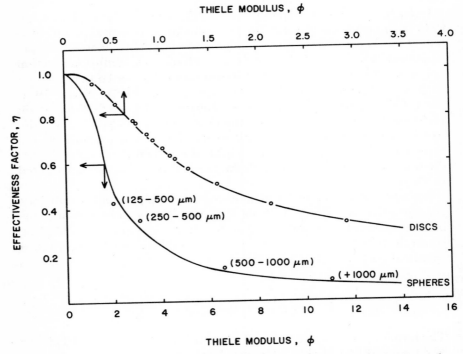

Fig. 6. Variation of effectiveness factor, η, with Thiele modulus (for pseudo first-order reaction kinetics) for spherical particles and disks. The solid line represents the theoretical prediction. Numbers in parentheses represent diameter of spheres. [Adapted from Bunting and Laidler (1972) and Rovito and Kittrell (1973).]

$$\phi_m = l\,[V'_m/K'_m D_e]^{0.5} \tag{66}$$

With this definition of ϕ_m, approximate analytical solutions have been obtained for the limiting case of large ϕ_m; i.e., the reaction is very rapid relative to diffusion (Petersen, 1965; Broun *et al.*, 1970). Bischoff (1965) suggested the use of a general modulus for describing the asymptotic behavior of η. Moo-Young and Kobayashi (1972) employed this suggestion to work out effectiveness factors for Michaelis–Menten kinetics as well as kinetic expressions involving competitive and non-competitive inhibitions; it was shown that the substrate-inhibited enzyme systems might have effectiveness factor values higher than unity. This is because under certain conditions the increase in rate caused by decreased substrate inhibition toward the center of the supported enzyme particle more than compensates for the decrease in rate caused by the concentration drop. Therefore, the overall reaction in these cases is faster than the case when the interior is at the same substrate concentration as in the bulk. Engasser and Horvath (1973) used the general modulus to treat Michaelis–Menten kinetics in spheres and spherical shells. Numerical solutions for Michaelis–Menten kinetics and effectiveness factor charts (η as a function of ϕ_m) have been developed by several authors recently (Marsh *et al.*, 1973; Miyamoto *et al.*, 1973; Engasser and Horvath, 1973; Blanch and Dunn, 1974; Hamilton *et al.*, 1974). The use of a different modulus has been advocated by Fink *et al.* (1973). Hultin *et al.* (1972) and Horvath and Engasser (1973) have used a modulus defined on the basis of observed reaction rate.

Davidson *et al.* (1974) approached the analysis of intradiffusional mass transfer using zero-order, first-order, and Michaelis–Menten kinetic models and compared the effectiveness factor values of the different kinetic models. Their analysis was done for a packed-bed reactor containing chips (approximately 6 mm × 6 mm × 0.05 mm) of collagen–enzyme membrane. Such a reactor is shown schematically in Fig. 7. The enzyme-membrane chips were viewed as semi-infinite slabs of thickness $2l$. This assumption is justified because the thicknesses of the chips were orders of magnitude smaller than the length and the width dimensions. x is the direction of diffusion of substrate into the membrane chip; diffusion occurs through two faces of the chip in the x direction such that the substrate concentration profile is symmetrical at the center. Consequently, the concentration gradient at the center of the chip is zero.

The reactor model based on Michaelis–Menten kinetics was divided into a membrane-phase model and a fluid-phase model. These sub-

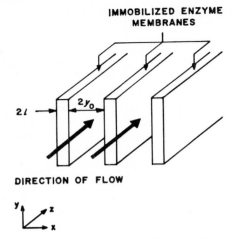

Fig. 7. Simplified parallel membrane model of the spiral-wound multipore reactor shown in Fig. 2. Membranes are considered to be semi-infinite slabs of thickness $2l$. Contiguous membrane layers are separated by a distance of $2y_0$. z is the direction of bulk flow along the reactor length. Substrate diffusion into the membrane occurs in the x direction.

models are coupled through the boundary conditions at the interface between the two phases. Eq. (62) expresses the membrane phase equation. In dimensionless form, it can be written as

$$\frac{\partial^2 Y_{\rm S}}{\partial L^2} - \phi_{\rm m}^2 \beta \left(\frac{Y_{\rm S}}{\beta + Y_{\rm S}}\right) = 0 \tag{67}$$

where

$$Y_{\rm S} = S/S_0; \qquad L = x/l; \qquad \beta = K_{\rm m}'/S_0; \qquad \phi_{\rm m} = l \, [kE_0/K_{\rm m}'D_{\rm e}]^{0.5}$$

Equation (67) has the following boundary conditions (B.C.) [separate detailed calculations for this system had established that there was no external mass transfer resistance, i.e., the concentration at the carrier surface ($L = 1$) was the same as that of the bulk]. The suffix F denotes that the concentrations are in the fluid phase

B.C. No. 1:

$$L = 0, \qquad \partial Y_{\rm S}/\partial L = 0 \qquad \text{(symmetry at the center)}$$

B.C. No. 2:

$$L = 1, \qquad Y_{\rm S} = Y_{\rm F} \qquad \text{(at solid surface, solid and bulk concentrations are equal)}$$

The concentration change along the length of the reactor (the flow direction is represented by z) is described by a fluid phase model. Ideal plug-flow conditions, i.e., no longitudinal or radial dispersion effects, are considered. Referring to the control volume shown in Fig. 8, the steady-state fluid phase mass balance can be written as

$$\begin{bmatrix} \text{rate of substrate} \\ \text{entering by convection} \end{bmatrix} - \begin{bmatrix} \text{rate of substrate} \\ \text{leaving by convection} \end{bmatrix} - \begin{bmatrix} \text{rate of transport} \\ \text{to catalyst surface} \end{bmatrix} = 0$$

i.e.,

$$(2\pi r \,\Delta r uS)_{\text{at } z} - (2\pi r \,\Delta r uS)_{\text{at } z+\Delta z} -$$
$$\left[\frac{(1-\epsilon)2\pi r \,\Delta r \,\Delta z}{2l} \right] D_e \left(\frac{\partial S}{\partial x} \right)_{\text{at } x=1} = 0 \quad (68a)$$

The control volume, $2\pi r \Delta r \Delta z$, is multiplied by $(1-\epsilon)$ to obtain the volume fraction of solids in the control shell. Dividing this quantity by the chip (catalyst) thickness, $2l$, results in the external surface area for mass transfer.

Dividing through by $2\pi r \Delta r \Delta z$ and taking the limit as $\Delta z \rightarrow 0$,

$$u(dS/dz) = (1-\epsilon)(D_e/2l)(\partial S/\partial x)_{\text{at } x=1} \quad (68b)$$

In dimensionless form, this becomes the following equation:

$$dY_F/dZ = \alpha(1-\epsilon)(\partial Y_S/\partial L)_{\text{at } L=1} \quad (68)$$

where (with B.C. No. 1: $Z = 0$, $Y_F = 1$) Z is the dimensionless reactor length $(= z/R_L)$, α is the dimensionless group $(= R_L D/2ul^2)$, R_L is the reactor length, z is the longitudinal distance in the reactor bed, u is the superficial fluid velocity, and ϵ is the void volume fraction. The membrane effectiveness factor, η, was expressed as

$$\eta = \frac{(\partial Y_S/\partial L)_{\text{at } L=1}}{[\phi_m^2 \beta Y_F/(\beta + Y_F)]} \quad (69)$$

It was determined that, at a particular value of Z, Eq. (69) can be expressed as a polynomial in Y_F:

$$\eta = a_0 + a_1 Y_F + a_2 Y_F^2 \quad (70)$$

With this simplification, Eq. (69) becomes

$$\frac{dY_F}{dZ} = \alpha(1-\epsilon)(a_0 + a_1 Y_F + a_2 Y_F^2) \left[\frac{\phi_m^2 \beta Y_F}{\beta + Y_F} \right] \quad (71)$$

The above mathematical formulation greatly simplifies the numerical iterative computation algorithm. From observed kinetic data (e.g., conversion as a function of reactor space time), the "best" kinetic constants for the system can be evaluated using this procedure. This is done as follows: For an assumed set of parameter values, K'_m and V'_m, which define a corresponding value of ϕ_m, the constants of Eq. (71) are determined numerically. Using these, the expected conversion values are predicted. If the value predicted by the model is different from the experimentally observed value, the recursive procedure is repeated with a newly guessed set of K'_m and V'_m values until the predicted and experimental values are the same. Kinetic constants estimated in this manner represent intrinsic parameter values free of any structural diffusion effects.

D. Combined External and Internal Diffusion Effects

In the preceding sections, the analysis of chemical reaction coupled with either external or internal transport resistance alone was discussed. Many practical reactor systems are likely to have both significant intra- and interparticle mass-transfer effects, which complicate even further the problem of reactor design. Several authors have attempted recently to model these cases; that is the subject matter of this section.

1. First-Order, Irreversible Reaction

For a fixed-bed reactor containing an enzyme-membrane, Vieth *et al*. (1973) developed expressions to describe different reaction kinetic schemes. The physical system modeled is the spiral-wound multipore reactor shown in Fig. 2. Considering the membrane to be a semi-infinite plate of thickness $2l$, the multipore reactor can be construed to be made up of a series of membranes. Thus, it can be approximated by a parallel membrane model as shown in Fig. 7. At steady state, the mass flux through the boundary layer is equal to that into the membrane, i.e.,

$$J = k_L(S_F - S_S) = \eta l k_f S_S \tag{72}$$

where J is the steady-state flux, S_f is the bulk substrate concentration, S_S is the substrate concentration at the membrane surface, and k_f is the pseudo first-order constant.

Rewriting Eq. (72),

$$(S_F - S_S)/J = 1/k_L \quad \text{and} \quad S_S/J = 1/\eta k_f l \tag{73}$$

which leads to

$$S_F/J = (1 \ k_L) + (S_S/J) = (1/k_L) + (1/\eta k_f l) \tag{74}$$

Defining a combined mass-transfer–kinetic coefficient, $K'(\equiv S_F/J)$, Eq. (74) can be written as

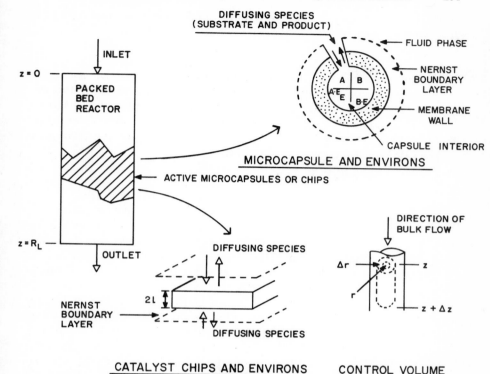

Fig. 8. Schematic diagram of a packed-bed reactor containing spherical microcapsules or collagen–enzyme membrane in the form of chips. z is the direction of fluid flow along the length of the reactor. Membrane chips are considered to be semi-infinite slabs of thickness $2l$. Substrate diffusion into the catalyst chips occurs in the x direction. The control volume shell, used in developing the fluid phase equation, Eq. (68), is also shown. Refer to page 259 for a discussion on model development. Transient analysis of a packed-bed reactor containing spherical microcapsules is discussed in detail in Section VII,B.

$$1/K' = (1/k_\text{L}) + (1/\eta k_t l) \tag{75}$$

Equation (75) represents a series of resistances, uncoupling the effects of diffusion and reaction. This is similar to the analysis of Aris (1957) for heterogeneous catalytic systems in which the reciprocal of an overall effectiveness factor is expressed as a series of resistances. In the case of spherical beads, Eq. (75) becomes

$$1/K' = (1/k_\text{L}) + (3/\eta k_t R) \tag{76}$$

where R is the radius of the particle. For microcapsules having an aqueous phase inside a semipermeable membrane wall, the wall permeability must also be accounted for. The wall resistance $(1/k_W)$ is included as shown below (Mogensen and Vieth, 1973):

$$1/K' = (1/k_L) + (1/k_W) + (3/\eta k_t R) \tag{77}$$

The results obtained above—even though based on an idealized solid phase geometry—can be readily extended to immobilized enzyme reactors of practical importance, such as the spiral wound, multipore biocatalytic modules discussed in an earlier section. Integrating the steady-state material balance on the substrate passing through a differential reactor element, we obtain:

$$\ln(1 - X) = -K'a\tau' \tag{78}$$

where a is the catalyst surface per unit of reactor fluid volume.

It is worth noting that the definition of K' is similar to that of k_t' in Eq. (43). It can be evaluated experimentally by knowing X and τ for steady-state reactor operation. The mass-transfer coefficient k_L is dependent on fluid velocity. Therefore, the combined coefficient K' would also be a function of fluid velocity. Experimental correlations of $(K'a)$ with flow rate (or linear velocity) can be developed easily. An example of such a correlation is shown in Fig. 9 for collagen–lactase (Eskamani, 1972) reactor system. The overall resistance $(1/K'a)$ was found to vary linearly with the reciprocal of the flow rate. From Eq. (75) it is seen that the effectiveness factor η can be evaluated from the intercept of Fig. 9. For instance, the effectiveness factor for the collagen–lactase system was thus evaluated to be 0.47. Correlations of this type are useful for scale-up purposes for this particular system. They can be used to design an immobilized enzyme reactor for other combinations of throughput and conversion.

2. First-Order, Reversible Reaction

Although a majority of enzyme-catalyzed reactions are irreversible in nature, there are some classes of reversible reactions effected by enzyme catalysis. Isomerization reactions typically fall under this category. Mass-transfer–kinetic models for such systems can be developed in a manner analogous to first-order, irreversible reaction cases. The first-order, reversible reaction considered here is

$$A \underset{k_{-1}}{\overset{k_{+1}}{\rightleftharpoons}} B$$

where k_{+1}, k_{-1} are the first-order rate constants for the forward and backward reactions, respectively.

Again considering the idealized parallel plate enzyme-membrane

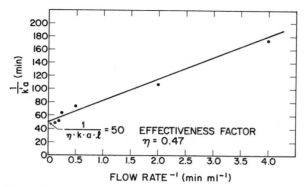

Fig. 9. Correlation of combined mass-transfer–kinetic coefficient with flow rate, based on Eq. (75). Experimental system studied was collagen–lactase multipore reactor of the type shown in Fig. 2. This system obeys pseudo first-order kinetics.

model which approximates the spiral-wound multipore reactor (refer to Figs. 2 and 8), the steady-state membrane phase equation [Eq. (62)] can be written as

$$D_e(\partial^2 S/\partial x^2) - [k_{+1}S - k_{-1}(S_0 - S)] = 0 \tag{79a}$$

The second term is the rate expression for the reversible reaction. $(S_0 - S)$ is the concentration of product, B, at any distance x along the direction of diffusion into the membrane. The above equation can be rewritten as

$$\frac{\partial^2 S}{\partial x^2} - \frac{k_{+1} + k_{-1}}{D_e}S + \frac{k_{-1}S_0}{D_e} = 0 \tag{79b}$$

In dimensionless form, this becomes

$$(\partial^2 Y_8/\partial L^2) - \phi^2 Y_8 + (k_{-1}l^2/D_e) = 0 \tag{79}$$

with the following boundary conditions:

B.C. No. 1:

$$L = 0; \quad \partial Y_8/\partial L = 0$$

B.C. No. 2:

$$L = 1; \quad (\partial Y_8/\partial L)_{\text{at } L=1} = N'_{\text{Sh}}(Y_\text{F} - Y_8)$$

where N'_{Sh} is the modified Sherwood number ($= k_L l/D_e$), and ϕ is the Thiele modulus, defined as $l[(k_{+1} + k_{-1})/D_e]^{1/2}$.

The fluid phase equation is represented [similar to Eq. (68)] by

$$-(\partial Y_\text{F}/\partial \tau') = -(D/ly_0)(\partial Y_8/\partial L)_{\text{at } L=1} \tag{80}$$

where y_0 is the half-distance between two flat plates in a compartment of the module (refer to Fig. 7) and

B.C. No. 1:

$$\tau' = 0, \qquad Y_F = 1$$

B.C. No. 2:

$$\tau' = 0, \qquad Y_S = 1$$

Equations (79) and (80) together with their boundary conditions can be solved analytically, yielding

$$Y_F = \exp A_1\tau'[1 + (A_2/A_1)] - (A_2/A_1) \tag{81}$$

where

$$A_1 = \frac{k_L}{y_0}\left[\frac{\phi \tanh \phi}{N'_{\text{Sh}} + \phi \tanh \phi}\right] \tag{82}$$

$$A_2 = \frac{k_L}{y_0}\left[\frac{\phi \tanh \phi(k_{-1}/k_{+1} + k_{-1})}{N'_{\text{Sh}} + \phi \tanh \phi}\right] \tag{83}$$

Unlike the case of irreversible reaction, it is not possible to uncouple the bulk and pore diffusional effects. Therefore, it is not possible to define a combined mass transfer-kinetic coefficient as in Eq. (75). However, these equations can be extended to actual enzyme reactor systems of the type mentioned earlier. A reactor balance for a differential element would be

$$dY_F/d\tau' = -k_La(Y_F - Y_S) \tag{84}$$

Equation (84) can be solved together with the material balance Eq. (79) for the membrane. The solution yields results similar to Eq. (81) except that the surface-to-volume ratio, a, replaces $1/y_0$ in Eqs. (82) and (83).

3. Michaelis–Menten Kinetics

Numerical computation has to be resorted to in the case of Michaelis–Menten kinetics for the solution of the mass transfer-kinetic model. The physical system modeled is the same as that for the previous two sections, i.e., a spiral-wound multipore enzyme-membrane reactor approximated by a parallel membrane model. The solid phase equation is the same as Eq. (66), but the presence of external mass transfer modifies the boundary conditions:

B.C. No. 1:

$$\text{at} \quad L = 0, \qquad \partial Y/\partial L = 0$$

B.C. No. 2:

$$\text{at} \quad L = 1, \qquad \partial Y/\partial L = N'_{Sh}(Y_F - Y)$$

where N'_{Sh} is the modified Sherwood number ($= k_l l/D_e$). Following Eq. (69), the effectiveness factor is given by

$$\eta = \frac{N'_{Sh}(Y_F - Y)}{[\phi_m^2 \beta Y_F/(\beta + Y_F)]} \tag{85}$$

Again, expressing η as a polynominal in Y_F, the fluid-phase equation can be expressed as in Eq. (70). Numerical solution of the model is presented graphically in Fig. 10. The variation of conversion with Sherwood number is shown in this figure.

Another approach to solve the combined external and internal mass transfer of substrate concomitant with the Michaelis–Menten reaction scheme has been advanced recently (Fink *et al.*, 1973). Considering an

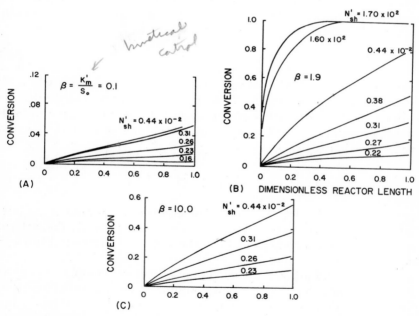

Fig. 10. Variation of fractional conversion, X, with dimensionless reactor length, Z, for different levels of mass-transfer resistance. Panels (A), (B), and (C) show the variation with respect to N'_{Sh} for three different β values; $\beta = k'_m/S_0$ is the dimensionless Michaelis–Menten constant. Larger values of N'_{Sh} correspond to lower resistances to mass transfer for a given value of diffusion coefficient, D_e. Profiles shown (A, B, and C) were generated based on the parallel membrane model of the multipore reactor system (refer to Figs. 2 and 8) for Michaelis–Menten kinetics. For a complete discussion, refer to the text.

idealized system of a permeable membrane entrapping within it a homogeneous enzyme solution, mass transfer-kinetic equations similar to Eqs. (67) and (68) are developed. A geometric correction factor is introduced in the system equations so that they are applicable to spherical, planar, or cylindrical geometries. (The practical systems of interest would be packed-bed reactors containing microcapsules or hollow fibers.) Following Satterfield (1970), an overall effectiveness factor, η_0, is defined as the ratio of the actual reaction rate to the uniform rate that would be observed at the external substrate concentration, i.e., in the absence of diffusional effects. Thus, η_0 lumps into it the effects of both bulk and solid phase mass-transfer constraints:

$$\eta_0 = \frac{[(p + 2)/R]J_8}{V'_m S_0/K'_m + S_0} \tag{86}$$

where p is the geometric factor, $p = -1, 0, +1$ for planar, cylindrical, and spherical membranes, respectively, R is the radius or depth of the enzyme-containing region, J_S is the substrate flux across the membrane into the enzyme solution, given by

$$J_8 = P'[S_0/h_e) - h_i S_i] \tag{87}$$

where h_e, h_i are the external and internal partition coefficients for the membranes, respectively, S_0, S_i are the substrate concentration in the bulk and in the capsule interior, respectively, and P' is the total substrate permeability coefficient for the membrane and external boundary layer, where

$$1/P' = (A_{R_0}/A_{R_0'})(1/h_e k_L) + (1/P'_8) \tag{88}$$

and where A_{R_0}, $A_{R_0'}$ are the area of the capsule membrane at the inner radius R_0, and the outer radius, R_0', respectively, k_L is the external mass-transfer coefficient, and P'_S is the substrate-permeability coefficient for the membrane.

With these definitions, and using the numerical computation procedure of Na and Na (1970), the two-point boundary value diffusion-reaction problem is transformed into an initial value problem, which facilitates the numerical procedure. Overall effectiveness factor charts, as a function of the modulus ϕ_m and design correlations for enzyme constrained within spherical membranes, have also been presented in the work of Fink *et al.* A similar approach has been employed by Hamilton *et al.* (1974); surface and the bulk substrate concentrations are related through a partition coefficient. Rony (1971) has also developed combined diffusion-reaction models through the use of partition coefficients for zero- and first-order reactions. It is important to note here that in this approach the bulk mass-transfer effect is accounted for by means of a partition coefficient (which is based on an equilibrium assumption) rather than by an external mass-transfer coefficient.

E. Analysis and Interpretation of Laboratory Data

From the preceding discussions, it is obvious that laboratory kinetic data may often be disguised by diffusional and/or electrostatic effects. Therefore, when these effects are present, recourse to simple graphical methods such as Lineweaver–Burk plots—which are chiefly applicable to homogeneous enzyme catalysis—to estimate kinetic parameters is likely to result in erroneous parameter values. It is necessary to exercise sufficient care in the design of experiments and in subsequent data analysis to obtain meaningful results. In this connection, several principles outlined by Churchill (1974) and Smith (1970) for heterogeneous catalytic systems can be extended to immobilized enzyme-catalyzed reactions.

Methods discussed in the preceding sections can be used to determine the rate-controlling step—if there really is a rate-limiting step—of the overall reaction. Experimental criteria for bulk-diffusion controlled, pore-diffusion controlled, and kinetic-controlled reactions have been spelled out by different authors (Horvath *et al.*, 1972, 1973b; Engasser and Horvath, 1973, 1974; Davidson *et al.*, 1974; Hamilton *et al.*, 1974; Gondo *et al.*, 1973; Kobayashi and Laidler, 1974a). If the reaction is kinetically controlled, graphical procedures similar to those for free enzyme kinetics can be employed. Under significant diffusional resistances, Lineweaver–Burk plots become nonlinear; the shape of the curve would depend on the value of the ratios of the Michaelis–Menten constant to the substrate concentration (β) and the Thiele modulus. At relatively low substrate concentrations, the plot is nonlinear only at the lowermost concentration values. As substrate concentration increases, the plot assumes a concavity with respect to the abscissa, the curvature regime increasing with increasing Thiele modulus. At very high concentrations, the effectiveness factor approaches unity, thus leading to an asymptotic straight line corresponding to diffusion-free kinetics. The concave curvature is diminished with a concomitant increase in the slope of the plot as a result of external diffusion (Hamilton *et al.*, 1974). The influence of electrostatic effects on the Lineweaver–Burk plot has been discussed by Hamilton *et al.* (1973). Different approaches for determination of intrinsic kinetic parameters under these conditions have been discussed by these authors.

A proper experimental design could eliminate many difficulties introduced by diffusional effects. If the experimental system is operated as a differential reactor, i.e., the fluid is passed over the bed at such high rates that only small conversions are obtained, then complications due to external diffusional gradients and/or nonideal flow patterns could be avoided. However, the assay procedures for the reaction system may not be sensitive enough to detect such small conversions. A possible way of circumventing this shortcoming is to operate the reactor as a recycle reactor. The use of recycle reactor systems in obtaining kinetic data and analysis of such data have been discussed in detail elsewhere (Ford *et al.*, 1972; Venkatasubramanian *et al.*, 1972; Constantinides *et al.*, 1973).

By operating a reactor as an integral reactor, i.e., in one-pass operation with appreciable conversion per pass, rate data can be gathered over a wide range of flow rates and concentrations. Kinetic parameters can be evaluated from an assumed reactor model incorporating both mass transfer and kinetic phenomena and the experimental data by minimizing the sum of squares of the difference between the model predictions and the observed data, i.e., an objective function of the form

$$\sigma = \sum_{i=1}^{n} [X_{i(\text{model})} - X_{i(\text{expt})}]^2 \tag{89}$$

where: $X_{i(\text{model})}$ is the conversion value predicted by the model, $X_{i(\text{expt})}$ is the observed conversion value, and n is the number of experimental points. The set of parameter values that minimize the function σ are the statistically "best" values. Several schemes are available for finding the maximum σ value. Statistical procedures involved in such parameter estimation techniques can be found in other reviews (Cleland, 1967; Kittrell, 1970; Hultin *et al.*, 1972). It may be noted that any nonlinear model can be used in Eq. (89).

F. Models for Kinetically Controlled Reactions

In the case of intrinsically slow biochemical reactions, the mass transfer resistances often become negligible. An important case in point is the isomerization of glucose to fructose by immobilized glucose isomerase. The isomerization reaction is reversible and has an equilibrium constant of approximately 1.0 at 70°C. In our laboratory we have recently completed a detailed kinetic study of this system (Saini and Vieth, 1975). Owing to the surging commercial interest in this system, this analysis is briefly discussed here.

A packed bed of chipped collagen–whole cell (containing glucose isomerase activity) was used in this study. Variation of fluid mass velocity over a wide range (15 to 760 gm/cm²hr) had no appreciable effect on the initial reaction rate. Using the (bulk) mass-transfer correlations outlined earlier, a percentage concentration drop between bulk and membrane surface was estimated to be only 0.005%. Therefore, external diffusional limitations can be ruled out. Using zero-order kinetics, the effectiveness factor was found to be 0.95, which was also corroborated experimentally. Thus, intramembrane transport resistance was also insignificant. Similar observations were also made by Giniger (1973).

The isomerization reaction $A \rightleftharpoons B$ may be considered to occur in three sequential steps: (a) formation of an enzyme–substrate complex, (b) isomerization of the complex, and (c) desorption of the product. No step is considered to be exclusively rate controlling in this mechanism. The rates of the individual steps can be written as follows for the postulated reaction mechanism:

$$A + E \underset{k_{-1}}{\overset{k_1}{\rightleftharpoons}} A \cdot E \underset{k_{-2}}{\overset{k_2}{\rightleftharpoons}} B \cdot E \underset{k_{-3}}{\overset{k_3}{\rightleftharpoons}} B + E \tag{90}$$

$$r_1 = k_1[A][E] - k_{-1}[A \cdot E] \tag{91}$$

$$r_2 = k_2[A \cdot E] - k_{-2}[B \cdot E] \tag{92}$$

$$r_3 = k_3[B \cdot E] - k_{-3}[B][E] \tag{93}$$

where k_1, k_2, k_3 are the rate constants for the forward reactions, k_{-1}, k_{-2}, k_{-3} are the rate constants for the backward reactions, $[A \cdot E], [B \cdot E]$ are the reaction intermediates, and $[A], [B]$ are the concentrations of glucose and fructose, respectively. If the enzyme E and the intermediate species $A \cdot E$ and $B \cdot E$ are assumed to be nondiffusing, the following stoichiometric invariance is obtained:

$$E_0 = E + A \cdot E + B \cdot E \tag{94}$$

where E_0 is the total enzyme concentration. At steady state, all steps should be mutually rate controlling, i.e.,

$$r_1 = r_2 = r_3 = r \tag{95}$$

The generalized rate expression r is obtained by solving simultaneously Eqs. (91)–(95):

$$r = \frac{m_4 E_0[A] + m_5 E_0[B]}{m_1 + m_2[A] + m_3[B]} \tag{96}$$

where the lumped constants m_1 through m_5 are given by

$$m_1 = (1/k_2) + (1/k_{-1}) + (k_{-2}/k_2k_3) \tag{97}$$

$$m_2 = (k_1/k_{-1})[(1/k_2) + (1/k_{-3}) + (k_{-2}/k_2k_3)] \tag{98}$$

$$m_3 = (k_{-3}/k_3)[(1/k_{-2}) + (1/k_{-1}) + (k_{-2}/k_{-1}k_2)] \tag{99}$$

$$m_4 = k_1/k_{-1} \tag{100}$$

$$m_5 = -(k_{-2}k_{-3}/k_2k_3) \tag{101}$$

The lumped, kinetic constants were evaluated experimentally through the use of Lineweaver–Burk plots and the following rate expression is obtained on substituting these constants:

$$r = \frac{0.128[A] - 0.098[B]}{0.096 + 0.383[A] + 0.25[B]} \tag{102}$$

Combining this with the plug-flow reactor equation [Eq. (7)], a reactor design equation is obtained:

$$[1.363\, A_0 + 0.424] \ln\left[\frac{0.128\, A_0}{0.128\, A_0 - 0.225\, X}\right] + 0.589\, A_0 X = \tau' \tag{103}$$

Here, the fractional conversion X is defined as $(A_0 - A)/A_0$. In order to make the application of Eq. (103) general, certain correction factors have to be incorporated. These are necessary because (a) each batch of immobilized preparation may differ in activity; (b) contact efficiency may be different depending upon many factors, such as spacing element, length/diameter ratio of the reactor; (c) packing density may be different.

Assuming linear dependence of reaction velocity on these three parameters, the right side of Eq. (103) can be modified as

$$(\tau')(P/0.227)(a'/235)(C'/1)$$

where $\bar{P}/0.227$ is the packing density correction factor, $a'/235$ is the activity correction factor, and $C'/1.0$ is the contact efficiency correction factor. The numbers 0.227, 235, and 1.0 are, respectively, the packing density (grams of catalyst per milliliter of fluid volume), activity (units per gram of catalyst), and contact efficiency (observed units of activity per available (total) units of activity). With this modification Eq. (103) becomes,

$$(1.363\, A_0 + 0.424) \ln\left[\frac{0.128\, A_0}{0.128\, A_0 - 0.225\, X}\right] + 0.589\, A_0 X = \frac{\tau' P a' C'}{53.4} \tag{104}$$

Based on the above design equation, digital computer simulations were performed to generate conversion–space time profiles at different feed concentrations. Experimental results agreed well with the predicted values, as shown in Fig. 11, for a particular combination of space time and inlet substrate concentration.

V. MASS TRANSFER-KINETIC MODELS FOR SPECIALIZED SYSTEMS: TUBULAR REACTORS WITH CATALYTICALLY ACTIVE WALLS/HOLLOW-FIBER REACTORS

Reactor tubes with enzymes attached to the inner tube wall, either directly or through a porous gel annulus, can readily be modeled. Such systems are of importance in analytical devices. Unlike the case in a packed-bed reactor, the chemical reaction occurs only at the tube wall or the porous annulus. In the case of tubular reactors with enzymes directly linked to the tube inner surface, the effectiveness factor may be considered to be unity; i.e., all the enzyme on the tube surface is active (Sundaram and Hornby, 1970; Hornby and Filippusson, 1970).

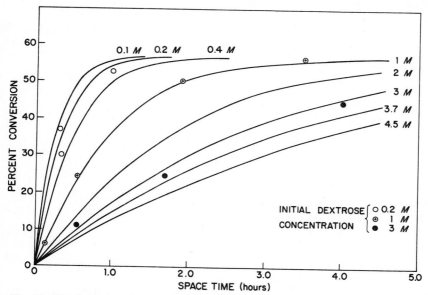

Fig. 11. Simulation profiles and experimental data for glucose isomerization model, based on Eq. (104). Numbers at the end of each curve represent the feed glucose concentration used in generating that curve. The physical system considered is the same as that for Fig. 10.

However, the diffusion of the substrate both in the bulk stream and the porous catalytic layer at the wall must be considered in the case where the enzyme is present in an annular porous matrix. Such reactor systems, called open tubular heterogeneous enzyme reactors (OTHER) have been extensively analyzed by Horvath and co-workers (Horvath and Solomon, 1972; Horvath et al., 1973c).

A. Reactors with Catalytic Inner Walls

Kobayashi and Laidler (1974b) have presented an analysis of a tubular reactor in which an enzyme is attached to the interior tube surface. The reactor model consists of a differential equation describing the substrate movement in the axial and radial directions with the enzymic reaction at the catalytic wall entering into the model as a boundary condition. Assuming a laminar velocity profile in the tube, the fluid-phase equation can be written as

$$D\left(\frac{\partial^2 S}{\partial \imath^2} + \frac{1}{\imath}\frac{\partial S}{\partial \imath}\right) - 2u_m\left(1 - \frac{\imath^2}{R'^2}\right)\frac{\partial S}{\partial z} = 0 \tag{105}$$

where $\imath$ is the radial position in the reactor, R' is the radius of the reactor, u_m is the maximum superficial velocity, and z is the axial position in the reactor. Equation (105) states that the diffusive flux in the radial direction is equal to the convective transport flux in the axial direction at steady state. The boundary conditions are

B.C. No. 1:

at $z = 0$,   all $\imath$; $S = S_0$

B.C. No. 2:

at $\imath = R'$,   all z; $D(\partial S/\partial \imath) = r(S_w)$

where $r(S_w)$ is the reaction rate based on substrate concentration at the catalytic wall, S_w.

For the limiting case of bulk diffusion-control, the substrate concentration would be essentially zero, i.e., boundary condition No. 2 becomes $S = 0$, throughout the reactor length. An analytical solution for the reactor model can be obtained for this condition for a kinetically controlled reaction, i.e., very slow enzyme reaction, the initially flat concentration profile does not change along the reactor. Therefore, the reaction rate would simply be based on bulk concentration value. Numerical solutions for the reactor model have been obtained for the intermediate region between these two limiting cases. Theoretical plots of the numerical solutions have been developed to delineate the dependence of the overall reaction rate on the substrate concentration, the Michaelis–Menten constant, and the Damköhler number. The latter is the ratio of the rate of reaction without diffusional effects to that for the diffusion-controlled reaction. Some of these theoretical predictions have been verified experimentally using a nylon tubular reactor with a catalytically active wall coated with asparaginase (Bunting and Laidler, 1974).

B. Reactors with Porous Annular Catalytic Walls

In the case of a tubular reactor with a porous annular catalytic shell at the tube wall, a solid-phase equation is also necessary in addition to the fluid-phase equation given by Eq. (105). The solid-phase equation, which equates the diffusive flux of substrate inside the porous annulus to the reaction rate of substrate conversion, is identical to Eq. (67). However, the modified Thiele modulus, ϕ_m, is now based on the thickness of the porous catalytic annulus rather than on the half-thickness of the catalyst pellet as in Eq. (67). An effectiveness factor similar to Eq. (69) can be defined and evaluated for different substrate concentration values. Numerical solutions of these equations have been obtained by Horvath *et al.* (1973c). In order to simplify the computational algorithm, the effectiveness factor was expressed as a weighted sum of the effectiveness factors for zero- and first-order reactions (limiting orders of Michaelis–Menten kinetic expression).

Simulation results from this model are presented in Fig. 12. The regimes of diffusion control and kinetic control are highlighted in terms of a dimensionless parameter called reactor modulus. It is a measure of the maximum possible rate in the catalytic annulus to the maximum possible rate of radial molecular diffusion. (Note that the reactor modulus resembles a Damköhler number.) The reactor performance can be described in terms of the reactor modulus and a dimensionless reactor length, ζ, which is proportional to residence time or reactor length and inversely proportional to the average rate of reaction to achieve a desired conversion level. As shown in Fig. 12, at reactor modulus values far greater than unity, the reaction is bulk diffusion-controlled; ζ is independent of reactor modulus in this region. At low modulus values, the reaction becomes kinetically controlled. These limiting functional dependencies of ζ on the reactor modulus for diffusion control and for kinetic control may be represented as ζ_{diff} and ζ_{kin}, respectively. The intermediate reactor modulus values characterize the transition regime. The reactor length

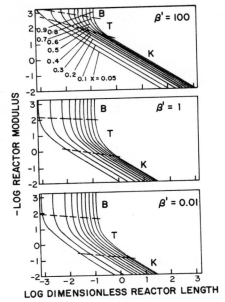

Fig. 12. Relationship between reactor modulus and dimensionless reactor length required to achieve conversions indicated at different values of dimensionless substrate concentration, $\beta'(= S_0/K'_m)$, and fractional conversion, x. Note the regimes of bulk diffusion control (B), kinetic control (K), and the transition regime (T). The physical system considered here is an enzyme reactor with porous annular walls. Refer to Section V,B for a detailed discussion. (From Horvath *et al.*, 1973c.)

ζ required to achieve a particular conversion level can be calculated by the approximate relation

$$\zeta \approx \zeta_{\text{diff}} + (\zeta_{\text{kin}}/\langle\eta\rangle) \tag{106}$$

where $\langle\eta\rangle$ is an average catalyst effectiveness factor. The second term on the right-hand side may be construed as ζ_{cat}, i.e., the axial distance required to achieve a desired conversion when the overall reaction is pore diffusion-controlled. This procedure of evaluating ζ values by simple additivity of the limiting ζ values permits a simplified treatment of the reactor data and problems of reactor design.

Based on the above definitions of the reactor coordinate, Horvath *et al.* (1973c) have also proposed a reactor effectiveness factor. It is expressed as the ratio of the actual reaction rate to some maximum reaction rate characteristic for that reactor both evaluated for a given conversion. Thus, the efficiency of the reactor can be represented by a kinetic effectiveness factor, E_{kin}, and a diffusional effectiveness factor, E_{diff}, thus:

$$E_{kin} = \zeta_{kin}/\zeta \tag{107}$$

$$E_{diff} = \zeta_{diff}/\zeta \tag{108}$$

E_{kin} is a measure of the degree of utilization of the enzyme because it is a ratio of the actual heterogeneous catalytic rate to the rate that would be obtained if the same amount of enzyme were used in a homogeneous plug-flow reactor under the same conditions. The actual reaction rate relative to the rate that would result if sufficient enzyme were present at the catalytic wall such that the surface concentration of substrate would be essentially zero is given by E_{diff}. Plots of the type shown in Fig. 13 have been presented which are useful to evaluate the reactor effectiveness factors. From Fig. 13 it may be noted that the dependence of E_{kin} on the reactor modulus is similar to that of the catalyst effectiveness factor on the modified Thiele modulus. Using these theoretical developments, reactor design equations for the limiting cases of bulk diffusion-control and kinetic-control have been presented (Horvath and Solomon, 1972). Correlations of Sherwood number with dimensionless reactor lengths from the chemical engineering literature have been used in developing these design equations. The foregoing analysis has also been extended to shell-structured enzyme reactors—known as pellicular enzyme reactors—in which the catalyst is confined to a porous outer shell surrounding an impervious inert core (Horvath and Engasser, 1973).

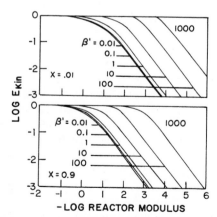

Fig. 13. Reactor effectiveness factor, E_{kin}, as a function of reactor modulus for various β' values at 1% and 90% conversion levels. The reactor system is the same as that for Fig. 12. (From Horvath *et al.*, 1973c.)

C. Hollow-Fiber Enzyme Reactors

Modeling of biochemical reactors containing enzymes trapped in hollow fibers has been carried out by Rony (1971) for zero- and first-order reactions. Lambert and Chambers (1973) have also proposed simple models for enzymes immobilized in hollow fibers. Theoretical treatment of asymmetric hollow-fiber membrane systems has been developed recently (Waterland *et al.*, 1974). As pointed out earlier, in this type of hollow-fiber membrane, the enzyme solution is contained within an annular porous shell structure surrounding the hollow fiber, which is impermeable to the enzyme. The substrate flowing through the fiber core, however, can diffuse through the hollow fiber to gain ingress to the catalytic site. The similarity of this system and the tubular reactor with a porous catalytic layer at the inner tube wall can be noted readily; therefore, the development of mathematical expressions for these two systems is very similar. However, to be fully rigorous, the asymmetric membrane model should also contain a consideration of the possible resistance of the finite thickness of the membrane separating the catalytic annulus and the cylindrical fiber lumen. Further, the curvature effects in the diffusive mass transport equation in the porous catalytic annulus must also be taken into account. The model developed by Waterland and associates in cylindrical coordinates includes these aspects.

While Horvath *et al.*(1973c) obtained numerical solutions to their model by uncoupling the system equations through the use of an effectiveness factor, the coupled differential equations in the asymmetric membrane model were solved directly. For first-order reaction, analytical solutions were obtained; numerical solutions for Michaelis–Menten kinetics were presented in graphical form in terms of the predicted profiles of dimensionless reactor length, required to yield a desired conversion level as a function of modified Thiele modulus, ϕ_m, for different β values. These profiles have a qualitative agreement with Horvath's model. For very small and very large values of the Thiele modulus, the overall reaction would be diffusion-controlled and kinetics-controlled, respectively. The transition region between conditions of diffusion control and kinetic control occurs over a narrower range of the modulus, as β is increased. It is also predicted from the simulations that as the reactor inlet substrate concentration is increased (i.e., a lower β value), the reactor might switch from a largely diffusion-controlled region to a kinetically controlled regime. This implies that a disproportionately longer reactor would be required to achieve the same level of conversion. Comparison of the profiles generated for the asymmetric hollow-fiber reactor system and a hollow-fiber reactor with the enzyme filled within the fiber indicates that the two configurations yield similar results under the kinetically controlled regime. However, for partial or total diffusion control, the reactor length required to effect a given conversion is much lower for the asymmetric hollow-fiber reactor system, thus alluding to its greater efficiency.

VI. NONIDEAL FLUID-FLOW PATTERNS IN ENZYME REACTORS

A. Analysis of Dispersion Effects

One of the important factors contributing to the deviation from an ideal plug-flow hydrodynamic condition in a packed-bed reactor is longitudinal dispersion. By axial or longitudinal dispersion is meant the combined turbulent and molecular diffusion of substrate and product molecules from one fluid element to another in the direction of flow. Since the substrate concentration decreases along the length of a packed-bed reactor, longitudinal dispersion provides a transport mechanism in addition to convective transfer by bulk flow. If the fluid flow through the reactor is laminar, only molecular diffusion would occur. This would be quite insignificant, provided the length of the reactor is far greater than its diameter. However, under turbulent flow conditions, the rate of eddy diffusive transport could be substantial because the eddy diffusion coefficient is normally orders of magnitude higher than molecular diffusivities. With longitudinal dispersion, the size of the reactor needed to effect a given conversion might be appreciably greater than that computed on the basis of plug-flow conditions. This disadvantage might be compounded if complex reaction schemes are involved (Levenspiel, 1972).

Most laboratory immobilized-enzyme reactors have poor length to diameter ratios, indicating the possibility of significant axial dispersion. Even in industrial-scale operation, if large particles are employed, the effect of backmixing may become substantial. Furthermore, the effect of longitudinal dispersion must be accounted for, when fluidized-bed reactors are employed, even if very small particles are used as solid supports. Recently, Kobayashi and Moo-Young (1971) analyzed theoretically the effects of backmixing and bulk mass transfer on the performance of immobilized enzyme reactors.

The axial dispersion model characterizing the material balance for the substrate is given by

$$D_a(d^2S/dz^2) - u(dS/dz) - k_La(S_F - S_S) = 0 \tag{109}$$

where D_a is the axial dispersion coefficient, z is the reactor axial position, S_F, S_S are the substrate concentrations in the bulk and at the carrier surface, respectively, and u is the superficial velocity. The boundary conditions for the dispersion model are given by Danckwerts (1953):

at $z = 0$,

$$S - (D_a/u)(dS/dz) = S_0 \tag{110}$$

at $z = R_L$,

$$dS/dz = 0 \tag{111}$$

where S_0 is the inlet substrate concentration and R_L is the reactor length. Assuming no intraparticle diffusional barrier, the external mass transfer rate would be equal to the reaction rate at steady state:

$$k_La(S_F - S_S) = \frac{k_2E_0S_S}{K'_m + S_S} \tag{112}$$

Combining Eq. (109) and Eq. (112), the dispersion model can be solved.

Kobayashi and Moo-Young (1971) have considered several subcases of varying Bodenstein number values. Bodenstein number (Bo), defined as

$$\text{Bo} = uR_L/D_a \tag{113}$$

is a measure of the degree of dispersion. When Bo is zero, the reactor operates under perfectly mixed conditions, i.e., as an ideal CSTR. The plug-flow reactor is characterized by a Bo value of ∞. When the external boundary layer resistance is very small (k_L very high), dispersion of substrate does not affect the degree of conversion because the reaction becomes zero order. For the cases of significant external diffusional effects, the dispersion model has been solved numerically and design charts have been developed. These charts allow comparison of reactor sizes based on plug-flow and dispersion models, as shown in Fig. 14 for a particular set of parameter values. It may be noted that as Bo values decrease, the reactor size required to achieve a given conversion level increases, approaching the maximum size corresponding to a CSTR, i.e., Bo = 0. Choi and Fan (1973) solved the axial dispersion model together with intrasolid phase mass transfer resistance (but boundary-

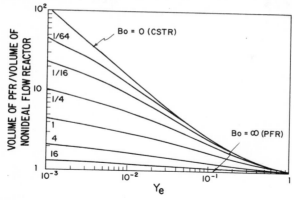

Fig. 14. Comparison of reactor performance according to relative reactor sizes at given conversion levels between plug-flow (PFR) and nonideal flow reactors for Michaelis–Menten kinetics for a (K'_m/HS_0) value of 0.5. H is the electrostatic partition coefficient, as defined by Eq. (47). Numbers on the curves are Bodenstein number, Bo [defined by Eq. (113)], values. Bo is a measure of the degree of axial disperson in the reactor. Bo values of zero and ∞ signify a perfectly mixed CSTR and an ideal plug-flow reactor, respectively. Y_e is the dimensionless exit concentration of substrate. External and internal diffusional resistances are considered to be negligible in this case. Refer to page 278 for a more detailed analysis. (Adapted from Kobayashi and Moo-Young, 1971.)

layer resistance was not considered) under transient conditions. The effects of several parameters, such as Michaelis–Menten constant, membrane diffusional resistance, linear velocity, on the extent of dispersion were investigated theoretically.

As indicated earlier, there are no temperature gradients in a packed-bed reactor under isothermal plug-flow conditions, so that radial dispersion effects can be neglected. Even though most practical reactor systems can be considered to operate isothermally, nonisothermal operation would introduce radial dispersion. Lin (1972) has presented an analysis of such a system. Another source of radial dispersion is from plug flow operation of a reactor. By using a heat-transfer analogy, the radial mixing in slug flow has been modeled (Horvath *et al.*, 1973b).

B. Analysis of Fluidized-Bed Reactors

Despite the well-documented advantages of fluidized-bed reactors in heterogeneous catalysis, their application as enzyme reactors is only beginning to emerge. Little surprising is it, therefore, that hardly anything is reported on the quantitative aspects of fluidized-bed reactors. Even in heterogeneous catalytic systems, most of the published literature pertains to gas–solid systems, whereas in immobilized enzyme catalysis, liquid–solid systems are of primary interest. Information available from the chemical engineering literature on liquid–solid fluidization can be extended to fluidized-bed enzyme reactor systems. Presented below is a brief summary of such information.

Most of the books published on fluidization touch upon liquid–solid systems briefly (Leva, 1959; Zenz and Othmer, 1960; Davidson and Harrison, 1963; Kunii and Levenspiel, 1969). Transport phenomena of liquid–solid fluidized systems have been studied for momentum (McCune and Wilhem, 1949), heat (Hamilton, 1970; Zahavi, 1971), and mass (Hanesian and Opalewski, 1967; Kato *et al.*, 1970). Correlations of the variables governing the process in fluidization have been reported (Bransom and Pendse, 1961). Fluidization has often been compared to sedimentation in order to predict minimum and terminal velocities (Richardson and Zaki, 1954; Bourgeois and Grenier, 1968). Bed expansion characteristics of annular liquid fluidized beds have been examined (Ramamurthy and Subbaraju, 1973). Methods to calculate the minimum fluidization voidage and fluid mass velocity (Leva, 1959) and pressure drops (Wilhem and Kwauk, 1948; Zenz and Othmer, 1960) are also available.

The few reports on fluidized–bed enzyme reactors have essentially been limited to the comparison of the performances of packed–bed and fluidized–bed reactors. Coughlin *et al.* (1973) found that a fluidized–bed reactor containing immobilized lactase yielded higher conversions of lactose (from cheese whey) hydrolysis than a packed-bed reactor containing the same amount of catalyst. The differences in conversion levels between the two reactor types increased with the degree of expansion of the bed. For the case of glucose isomerization by immobilized whole cells (containing glucose isomerase activity), no definite advantage of a fluidized–bed reactor over a packed–bed reactor was observed (Goldberg, 1974). Experimental comparisons of the pressure drops in the two reactor types have been made in both these cases; the pressure drop values were substantially lower for the fluidized–bed reactor, as predicted from theory.

Since the mixing pattern in a fluidized–bed reactor falls between the limiting cases of perfect mixing and plug flow, the axial dispersion model discussed earlier [Eqs. (109) and (112)] can be applied to describe the hydrodynamic effects in the reactor. A modified dispersion model has recently been proposed (Emery and Revel-Chion, 1974). Based on that model and data obtained from an experimental column, correlations for dispersion have been developed.

VII. TRANSIENT ANALYSIS OF IMMOBILIZED-ENZYME REACTORS

Most of the studies published on enzyme-reactor modeling have been confined to steady-state reactor operation. Although many practical systems of interest would fall under this category, it is useful to probe into their transient behavior as well for several reasons: (a) to understand the start-up and shut-down performances of a reactor; (b) to delineate reactor dynamics and control characteristics; (c) to estimate the time required to reach a new steady state when a disturbance is introduced into a reactor system in steady state; (d) to explore the possibilities of multiple steady states of a given reaction scheme; (e) to study multienzyme-catalyzed sequential reaction schemes where intermediate products could have an inhibiting or an activating effect on the overall reaction; (f) to analyze regulation and control of reactors by injected activators, inhibitors or cosubstrates; and (g) to examine the loss of reactor catalytic potency by mechanisms such as enzyme inactivation and elution from the carrier matrix which are associated with the intrinsic process reaction dynamics. Further, batch reactor operation is phenomenologically an unsteady-state process.

A. Idealized Reactor Systems

Michaelis–Menten reaction kinetics, which are based on an enzyme-substrate steady-state hypothesis, have been used by almost all the authors to analyze transient reactor performance. This stems from the assumption that the time scale of function of the enzyme is considerably smaller than the macroscopic time scale of the reactor system dynamics. For an ideal enzyme reactor, i.e., without mass transfer and/or electrostatic partitioning effects, the following equations describe the reactor unsteady-state behavior:

CSTR:

$$V_R(dS/dt) = QS_0 - QS - V_R r \tag{114}$$

PFR:

$$(\partial S/\partial t) + u(\partial S/\partial z) = \epsilon V_R r \tag{115}$$

where z is the axial position in a packed-bed reactor, V_R is the reactor volume, Q is the flow rate, u is the linear velocity, and r is the reaction rate obtained by invoking the steady-state hypothesis. It may be noted that the linear velocity u may be a function of radial position in a packed-bed reactor under certain hydrodynamic conditions, e.g., laminar flow through the column, in which case the velocity is described by a parabolic variation with radial position.

The transient plug-flow reactor Eq. (115) has been solved numerically for Michaelis–Menten kinetics by using kinetic parameters (K'_m and V'_m) obtained separately from steady-state measurements (Gellf et al., 1974). The special case of varying inlet substrate concentration with time was considered by these authors. Consequently, the reactor would not reach a steady state. This special case simulates the kinetic behavior of multienzyme systems with the intermediate enzymes in the sequence experiencing time-dependent substrate concentrations. The following boundary conditions and initial conditions (I.C.) were used:

$$\text{I.C.} \quad \text{at} \quad t = 0, \quad z \geqslant 0, \quad S = 0 \tag{116}$$

$$\text{B.C.} \quad \text{at} \quad z = 0, \quad t > 0, \quad S_0 = mt \tag{117}$$

Equation (117) describes the variation of substrate concentration at the reactor inlet with time. S is assumed to vary linearly with time, the constant of proportionality being $\bar{m}$. Time-varying concentration profiles at reactor exit were generated for three representative values of $\bar{m}$. Experimental data obtained from an immobilized α-chymotrypsin column were in agreement with the numerical solutions.

Transient behavior of a CSTR and a PFR was analyzed by Ryu et al. (1972) for the hydrolysis of penicillin by penicillin amidase. The reaction kinetics of this system is very complex; the enzyme is subjected to competitive product inhibition by one of the products of the hydrolysis reaction and noncompetitive inhibition by the other reaction product. The fractional conversion as a function of reactor operating time of the CSTR system is portrayed in Fig. 15. Depending on the feed concentration of the substrate and the reactor space time, different steady-state conversion levels are reached. The final steady-state conversion value approaches asymptotically a single value, irrespective of the initial conversion value used in the simulation for a given combination of inlet concentration and space-time values. The transient period is relatively short for low reactor space times but increases with increasing space-time values. Similar results on the simulation of the hydrolysis of sucrose by invertase in a CSTR have been reported (Bowski et al., 1972). The transient plug-flow reactor equation was solved by Ryu et al. (1972) by treating it as a multistage CSTR system

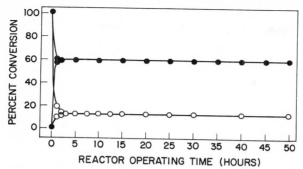

Fig. 15. Percent conversion, x, as a function of reactor operating time of a continuous-flow stirred-tank reactor (CSTR) under the given conditions of reactor space-time (τ) and substrate inlet concentration (S_0). Reaction system studied was the hydrolysis of benzyl-penicillin by bentonite-immobilized penicillin amidohydralase kept in suspension in a CSTR. Note that the transient period is relatively short irrespective of the initial conditions. ●, $\tau = 0.5$ hr, $S_0 = 0.01\ M$; ○, $\tau = 0.5$ hr, $S_0 = 0.10\ M$. (From Ryu *et al.*, 1972.)

described by

$$dX_i/dt = (1/\tau_i)(X_{i-1} - X_i) + (1/S_0)(r_i) \tag{118}$$

$$r_i = r(X_i) \tag{119}$$

where the subscript i refers to the ith stage of the multistage CSTR system. τ_i and X_i are the space-time and conversion values, respectively, at the ith stage. Simulation results of the transient condition were in agreement with experimental results.

Choi and Fan (1973) have developed mathematical models for depicting the transient behavior of a CSTR and a packed-bed tubular reactor containing enzyme microcapsules. They consider the wall resistance of the microcapsule in their analysis, even though the bulk transport resistance is ignored. The reaction rates are therefore modified as discussed in Section IV,C. The modified rate expressions when substituted in Eqs. (114) and (115) form the system equations for transient reactor operation. Numerical solutions have been presented, for the limiting case of negligible membrane resistance to fluid transport, and an analytical solution has been obtained for the CSTR system. The effects of several parameters, such as the enzyme reaction rate constant, the Michaelis–Menten constant, and diffussional resistance of the membrane phase on the substrate concentration profile as a function of reactor operating time are examined theoretically. The packed-bed reactor is described by the axial dispersion model in this analysis (cf. Section VI,A); i.e., the axial dispersion of the substrate is taken into account rather than assuming an ideal plug-flow condition.

B. Unsteady-State Mass Transport-Kinetic Models

1. First-Order Reactions

When solid–fluid interfacial and intrasolid mass-transfer impedances are significant, it becomes difficult to solve the unsteady-state transport-kinetic models. However, if the enzymic reaction can be approximated as a pseudo first-order process, analytical solutions can be obtained (Vieth *et al.*, 1972c, 1973). The system equations are similar to Eqs. (79) and (80), except that their right-hand sides are now replaced by a term describing the variation of substrate concentration with time. Analytical solutions for both first-order reversible and irreversible reactions have been obtained. Mogensen and Vieth (1973) have solved analytically similar transient equations for a batch reactor containing spherical microcapsules. It is presented here as an example of an explicit solution to the transient mass transfer-kinetic model. The system equations in dimensionless form are

Capsule interior:

$$(\partial^2 Y_S/\partial \imath^2) + (2/\imath)(\partial Y_S/\partial \imath) - \phi^2 Y_S = \partial Y_S/\partial \theta \tag{120}$$

B.C. No. 1:

$$(\partial Y_S/\partial \imath)_{\text{at } \imath=1} = N'_{\text{Sh}}(Y_F - Y_S) \tag{121}$$

B.C. No. 2:

$$\imath = 0, \quad (\partial Y_S/\partial \imath) = 0 \tag{122}$$

I.C.:

$$\theta = 0, \quad Y = 0 \tag{123}$$

Bath:

$$\partial Y_F/\partial \theta = -(3/\bar{V})(\partial Y_S/\partial \imath)_{\text{at } \imath=1} \tag{124}$$
$$\text{I.C.}_f: \quad \theta = 0, \quad Y_F = 1 \tag{125}$$

where $\bar{V}$ is the ratio of the bath volume to the capsule volume, $\imath$ is the radial distance from the capsule center, θ is the dimensionless time ($= Dt/R^2$),

$$\phi^2 = R^2 k_t E_0/D_e$$
$$N'_{\text{Sh}} = k_{\text{L+M}}R/D_e; \quad 1/k_{\text{L+M}} = (1/k_{\text{L}}) + (1/k_{\text{Mem}})$$

$k_{\text{L+M}}$ is the combined fluid and membrane mass-transfer coefficient.

Equations (120) and (125) coupled through the indicated boundary and initial conditions can be solved by the use of Laplace transform to obtain Y_F, the dimensionless bath substrate concentration, as

$$Y_F = 6(N'_{Sh}) \bar{V} \sum_{j=1}^{\infty} - \left[\frac{(S_j + \phi^2) \exp (S_j \theta)}{Q_j} \right] \tag{126}$$

where S_j's are the eigenvalues of the solution which satisfies Eq. (127).

$$\frac{\tanh (S_j + \phi^2)^{1/2}}{(S_j + \phi^2)^{1/2}} = \frac{3N'_{Sh} + \bar{V}S_j}{3N'_{Sh} - \bar{V}S_j(N'_{Sh} - 1)} \tag{127}$$

and

$$Q_j = \bar{V}_j^2 S_j^2 \{ N'_{Sh}(N'_{Sh} - 1) - (S_j + \phi^2) \} - 6\bar{V}N'_{Sh}(S_j + \phi^2)S_j \\ + 3\bar{V}(N'_{Sh})^2 \phi^2 - 9(N'_{Sh})^2(\bar{V} + 1)(S_j + \phi^2) \tag{128}$$

The full series expansion given by Eq. (126) can be approximated for short times by

$$Y_F = \exp (S_1 \theta) \tag{129}$$

with only the first term from Eq. (127). The above solution has been used to study the batch decomposition of hydrogen peroxide by encapsulated catalase (Mogensen and Vieth, 1973).

2. Transient Model with Detailed M-M Kinetics

The most detailed analysis of an enzyme reactor system has been completed recently at our laboratory (Shyam *et al.*, 1975). This generalized reactor dynamic model encompasses (a) the transient state behavior; (b) models for bulk and pore diffusional substrate transport; (c) detailed reaction rate model relaxing the steady-state assumption of Michaelis–Menten kinetics; (d) mass balances for substrates, reaction intermediates, products, and unoccupied enzyme active sites; and (e) models for enzyme inactivation and elution from the carrier. This analysis will be considered in depth here because it is useful as a well-structured general model providing an overview of almost all the mechanistic processes occurring in an immobilized enzyme reactor system. Although the analyses presented here are limited to packed beds of spherical catalyst particles (refer to Fig. 8), the methodology is perfectly general; extensions to other geometries such as slabs, rods, chips, etc., or to other reactor configurations such as spiral-wound enzyme-membrane biocatalytic modules are readily attainable.

Considering the reversible, monomolecular reaction represented by

$$A \rightleftharpoons B \tag{130}$$

which involves individual mechanistic steps as given by Eq. (90), the rates of the individual reaction steps are as outlined in Eq. (91) through Eq. (93). The overall steady-state reaction rate is expressed in Eq. (96). It is worth recalling that Eq. (96) was derived by assuming that the three individual reaction steps are equal at steady state. Under transient reactor operating conditions, this may not be so. Therefore, in the transient model, material balances should be written for each reaction species together with the appropriate rate expressions for the mechanistic steps involving that species. Both these kinetic formulations have been solved in Shyam's (1974) work, and the results have been compared.

The mass transfer-kinetic model for a packed-bed reactor (illustrated in Fig. 8) with spherical catalyst particles under transient conditions with a rate expression based on the enzyme–substrate steady-state hypothesis [i.e., Eq. (96)] is given below in dimensionless form:

Solid phase:

$$\frac{\partial Y_A}{\partial \theta} = \frac{\partial^2 Y_A}{\partial L^2} + \frac{2}{L} \frac{\partial Y_A}{\partial L} - \frac{E_0(M_4 Y_A + M_5 Y_B)}{(M_1 + M_2 Y_A + M_3 Y_B)} \tag{131}$$

$$\frac{\partial Y_B}{\partial \theta} = \frac{D_{eB}}{D_{eA}} \left(\frac{\partial^2 Y_B}{\partial L^2} + \frac{2}{L} \frac{\partial Y_B}{\partial L} \right) + \frac{E_0(M_4 Y_A + M_5 Y_B)}{(M_1 + M_2 Y_A + M_3 Y_B)} \tag{132}$$

Fluid phase:

$$\frac{\partial Y_{AF}}{\partial \theta} = - \frac{\partial Y_{AF}}{\partial Z} - 3 \left(\frac{1 - \epsilon}{\epsilon} \right) N_{Sh}^A (Y_{AF} - Y_A) \tag{133}$$

$$\frac{\partial Y_B}{\partial \theta} = - \frac{\partial Y_{BF}}{\partial Z} - 3 \left(\frac{1 - \epsilon}{\epsilon} \right) \frac{D_B}{D_A} N_{Sh}^B (Y_{BF} - Y_B) \tag{134}$$

where M_i ($i = 1$ to 5) are the dimensionless composite reaction-rate constants, Y is the dimensionless concentration ($= S/S_0$), subscripts A and B refer to species A and B, subscript F signifies the fluid phase, θ is dimensionless time ($= D_{eA} t/R^2$), D_{eA}, D_{eB} are the effective diffusivities of A and B, respectively, ϵ is the ratio of the fluid volume in the reactor to the total reactor volume, Z is the dimensionless reactor length, and L is the dimensionless membrane thickness ($= x/l$). These equations are basically similar to Eqs. (67) and (68), which are steady-state equations. The initial conditions for reactor start-up are

$$\theta = 0, \quad L \geqslant 0, \quad Z > 0; \qquad Y_{AS} = 0, \quad Y_B = 0 \tag{135}$$

$$\theta = 0, \quad Z \geqslant 0; \qquad Y_{AF} = 0, \quad Y_B = 0 \tag{136}$$

For the case where the system is about to be perturbed from an existing steady-state operation, the initial conditions are

$$\theta = 0, \quad L \geqslant 0, \quad Z \geqslant 0; \qquad Y_A = \bar{Y}_A, \quad Y_B = \bar{Y}_B \qquad (137)$$

$$\theta = 0, \quad Z \geqslant 0; \qquad Y_{AF} = \bar{Y}_{AF}, \quad Y_B = \bar{Y}_B \qquad (138)$$

where $\bar{Y}_i$ is the dimensionless steady-state concentration of the ith species.

To analyze the transient process, a disturbance is imposed on the reactor system. The forcing function which perturbs the system from its initial steady-state condition is given by

$$\theta = 0^+, \quad Z = 0; \qquad Y_{AF} = Y_{A0} + \Delta Y_{AF} \qquad \text{(a step change)} \qquad (139)$$

The boundary conditions are

$$\theta \geqslant 0, \quad Z \geqslant 0, \quad L = 0; \qquad \partial Y_A/\partial L = 0, \qquad \partial Y_B/\partial L = 0 \qquad (140)$$

$$\theta \geqslant 0, \quad Z \geqslant 0, \quad L = 1; \qquad \partial Y_A/\partial L = N_{\text{Sh}}^A(Y_{AF} - Y_A) \qquad (141)$$

and

$$\partial Y_B/\partial L = N_{\text{Sh}}^B(Y_{AF} - Y_A) \qquad (142)$$

$$\theta \geq 0, \quad Z = 0; \qquad Y_{AF} = Y_{A0}, \quad Y_{BF} = Y_{B0} \qquad (143)$$

It may be noted that in the limiting first-order case, the solution of the system equations presented here tallies with the analytical solution for the first-order case, Eq. (126).

3. Complete Transient Model without Invoking the Steady-State Hypothesis

When the steady-state hypothesis is relaxed, the following equations describe the reactor transient model.
Solid phase:

$$\frac{\partial Y_A}{\partial \theta} = \frac{\partial^2 Y_A}{\partial L^2} + \frac{2}{L}\frac{\partial Y_A}{\partial L} - E_0 K_1 \left(Y_A Y_E - \frac{Y_{A \cdot E}}{K_4} \right) \qquad (144)$$

$$\frac{\partial Y_B}{\partial \theta} = \frac{D_{eB}}{D_{eA}} \left(\frac{\partial^2 Y_B}{\partial L^2} + \frac{2}{L}\frac{\partial Y_B}{\partial L} \right) - E_0 K_3 \left(Y_B Y_E - \frac{Y_{B \cdot E}}{K_6} \right) \qquad (145)$$

$$\frac{\partial Y_{A \cdot E}}{\partial \theta} = K_1 \left(Y_A Y_E - \frac{Y_{A \cdot E}}{K_4} \right) - K_2 \left(Y_{A \cdot E} - \frac{Y_{B \cdot E}}{K_5} \right) \qquad (146)$$

$$\frac{\partial Y_{B \cdot E}}{\partial \theta} = K_3 \left(Y_B Y_E - \frac{Y_{B \cdot E}}{K_6} \right) + K_2 \left(Y_{A \cdot E} - \frac{Y_{B \cdot E}}{K_5} \right) \qquad (147)$$

where K_1 to K_5 are dimensionless kinetic constants.

The fluid phase equations and the boundary conditions are identical as in the previous model based on the steady state hypothesis. In addition, the following initial conditions are needed:

For reactor start-up:

$$\theta = 0, \quad L \geqslant 0, \quad Z \geqslant 0; \qquad Y_{A \cdot E} = 0, \quad Y_{B \cdot E} = 0 \qquad (148)$$

For perturbation of steady state:

$$\theta = 0, \quad L \geqslant 0, \quad Z \geqslant 0; \qquad Y_{A \cdot E} = \bar{Y}_{A \cdot E}; \qquad Y_{B \cdot E} = \bar{Y}_{B \cdot E} \qquad (149)$$

The transient reactor model, represented by a system of nonlinear, coupled, partial differential equations is solved numerically. Results of the comparison made between the model based on the steady-state hypothesis and the complete transient model are presented graphically. The response of the product outlet concentration to reactor start-up (with substrate entering a regenerated reactor at time = 0), for two sets of kinetic parameters and two values of the modified Sherwood number, is shown in Fig. 16. As seen in this figure, there is a marked difference in the outlet concentrations predicted by the two models. When the model based on the steady-state hypothesis predicts the reactor outlet concentration to reach almost the final steady-state value at a certain time, the complete transient model predicts the concentration to be almost zero and just starting to increase. This is found to be true, regardless of the value of k_1/k_{-3} ratio for a particular set of kinetic parameters and the modified Sherwood number. The process time lags are due to pure transportation delays plus delays associated with the time to form the reaction intermediates. Thus, during reactor start-up, the substrate is complexed at the active sites within the catalyst particles at a high rate until a steady state is reached. Although chemical transformations begin when the intermediate complexes are formed, in the initial transient stages, most of the substrate molecules and any product molecules formed are being used up for the production of the intermediate complexes. The concentration of each of these intermediates reaches a steady-state value only after a finite time lag, while the product formed must diffuse through the solid phase to the surface and from the surface into the fluid phase. This explains the time lag of several residence times before any product appears in the reactor fluid, as predicted by the complete transient model. By contrast, the model based on the steady-state hypothesis, because of the a priori imposed conditions that the concentrations of intermediates reach the steady-state levels rapidly, predicts time lags and concentration levels that are grossly inaccurate for the reactor start-up case.

The responses of the product and substrate outlet concentrations to different step changes in substrate inlet concentration were investigated. It was shown that the level of "error" involved in invoking the

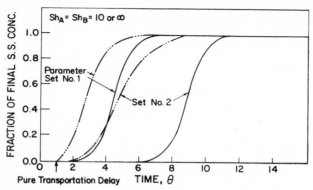

Fig. 16. Transient response of product outlet concentration to reactor start-up in packed-bed reactor with spherical particles (refer to Fig. 8 for pictorial description of the system), catalyzing a reversible reaction $A \rightleftharpoons B$. Fraction of final steady state (S.S.) concentration of substrate is plotted as a function of reactor operating time. Final steady state (S.S.) is characterized by the time-invariance of the enzyme–substrate complex concentration. Note that S.S. concentration values plotted are normalized values, i.e., not absolute concentration values. Equations (133) through (149) form the basis of this simulation. Two Sherwood number values (10 and ∞) have been used for the simulation. $N'_{Sh} = \infty$ and $N'_{Sh} = 10$ indicate negligible and significant diffusional resistances, respectively. Transient response curves are independent of Sherwood number; i.e., the severity of mass-transfer resistance. (Again, it must be noted that we are considering normalized responses.) Parameter values have been chosen carefully to highlight the differences between the model based on enzyme–substrate steady-state hypothesis ($—\cdot\cdot—$) and the complete transient model ($—$) without invoking the above hypothesis. Among those examined, the ratio of the kinetic constants, k_1/k_{-3}, exerted the maximum influence. Therefore, in this figure, the effect of variation of this ratio is shown. At low values of this ratio (set No. 2: $k_1/k_{-3} = 0.1$), the difference between the two models is less compared to high k_1/k_{-3} values (set No. 1: $k_1/k_{-3} = 2$). This ratio appears in the kinetic expression for the reaction under consideration [Eq. (96)]. Transportation delay indicates the time elapsed between the time the feed enters the reactor inlet and the time it leaves the reactor. Under plug-flow conditions, this delay is equal to one reactor space time. Refer to page 287 for a more detailed discussion. Other parameter values used in generating the profiles in this figure are: $k_1, k_{-1} = 10 \times 10^7$ ml/mole sec; $k_2, k_{-2} = 2.0$ sec^{-1}; $k_3 = 0.5$ ml/mole (set 1); $k_3 = 1.0$ ml/mole (set 2).

steady-state hypothesis depends on the relative magnitude of the kinetic parameters and also on the level of "disturbance" at the reactor inlet (i.e., percent change in substrate inlet concentration). The "error," however, did appear to be strikingly insensitive to the magnitude of the resistances to mass transfer, as characterized by the modified Sherwood number. It was concluded that, given any complete set of kinetic parameters, a transient, heterogeneous, isothermal reactor model based on the steady-state hypothesis may be used for predicting time-varying concentration profiles for minor (i.e., less than 5% change in substrate inlet concentration) "disturbances" at the reactor inlet. The corresponding "errors" would be at an acceptable level (i.e., less than 2% in the concentration and less than 10% in the time lag) under these conditions.

4. Model for Enzyme Denaturation

As noted before, enzyme denaturation may occur owing to one or more of several phenomena. One of the more mathematically tractable ways of expressing denaturation in kinetic terms is by assuming that the enzyme denatures reversibly, and that the rate of denaturation is proportional to the enzyme concentration only. Hence,

$$\text{E} \underset{k_{E'}}{\overset{k_E}{\rightleftharpoons}} \text{I} \tag{150}$$

The corresponding equation for the rate of enzyme inactivation is

$$r = k_E[\text{E} - (\text{I}/K_E)] \tag{151}$$

where

$$K_E = k_E/k'_E. \tag{152}$$

If E, I, A · E, and B · E are considered as nondiffusing species, then the following stoichiometric invariance is obtained:

$$\text{E} + \text{I} + \text{A} \cdot \text{E} + \text{B} \cdot \text{E} = \text{E}_0 \tag{153}$$

In the solid-phase model Eqs. (144)–(147) an additional equation should be incorporated to account for enzyme inactivation.

$$\frac{\partial Y_E}{\partial \theta} = -K_1\left(Y_A Y_E - \frac{Y_{A \cdot E}}{K_4}\right) - K_3\left(Y_B Y_E - \frac{Y_{B \cdot E}}{K_6}\right) - K_7\left(Y_E - \frac{Y_I}{K_8}\right) \tag{154}$$

The fluid phase equations and the boundary conditions are identical, as before. The following additional initial conditions need to be imposed:

$$\theta = 0, \quad L \geqslant 0, \quad Z \geqslant 0; \quad Y_I = 0 \tag{155}$$

$$\theta = 0, \quad L \geqslant 0, \quad Z \geqslant 0; \quad Y_E = \tilde{Y}_E \tag{156}$$

These conditions are based on the assumption that the reactor system is at a given steady state when denaturation sets in.

5. Model for Enzyme Elution from the Matrix

The model for the elution of a loosely held portion of the initial enzyme in the solid particles is developed, based on the following assumptions:

i. The initial enzyme concentration in the carrier particle consists of firmly bound ("nondiffusing fraction") and an elutable ("diffusing fraction") portions.

ii. In addition to the enzyme, intermediate-enzyme species can also be eluted.

iii. The substrate and product react with the eluting portion of the enzyme to produce only eluting species.

iv. The substrate and product react with the noneluting portion of the enzyme to produce only noneluting species.

v. Diffusion of enzyme and the intermediate species into the bulk fluid is modeled in the same manner as that of the free molecular substrate and the product.

vi. All the eluted species react in the fluid phase also.

The diffusing fraction, i.e., the "unbound" fraction, λ, of the enzyme can be defined as the ratio of the initial concentration of unbound enzyme E_0' to the total initial enzyme concentration E_0, i.e.,

$$\lambda = E_0'/E_0 \tag{157}$$

Therefore, the stoichiometric balance for the nondiffusing fraction of the enzyme becomes

$$(1 - \lambda)E_0 = E + A \cdot E + B \cdot E \tag{158}$$

Accordingly, the solid- and the fluid-phase equations would be changed to discriminate between the diffusing and the nondiffusing portions of the enzyme. For instance, the solid phase Eq. (144) becomes,

$$\frac{\partial Y_A}{\partial \theta} = \frac{\partial^2 Y_A}{\partial L^2} + \frac{2}{L} \frac{\partial Y_A}{\partial L} - E_0 K_1 \left[Y_A(Y_E + Y_E') - \left(\frac{Y_{A \cdot E} + Y_{A \cdot E}'}{K_4} \right) \right] \tag{159}$$

The primed quantities refer to the unbound fraction of the corresponding species. Equations similar to the above can be deduced for other species B, A · E, and B · E in solid and liquid phases. In addition to these equations, material balances must be written for the eluting portions of E, A · E, and B · E. For the sake of brevity, the equation for only one of the species (species E) is shown here. The equation for the others are straightforward; they can be found elsewhere (Shyam, 1974).

Solid phase equation for eluted species E:

$$\frac{\partial Y'_E}{\partial \theta} = \frac{D_E}{D_A}\left[\frac{\partial^2 Y'_E}{\partial L^2} + \frac{2}{L}\frac{\partial Y'_E}{\partial L}\right] - K_1\left(Y_A Y'_E - \frac{Y'_{A\cdot E}}{K_4}\right)$$
$$- K_3\left(Y_B Y'_E - \frac{Y'_{B\cdot E}}{K_6}\right) \quad (160)$$

Fluid phase equation:

$$\frac{\partial Y'_{EF}}{\partial \theta} = -\frac{\partial Y'_{EF}}{\partial Z} - 3\left(\frac{1-\epsilon}{\epsilon}\right)\frac{D_E}{D_A}(N_{Sh}^E)(Y'_{EF} - Y'_E)$$
$$- K_1\left[Y_{AF}Y'_{EF} - \frac{Y'_{A\cdot EF}}{K_4}\right] - K_3\left[Y_{BF}Y'_{EF} - \frac{Y'_{B\cdot EF}}{K_6}\right] \quad (161)$$

The following additional initial conditions are imposed:

$$\theta = 0, \quad L \geqslant 0, \quad Z \geqslant 0; \qquad Y'_E = \lambda \quad (162)$$
$$\theta = 0, \quad L \geqslant 0, \quad Z \geqslant 0; \qquad Y'_{A\cdot E} = Y'_{B\cdot E} = 0 \quad (163)$$

Additional boundary conditions similar to Eq. (140) through Eq. (143) are also necessary. These conditions state that (a) at the center of the catalyst particle, the concentration gradient of each of the species discussed above is zero, for all values of Z, and (b) at the particle surface the radial mass flux of the species is equal to its mass-transfer rate from the solution bulk.

6. Denaturation and Elution Model

Enzyme denaturation and elution may occur simultaneously in an enzymic packed-bed reactor. Then the two models outlined above may be combined into a single model. The stoichiometric invariance for the nondiffusing fraction of the enzyme therefore becomes,

$$(1 - \lambda)E_0 = E + A\cdot E + B\cdot E + I \quad (164)$$

Considering that the eluted enzyme may be in active and denatured forms, system equations are developed in an identical manner as before. The solid- and fluid-phase equations are identical to the elution model for A, B, A · E, B · E, A · E', B · E'. Additional equations are necessary for species E and I', which are obtained in a similar manner, as discussed before. (The material balance for I is given by the stoichiometric invariance.) The equation for E is the same as Eq. (154). The system equation for E' has an additional term given by

$$-K_7[Y'_E - (Y'_I/K_8)] \quad (165)$$

The same initial and boundary conditions as before apply.

The transient reactor model together with the enzyme denaturation and/or elution model was solved in an analogous manner. The integration with respect to time is started at a given steady-state condition of the reactor. Various values of K_E (for the denaturation model), λ (for the elution model), and a combination of K_E and λ (for the combined denaturation and elution model) were chosen. Results of computer simulation runs for the denaturation and elution models are presented graphically in Figs. 17–19 for a representative set of parameters.

It is seen that simulated data (percent initial conversion vs time) (Figs. 17–19), obtained for the three models, cannot be differentiated in qualitative terms. This is true for any given set of kinetic parameters and any given value of K_E or λ or a combination of K_E and λ. It follows that clear discrimination between the various forms of the overall process models is not possible, given the usual conversion vs process time data. It must be emphasized here that the primary objective in developing these models is not to discriminate the different mechanisms of enzyme activity loss but to gain some insight to the overall process. The above conclusions simply mean that we need additional information (e.g., concentration of bound active enzyme as a function of time) to obtain qualitative distinction between the different mechanisms.

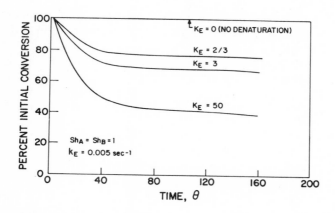

Fig. 17. Transient solution to denaturation model. Physical system is the same as that for Fig. 16. Parameter values used: $k_1 = 1.0 \times 10^7$ ml/mole sec; $k_2 \times 2.0$ sec^{-1}; $k_3 = 1.0$ ml/mole sec; $k_{-1} = 0.1$ sec^{-1}; $k_{-2} = 1.0$ sec^{-1}; $k_{-3} = 1.0$ ml/mole sec. Profiles shown are generated from the solutions of Eqs. (144) through (147), (153) through (156), and (133) and (134). K_E is the equilibrium constant characterizing the reversible denaturation of the supported enzyme, as defined by Eq. (152). The constants k_E and K_E appear in Eq. (154).

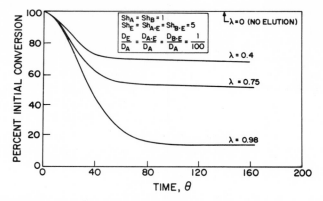

Fig. 18. Transient solution to elution model. Physical system and parameter values used are the same as those for Fig. 17. Equations (159) through (163) are used in generating these profiles. Refer to pages 291 and 292 for a more detailed discussion. Sherwood number and diffusion coefficient values used are shown on the figure. λ is the fraction of initial "unbound" enzyme, as defined by Eq. (157). Profiles are shown for different λ values.

Experimental data for loss in activity of various reactor systems is given in Fig. 20. The qualitative similarity of the simulated and the experimental data for the various models is evident from these figures. Hence, the loss in activity of immobilized enzyme with respect to process time may be a result of slow denaturation of enzyme or elution of unbound enzyme from the host matrix or a combination of denaturation and elution.

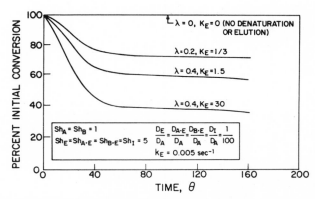

Fig. 19. Transient solution to combined denaturation-elution model. Physical system and parameter values used are the same as those for Fig. 17. Simulations are based on a combination of the equations used for Figs. 15 and 16. Refer to page 292 for a complete discussion. Profiles are shown for different values of λ and K_E.

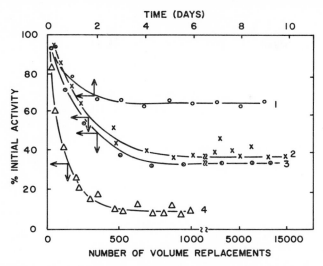

Fig. 20. Experimental data for the stability and reusability of different collagen–enzyme membranes housed in spiral-wound multipore reactor systems and operated as recycle reactors. Reactor volumes were between 100 and 400 ml. For reactor operating conditions, refer to the original papers. Note the qualitative similarities between these profiles and those in Figs. 17–19, the significance of which is discussed on page 293. Curve 1: *l*-asparaginase (Venkatasubramanian *et al.*, 1974b); curve 2: lysozyme (Venkatasubramanian *et al.*, 1972); curve 3:invertase (Vieth *et al.*, 1972a); curve 4: glucose oxidase (Constantinides *et al.*, 1973).

Thus, even though discrimination between different mechanistic models is not possible, the usefulness of this analysis in conceptual understanding of these mechanisms is obvious. Shyam *et al.* conclude appropriately, "It is beyond the state-of-the-art to expect that experimental values of the reaction intermediates in mixed-gradient, heterogeneous reactors are available for even the most elementary systems. For this reason alone, the generalized map of the validity of the Steady-State Hypothesis for mixed-gradient reactor systems may well prove to be an important research result for the experimentalist and design engineer alike. This analysis provides a theoretical rationale for invoking with impunity the Steady-State Hypothesis under certain conditions."

C. Analysis of Multiple Steady States

Continuous-flow enzyme reactors can attain more than one steady state under certain conditions. Not all these steady states may be stable. The existence and stability of multiple steady states in chemical reactors have been extensively studied (Aris, 1969). In exothermic chemical reaction systems, multiple steady states arise because the reaction rate goes through a maximum with increasing temperature. Enzymic reactions are essentially isothermal, but they might exhibit a maximum in reaction rate with increasing concentration, as in the case of substrate-inhibited enzyme systems. Therefore, multiple steady states are possible in these instances (O'Neill *et al.*, 1971c).

The material balance equation for a CSTR under transient conditions is given by

$$V_R(dS/dt) = Q(S_0 - S) - V_R(r) \tag{166}$$

where r is the reaction rate. When the reaction rate equals the product removal rate $(QS_0 - QS)$, the reactor system reaches a steady state. For simple Michaelis–Menten kinetics, only one steady state is attained for each set of Q, V_R, S_0, and kinetic constants.

For substrate-inhibited enzyme systems, the above equation becomes

$$V_R \, dS/dt = Q(S_0 - S) - \frac{k_2 E_0}{1 + (K'_m/S) + (S/K_S)} \tag{167}$$

where K_S = substrate inhibition constant.

Since the reaction rate goes through a maximum with substrate concentration, the product removal rate is found to be equal to the reaction rate at three different product concentration values, signifying three possible steady states for the same values of Q, S_0, and V_R, as shown in Fig. 21. Numerical solution of the transient reactor Eq. (167) also shows that three steady states can be reached (O'Neill, 1971). The middle steady state B in Fig. 21 is unstable, since a slight fluctuation in concentration will result in a shift to new steady states A or C. The occurrence of unstable steady states depends upon the ratios K'_m/S and K_S/S. The regions of unstable steady-state conditions occur mainly at small values of these ratios and in the conversion ranges of 50–100% (O'Neill *et al.*, 1971c). Experimental results on a combined CSTR/UF membrane reactor with the sucrose-invertase system were found to agree with the above criteria for single steady states (Bowski *et al.*, 1972).

Shuler *et al.* (1973) have recently extended this analysis to incorporate bulk diffusive and electrostatic effects in a nonporous particle-immobilized enzyme system. The conditions under which multiple steady states will occur have been determined. They observe that in the absence of electrostatic effects, multiple steady states result only when the external mass transfer coefficient is very low. However, at normal conditions of CSTR operation, the bulk mass transfer rates are

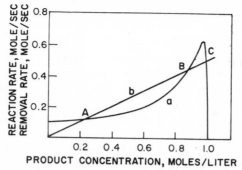

Fig. 21. Existence of multiple steady states for substrate-inhibited enzyme systems. a: Reaction rate curve; b: product removal curve. A, B, C are the three possible steady states. B is an unstable steady state. (From O'Neill *et al.*, 1971c.)

likely to be high enough to exclude the possibility of multiple steady states. When electrostatic effects are present, the possibility of multiple steady states is increased with increasing electrostatic attraction between the carrier and the substrate. Again, if sufficiently high mass-transfer rate to the carrier surface can be maintained, the chances of attaining multiple steady states are diminished. When the enzyme is immobilized in a porous matrix, the pore diffusional effects must also be considered in evaluating reactor stability. Moo-Young and Kobayashi (1972) have found that at low (modified) Thiele modulus values, there are no possibilities of multiple steady states. In other words, when the carrier-enzyme particle is small, the instability problem does not occur. As the particle becomes larger, an unstable region appears and this region increases with increasing particle size.

A more rigorous and complete stability analysis using the theory of nonlinear mechanics has been reported for the following reaction (Bruns *et al.*, 1973):

$$aA + sS \rightarrow pP \tag{168}$$

with the reaction rate given by

$$r(C_S, C_A) = \frac{V'_m C_S}{(K'_m + C_S)[1 + (C_A/K) + (K'/C_A)]} \tag{169}$$

This type of reaction rate is applicable to systems where the reaction is inhibited by high concentrations of the substrate A. Results from stability analysis based on a linearized dynamic model and a nonlinear model are compared. It is shown that linearized models can sometimes give misleading results.

VIII. MODELING OF MULTIENZYME REACTOR SYSTEMS

Immobilized multienzyme systems, in which two or more enzymes are fixed on the same carrier matrix, simulate the way in which several enzymes are compartmentalized in the living cell. An intermediate produced by one type of enzyme would encounter the next enzyme in the sequence more rapidly if the two enzymes were in close spatial proximity than if they were randomly distributed. This process of constrained metabolic routing enhances local concentrations of the reaction intermediates in the microenvironment of the enzyme, resulting in higher overall reaction rates. In this way, partial regulation of the overall reaction by diffusion of intermediate reaction species within the compartmentalized structure is achieved. There are several well documented examples of such concerted cellular processes mediated by multienzyme complexes *in vivo*. Even though some of these processes might further be regulated by a sophisticated metabolic control machinery, it is obvious that introduction of the transport constraints for other, simpler biochemical sequences *in vitro* would augment reaction velocities compared to arbitrarily mixed, bulk phase-mediated catalysis by soluble enzyme mixtures.

The process of enzyme immobilization results in a microstructure with diffusional impedances similar to those present in cellular organelles. Thus, these systems are quite useful in analyzing the metabolic processes effected by naturally occurring multienzyme complexes. Furthermore, there are many enzymic reactions of practical interest that could be catalyzed by immobilized multienzymes. Transformation of steroids (Mosbach, 1971), production of gluconic acid from sucrose (Fernandes, 1974), starch conversion to maltose (Martensson, 1974), desugaring of egg products (Hultin *et al.*, 1974), and production of fructose from starch (Kent and Emery, 1974) are but a few examples.

Two types of enzyme arrangements are possible in multienzyme reactors. In the first type, the enzymes are immobilized separately on different membranes (or particles) and then mixed in the desired ratio in the reactor (physical admixture). In the second type, the enzymes are coimpregnated simultaneously on the same membrane (or particle) and packed in the reactor. Type I formulations are advantageous for systems having negligible mass-transport resistances since it is possible to sequentially pack the enzymes in the reactor. Mosbach (1971) used a sequentially packed two-enzyme reactor to study a two-step steroid conversion process. A two-enzyme reactor containing immobilized pyruvate kinase and lactate dehydrogenase has also been reported (Wilson *et al.*, 1968). It should be recognized, however, that in most practical systems of interest, mass transport would play a dominant role and it will be beneficial to use the type II arrangement.

Theoretical and experimental analyses of supported multienzyme systems have been confined primarily to isolated membranes, pellets, and particles. The spatial proximity of hexokinase and glucose-6-phosphate dehydrogenase covalently bound to a polymer matrix augmented the overall reaction by 100% compared to the rate obtained in free solution (Mosbach and Mattiasson, 1970). For a three-enzyme system, the activity enhancement was even higher (Mattiasson and Mosbach, 1971). Goldman and Katchalski (1971) analyzed theoretically the kinetic behavior of an isolated impermeable membrane with two randomly distributed enzymes attached to the membrane surface, in contact with a substrate solution of finite volume. They demonstrated that the main distinction between the two-enzyme membrane and the same pair of enzymes in free solution was the existence of a diffusional barrier to the transport of the intermediate product.

Models for an isolated bienzymic membrane containing two uniformly distributed enzymes and for two membranes each containing an individual enzyme and acting in tandem have been developed (Selegny *et al.*, 1971; Broun *et al.*, 1972). The application of such a model to explain the mechanisms of (1) active, facilitated, and retarded transport of metabolites through cell walls and (2) feedback or allosteric control mechanisms in biological systems has been discussed (Thomas *et al.*, 1972; DeSimone and Caplan, 1973b).

The effects of combined internal diffusion and catalytic action in a pellet containing two enzymes have been quantitated theoretically by a mathematical model (Laurence and Okay, 1973). Effectiveness factors for a two-enzyme system have been defined similarly to those for the monoenzymic reactions. By assuming first-order kinetics, effectiveness factors for the individual reactions were evaluated and the effects of enzyme loadings on catalyst activity were examined. Ford *et al.* (1973) and Lambert and Chambers (1973) have reported the analysis of a hollow-fiber reactor system containing two enzymes. Their study was on the continuous oxidation of ethanol to acetic acid while retaining and enzymically regenerating the coenzyme NAD. The reaction scheme and membrane permeability properties are quantified by mathematical models that are compared with experimental results. Multiple immobilized enzyme reactors, involving monodisperse enzymes cross-linked to porous glass have been studied with respect to external mass-transfer limitations, the measured kinetic parameters, and reactor design (Kallen *et al.*, 1973; Newirth *et al.*, 1973).

Although some attempts have been made in these models to incorporate transport constraints, they lack an integrated approach in developing a meaningful multienzyme reactor model. It is important to take into account both solid and fluid phase mass-transfer resistances together with a consideration of an appropriate reactor flow environment. Such a model has recently been developed and tested experimentally at our laboratory (Fernandes *et al.*, 1975; Fernandes, 1974). This mass transfer-kinetic model can be applied to predict the performance characteristics of an immobilized multienzyme system. The generality of this model makes it applicable for use in industrial reactors, in addition to increasing our understanding of the kinetics of multienzyme sequences in cellular processes.

The system considered consists of a series of permeable membranes, each impregnated with two enzymes catalyzing the sequential reaction:

$$A \xrightarrow{\text{enzyme 1}} B \xrightarrow{\text{enzyme 2}} C \tag{170}$$

The plane membrane geometry is a reduction of a multipore, spiral biocatalytic module, as pointed out earlier. The two enzymes are immobilized together on a single porous membrane (e.g., collagen) as distinct from a physical mixture of two enzymes immobilized separately on different membranes. The substrate (A) diffuses from the bulk fluid to the membrane through a boundary diffusion layer adjacent to the membrane. The substrate then diffuses into the membrane, where the coupled reaction takes place forming the intermediate (B) and the final product (C), which diffuses back to the bulk fluid, together with any unreacted substrate. No inactivation of the enzymes occurs, and it is assumed that the enzymes behave independently of each other. To simplify the model, the fluid phase mass-transfer coefficients, k_L, and the diffusion coefficients, D_e, are assumed the same values for the components, A, B, and C. It is clear that this assumption can be later modified without affecting the validity of the model.

Since the membrane can accommodate only a finite number of enzyme molecules, it is necessary to introduce a control parameter, ω, to distinguish between the amounts of enzymes 1 and 2 present. If E_t is defined as the sum of the total molar concentration of both the enzymes, then when only one of the two enzymes is present, the maximum reaction velocity, V'_m, for a unit weight of membrane is given by

$$V'_{mi} = k_i E_t$$

where V'_{mi} is the maximum reaction velocity for enzyme 1 or 2, k_i is the turnover number of enzyme 1 or 2, and E_t is the total molar concentration of the two enzymes. When both enzymes 1 and 2 are present, the maximum reaction velocities for the enzymes reduce to ωV_1 and $(1 - \omega)V_2$, where

$$\omega = \frac{\text{molar concentration of enzyme 1}}{\text{total molar concentration of enzymes 1 and 2}}$$

The control ω is thus constrained between the limits of 0 (only enzyme 2 present) and 1 (only enzyme 1 present). On the basis of these assumptions the material balances for the species A, B, and C, assuming Michaelis–Menten kinetics, can be defined by the following differential equations:
In the solid phase:

$$\frac{\partial^2 Y_{Am}}{\partial L^2} - \omega \frac{L^2}{D_e}\left[\frac{V'_{m1}Y_{Am}}{K'_{m1} + S_{A0}Y_A}\right] = 0 \tag{171}$$

$$\frac{\partial^2 Y_{Bm}}{\partial L^2} - \omega \frac{L^2}{D_e}\left[\frac{V'_{m1}Y_{Am}}{K'_{m1} + S_{A0}V_{Am}}\right] + (1 - \omega)\frac{L^2}{D_e}\left[\frac{V'_{m2}Y_{Bm}}{K'_{m2} + S_{A0}V_{Bm}}\right] = 0 \tag{172}$$

$$\frac{\partial^2 Y_{Cm}}{\partial L^2} + (1 - \omega)\frac{L^2}{D_e}\left[\frac{V'_{m2}Y_{Bm}}{K'_{m2} + S_{A0}Y_{Bm}}\right] = 0 \tag{173}$$

where S_{A0} is the feed concentration of A, K'_{m1}, K'_{m2} are the Michaelis–Menten constants for enzymes 1 and 2, L is the dimensionless thickness of the membrane, and subscript m signifies that the concentrations referred to are in the membrane phase.
In the bulk fluid phase:

$$(dY_{iF}/d\tau') + (k_L/x_0)[Y_{iF} - Y_{is}] = 0 \tag{174}$$

where i = A or B or C; x_0 is the half-distance between two consecutive membranes, τ' is the space time (or axial distance in the reactor) based on fluid volume, and subscripts F and S represent bulk and surface concentrations, respectively.
Boundary conditions:

$$\text{at} \quad L = 0, \quad \partial Y_{im}/\partial L = 0, \quad i = A, B, C \tag{175}$$

$$\text{at} \quad L = 1, \quad Y_{im} = Y_{is}, \quad i = A, B, C \tag{176}$$

where the surface concentrations Y_{AS}, Y_{BS}, and Y_{CS} are determined by equating the surface flux to the bulk transport flux, i.e.,

$$(D_e/l)(\partial Y_{im}/\partial L)_{at\, L=1} = k_L[Y_{iF} - Y_{is}] \qquad i = A, B, C \tag{177}$$

where l is the half-thickness of the membrane. At the entrance of the reactor, i.e.,

$$\text{at } \tau' = 0, \quad Y_{iF} = 0, \quad Y_{AF} = 0 \qquad i = B, C \tag{178}$$

The above equations are not amenable to analytical solution and require numerical techniques. Defining the two Thiele moduli as

$$\phi_1 = L\left[\frac{V'_{m1}\omega}{K'_{m1}D_e}\right]^{0.5}$$

$$\phi_2 = L\left[\frac{V'_{m2}(1-\omega)}{K'_{m2}D_e}\right]^{0.5} \tag{179}$$

The membrane material balance equations for species A and C [Eqs. (171) and (173)] at any cross section of the reactor reduce to

$$\frac{d^2 Y_{Am}}{dL^2} - (\phi_1^2)\frac{Y_{Am}}{[1 + (S_{A0}/K'_{m1})Y_{Am}]} = 0 \tag{180}$$

$$\frac{d^2 Y_{Cm}}{dL^2} + (\phi_2^2)\frac{Y_{Cm}}{[1 + (S_{A0}/K'_{m2})Y_{Bm}]} = 0 \tag{181}$$

The integration of the above equations is started at $L = 0$ with initial values for species A and C equal to conditions given in Eq. (178). The initial estimates are corrected by matching the surface concentrations generated with those derived through the use of boundary conditions [Eqs. (175) and (176)]. The concentration profile for the species A, B, C in the bulk fluid can then be obtained by evaluation of the above Eqs. (180) and (181) together with the fluid-phase equations over an interval, $\Delta\tau$, thence repeating the procedure until the desired space time is reached.

Although detailed numerical solutions of the type shown above for nonlinear, mass transport-kinetic models are possible and in many cases warranted, approximate solutions are probably more useful for design purposes. As noted earlier, first-order kinetic equations will provide good first approximations to describing the performance of many immobilized reactor systems. With this assumption, an analytical solution can be readily obtained. Defining two pseudo first-order rate constants, k_{f1} and k_{f2}, as the ratio of the respective V'_m and K'_m, and substituting these rate constants in the material balance equations [Eqs. (171) through (173)], the concentration profiles in the membrane and fluid phases can be arrived at by analytical solutions of the differential equations. The solutions are given below:

$$Y_{AF} = \exp(-p_1\tau') \tag{182}$$

$$Y_{Am} = \frac{N'_{Sh} \cosh(\phi_1 L)}{N'_{Sh} \cosh \phi_1 + \phi_1 \sinh \phi_1} Y_{AF} \tag{183}$$

$$Y_{BF} = h[\exp(-p_1\tau') - \exp(-p_2\tau')] \tag{184}$$

$$Y_{Bm} = \left[\frac{Y_{BS} - 2g(\tau')\cosh\phi_1}{\cosh\phi_2}\right]\cosh(\phi_2 L) + 2g(\tau')\cosh(\phi_1 L) \tag{185}$$

$g(\tau')$, p_1, p_2, h, and Y_{BS} are complex functions of the transport and kinetic parameters shown in the Appendix. The concentrations of species C in the membrane and fluid phases are obtained from a species component balance.

It can be seen from the model developed that the concentration profiles generated depend to a large extent on the value of the control, ω. When $\omega = 1$, the product C cannot be produced, and when $\omega = 0$, no A can be converted. For maximum production of C, ω should vary optimally from a value of 1 at the entrance of the reactor to a value of 0 at the exit. (This optimization problem will be considered in a later section.) The construction of such a reactor with local optimal enzyme compositions may not be practical and in most systems a constant value of the control ω is preferable. Figure 22 shows the overall conversion with respect to different constant values of the control ω for three sets of kinetic rate data. The values of the control for maximum production of C for the curves I, II, and III are 0.75, 0.375, and 0.2, respectively. The skewness of the profiles is a reflection of the particular choice of rate constants k_1 and k_2—these being in turn dependent on the potency of the individual enzymes. Another important variable is the membrane thickness. Simulations based on the above model predict that the overall conversion increases with membrane thickness initially reaching a maximum asymptotically. This results from the fact that as the membrane thickness is increased, the number of potential active sites per unit surface area is increased thus resulting in an initial increase in the overall conversion level. Simultaneously, however, diffusion of the reactants and products is restricted by increasing the membrane thickness. Thus, the combination of these opposing factors is reflected in these results.

The evolution of the concentration profile for the intermediate inside the membrane is shown in Fig. 23. The concentration inside the membrane is higher than the bulk concentration, and this effect may be responsible for the efficient biofeedback regulation observed in cellular substances, where numerous enzymes may be immobilized on a

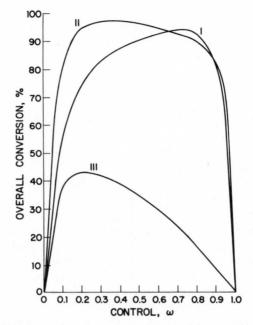

Fig. 22. Effect of enzyme composition, ω, on overall conversion of A in a spiral-wound multipore two-enzyme reactor system. The consecutive reaction considered here is $A \rightarrow B \rightarrow C$. The reactor system is approximated as a parallel membrane model (Figs. 2 and 7). The control, ω, is the concentration ratio of enzyme 1 to enzyme 2 in the reactor. Profiles shown were generated based on Eqs. (182) through (185), and Appendix I. For a detailed discussion on the significance of the simulation results, refer to page 303. Three sets of reaction rate parameters have been used for the simulation. Curve I: $k_{f1} = 0.02$ sec^{-1}, $k_{f2} = 0.12$ sec^{-1}; curve II: $k_{f1} = 0.1$ sec^{-1}, $k_{f2} = 0.06$ sec^{-1}; curve III: $k_{f1} = 0.05$ sec^{-1}, $k_{f2} = 0.002$ sec^{-1}. Membrane thickness $(2l) = 3$ mils; space time = 7.5 hr.

single matrix. The concentration profiles for A and C using Michaelis–Menten kinetics have also been simulated. As the parameter β (K_{m1}/S_{A0}) approaches the value of 0.1, the nonlinear model reduces to the first-order model. Experimental results have been found to be in agreement with these theoretical predictions (Fernandes, 1974).

IX. COMBINED REACTION AND SEPARATION PROCESSES

In combined CSTR/UF membrane reactor systems, the unused substrate and/or the enzymic reaction product is removed continuously as the reaction progresses. While the free enzyme-catalyzed reaction can

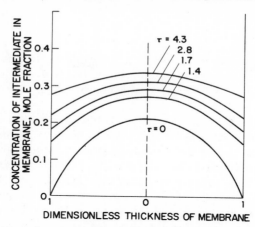

Fig. 23. Evolution of concentration profiles of B along the length of reactor in a two-enzyme reactor system catalyzing the sequential reaction A → B → C. Reactor type is the same as that for Fig. 22. The effect of variation of reactor space time, τ, is also shown. Refer to pages 302 and 303 for a more detailed discussion. Parameter values used: $k_{f1} = 0.01$ sec^{-1}; $k_{f2} = 0.002$ sec^{-1}; $\omega = 0.2$; membrane thickness $(2l) = 0.075$ mm.

be described easily by an ideal CSTR equation, a more detailed model for the system should include the separation process as well. The phenomenon of concentration polarization together with the membrane permeability and rejection characteristics should be taken into account. Concentration polarization is the process of slow buildup of the enzyme at the membrane surface. The turbulence at the membrane surface created by the agitation of the fluid is not usually sufficient to sweep the enzyme back to the solution bulk, resulting in the formation of a gel at the surface.

Bowski *et al.* (1972) and Ryu *et al.* (1972) modeled the CSTR/UF membrane reactor system by considering the ultrafiltration membrane to be ideal; i.e., there is no concentration polarization effect and that the membrane rejection efficiency is 100%. But Bowski *et al.* (1972) interpreted their experimental results in the light of the concentration polarization effect. Detailed models for the effect of concentration polarization in ultrafiltration membrane units have been developed (Michaels, 1968; Porter, 1972) but they are yet to be extended to combined CSTR/UF membrane reactor systems.

Assuming that the rate of ultrafiltration is directly proportional to the product of the membrane area and the pressure difference (which is the driving force) and inversely proportional to the resistance of the membrane and the deposited gel layer, the following simple expression can be derived for the rate of ultrafiltration (Wang *et al.*, 1970):

$$\frac{dV_f}{dt} = \frac{\bar{a}\,(\Delta P)}{R_M + k_c V_f} \tag{186}$$

where V_f is the volume of filtrate, $\bar{a}$ is the membrane surface area, ΔP is the pressure drop across the membrane, R_M is the resistance of the membrane, and k_c is the constant characteristic for the material deposited.

It may be noted that $(k_c V_f)$ represents the cake resistance of the gel. For steady-state operation of the CSTR/UF membrane system, the overall flow rate would be the ultrafiltration rate. Now, the reactor performance can be described by the ideal CSTR equation [Eq. (6)], with the space time value τ based on the filtration rate. In the derivation of Eq. (186), the rate of ultrafiltration remains constant with time, and the constants R_M and k_c are also time-invariant. But in many practical systems, the permeation rate does not reach a steady state (Closset *et al.*, 1974). Further, the enzyme activity can also decline with time (Butterworth *et al.*, 1970).

While ultrafiltration cells have been used widely in conjunction with CSTRs in experimental investigations, they suffer from low surface-to-volume ratios. A commercial process is more likely to employ a configuration where surface-to-volume ratios are higher, such as the spiral-wound ultrafiltration modules or batteries of tubular membranes. The analysis of a continuous, reactive-flow system in which the enzyme and the substrate flow past ultrafiltration membrane with the microsolutes permeating through the membrane walls has recently been attempted (Closset *et al.*, 1973; Tachauer *et al.*, 1974). The analysis is similar to that of a hollow-fiber reactor but differs significantly from it since the enzyme is also transported along with substrate in this case. Considering a plane membrane reactor, in which the enzyme-substrate solution flows past two plane membranes, the mass balance equations for the enzyme and the substrate are

$$[\partial(uS)/\partial z] + [\partial(vS)/\partial y] + r = D(\partial^2 S/\partial y^2) \tag{187}$$

$$[\partial(uE)/\partial z] + [\partial(vE)/\partial y] = D_E(\partial^2 E/\partial y^2) \tag{188}$$

with the following boundary conditions:

B.C. No. 1:

$$z = 0, \quad \text{any} \quad y; \quad S = S_0; \quad E = E_0 \tag{189}$$

B.C. No. 2:

$$y = 0, \quad \text{any} \quad z; \quad \partial S/\partial Y = 0; \quad \partial E/\partial Y = 0 \tag{190}$$

B.C. No. 3:

$$y = y_0, \quad \text{any} \quad z; \qquad D(\partial S/\partial y) = R_S V_W S \tag{191}$$

B.C. No. 4:

$$y = y_0, \quad \text{any} \quad z; \qquad D(\partial E/\partial y) = R_E V_W E \tag{192}$$

where z is the axial position in the membrane reactor, y is the vertical position, y_0 is the half-width of the channel, S is the substrate concentration, E is the enzyme concentration, u is the longitudinal velocity, v is the vertical velocity, D and D_E are the diffusivities of the substrate and enzyme, respectively, R_S and R_E are the rejection coefficients for enzyme and substrate, respectively, and V_W is the vertical velocity at the membrane wall.

By assuming laminar flow conditions and first-order reaction kinetics, these equations have been solved numerically. Simulation profiles for the effect of longitudinal velocity, vertical velocity, and enzyme concentration have been generated. Experimental results from a tubular membrane reactor show that the performance of a membrane reactor is much better than an equivalent solid wall reactor. Concentration polarization contributed to the decline in permeation and reaction rates (Closset *et al.*, 1974).

X. PROCESS OPTIMIZATION OF ENZYME REACTORS

A. Monoenzyme Systems

The overall efficiency of an immobilized enzyme-catalyzed reaction is determined by many—often opposing—process variables. For instance, if the reaction is carried out at an elevated temperature, the reaction rate is increased with concomitant decline in the half-life of the enzyme. Similarly, increasing the space time in a reactor results in higher conversions but lower productivities. In order to harness completely the potential of the immobilized-enzyme system, the optimal combination of the process variables must be employed. Since there are so many different variables that influence the reaction, a pragmatic approach to optimization is to identify a control variable(s) and carry out the optimization process with respect to the chosen control variable. What constitutes a control variable would depend on the process and the optimization objective. The optimization procedure itself is often complicated and requires a detailed knowledge of the process.

Of particular significance to immobilized enzyme reactor design is the knowledge of the response of a given reactor system to changes in different variables. This would be useful in performing limited optimizations within a realm of practical interest. Simulation studies based on a sound mathematical model can predict the performance of an enzyme reactor over a wide range of operating conditions so that the optimal productivity can be determined. This approach reduces considerably the amount of experimental work required. Simulation studies would be even more relevant for systems involving complex reaction kinetics since the complexities of the overall process do not permit an intuitive guess as to the best combination of operating parameters. Even rather simplistic models would be useful as a first approximation. Ryu *et al.* (1972) employed such an approach to optimizing reactor productivity.

The variation of productivity with inlet substrate concentration and reactor space time for a CSTR with immobilized penicillin amidase is presented in Fig. 24. This enzyme undergoes double inhibition by both the hydrolysis products of benzylpenicillin. In Fig. 24, the isoconversion lines (- - -) are superimposed on the productivity map. From these curves, it is possible to determine the productivity as a function of space-time for any desired level of conversion of any inlet substrate concentration. Each isoconversion line traverses a maximum

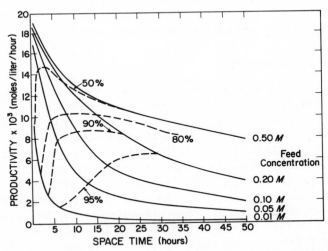

Fig. 24. Computer simulation of reactor productivity, Pr (moles of product produced per liter per hour) of a continuous-flow stirred-tank reactor containing immobilized penicillin amidase in suspension. The effect of reactor space-time, τ, on productivity at varying substrate feed concentration is shown. The dashed lines indicate the level of percent conversion, X, that can be achieved under varying conditions of substrate feed concentration, reactor space-time, and productivity.

that corresponds to the maximum productivity for a given conversion. Based on the information in Fig. 24, one can also determine the best combination of substrate concentration and the space-time that correspond to the maximal productivity for a desired conversion. These results represent, to a limited extent, a process optimization in terms of key operating variables such as space-time, feed concentration, and conversion level. Similar optimal process conditions have been worked out for sucrose–invertase system operating in a CSTR/UF membrane reactor (Bowski *et al.*, 1972) and in a packed-bed reactor (Kobayashi and Moo-Young, 1973).

Optimal temperature and pH control policies for a batch process using a free enzyme that is subject to inactivation have been formulated (Ho and Humphrey, 1970). Again, the system considered by these authors is penicillin amidase–benzylpenicillin. Assuming that at low benzylpenicillin concentrations the reaction obeys Michaelis–Menten kinetics, the system equations may be described as follows for the substrate (S), product (P), and the enzyme (E) concentrations:

$$\frac{dS}{dt} = -\left[\frac{k_2 E_0 S}{K_m + S}\right] - K_1' S \tag{193}$$

$$\frac{dP}{dt} = +\left[\frac{k_2 E_0 S}{K_m + S}\right] - K_2' P \tag{194}$$

$$\frac{dE}{dt} = -K_3' E \tag{195}$$

The terms $(K_1' S)$ and $(K_2' P)$ signify the degradation of the substrate and product in free solution. The constants K_1', K_2', K_3', and k_2 are expressed as functions of the control variables, i.e., temperature and pH. These dependencies were formulated empirically from separate experiments. Using this mathematical model, the process was optimized with respect to temperature and pH such that the yield of product in a given fixed reactor operating time is maximized while at the same time minimizing loss of enzyme in the operation. This objective function for the process is mathematically stated as

$$\text{Maximize } N = P(t) - W[1 - E(t)] \tag{196}$$

where $\bar{t}$ is the given duration of reactor operation time and W is a weighting constant that represents the cost of the enzyme loss in terms of the product yield. The solution to Eq. (196) is obtained by a common optimization technique, which yields the optimal temperature and pH profiles. These profiles give the best values of the control variables that must be used at any given time so as to achieve the objective function.

In a similar manner, optimal temperature policy for the enzymic isomerization of glucose to fructose by free glucose isomerase in a batch reactor has been formulated (Haas *et al.*, 1974). The system was assumed to follow first-order reversible reaction kinetics. Laboratory reactor data were used to determine the reversible reaction rate constants and the enzyme inactivation rate constants as functions of temperature. Results obtained by solving the optimization problem by using calculus of variation indicate that the catalyst inactivation can be reduced by 10% by adopting the optimal temperature policy compared to operation at an optimal isothermal temperature. This study is being extended to generate optimal temperature policies for plug-flow and fixed-bed reactors (Tavlarides *et al.*, 1974).

B. Multienzyme Systems

Multienzyme systems raise interesting optimization problems regarding pH, temperature and relative enzyme amounts in the mixed bed. Since any two enzymes are likely to have different temperature and pH optima for maximal activity, an optimization problem exists— even if isothermal or iso-pH operation is contemplated. Similarly, the packing policy of the enzymes in the reactor should be varied optimally along the length of the reactor.

For an immobilized two-enzyme system, Laurence *et al.* (1973) have worked out the best constant enzyme fraction distribution in the reactor for maximizing the yield of product, assuming no mass transfer restrictions. This represents a suboptimal policy since the optimization is accomplished with an a priori imposed constraint; viz., a constant ratio of the two enzymes throughout the reactor. This constant blending ratio was shown to be

$$\omega = \frac{1 + (b - 1)(b/2)^{1/2}}{1 + b^2} \tag{197}$$

where b is the ratio of the first-order kinetic constants for the reactions $A \rightarrow B$ and $B \rightarrow C$, respectively.

A more rigorous optimization problem is to consider the two-step reaction with mass transport impedances. In our laboratory such a case has recently been analyzed both theoretically and experimentally, based on the mass transfer-kinetic model for the two-enzyme system discussed earlier (Fernandes *et al.*, 1975). The problem is to choose the control, the packing policy $\omega(\tau)$ such that the concentration of C in the reaction scheme $A \rightarrow B \rightarrow C$ is maximized at each point in the reactor. It may be noted that this is a continuous optimization problem as opposed to evaluating the best constant-packing policy. The state equations are formulated by combining the solutions for the concentrations of A and B in the membrane phase [solutions to Eqs. (183) and (185), respectively] with the fluid phase equation [Eq. (174)]. These are given by

$$dY_{AF}/d\tau' = -f_1 Y_{AF} \tag{198}$$

$$dY_{BF}/d\tau' = f_2 Y_{AF} - f_3 Y_{BF} \tag{199}$$

with the boundary conditions

B.C. No. 1:

$$\tau' = 0, \quad Y_{AF} = 1.0 \tag{200}$$

B.C. No. 2:

$$\tau' = 0, \quad Y_{BF} = 0 \tag{201}$$

f_1, f_2, and f_3 are nonlinear functions of the control ω containing all the transport and kinetic parameters as presented in Appendix I. The objective function is to maximize $Y_{CF}(\tau)$. It is represented by

$$\text{Max}\,[Y_{CF}(\tau')] = \text{Max}\left[\int_0^{\gamma'} (f_1 - f_2) Y_{AF} + f_3 Y_{BF} \right] \tag{202}$$

with the boundary condition

$$\text{at} \quad \tau' = 0, \quad Y_{CF} = 0 \tag{203}$$

The optimization problem is solved numerically through the use of a sophisticated algorithm called the Continuous Maximum Principle (Pontryagin *et al.*, 1962).

The profiles for the control ω and the resulting overall conversions for a given set of kinetic and transport parameters are shown in Fig. 25. From a practical point of view, the construction of a reactor with continuously varying value of the control ω, along the length of the reactor may be difficult. It may be necessary to compromise some of the yield by employing a uniform packing policy throughout the reactor. For the conditions given on Fig. 25, the best constant value of ω is 0.35 with a resulting yield of 56.2% for a τ' value of 1. However, the corresponding yield for the case of a reactor packed with the optimal enzyme policy is 82.2%. The difference, which is significant in this case, is an important criterion in deciding whether local optimal control for $\omega(\tau')$ is warranted for a given reactor.

When low or negligible interphase mass-transport conditions are

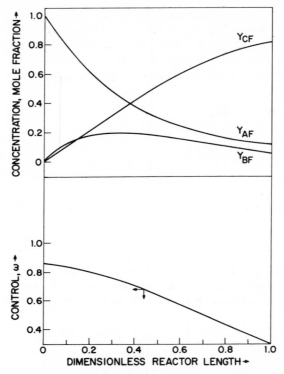

Fig. 25. Concentration profiles of species A, B, and C for locally optimal distribution of enzymes in a two-enzyme reactor. The physical system considered is the same as that for Figs. 23 and 24. Optimal profile for the control, ω, along the reactor length is also shown. Profiles were generated based on Eqs. (198) through (203). Parameter values used are: $k_{f1} = 0.4 \text{ sec}^{-1}$, $k_{f2} = 1.2 \text{ sec}^{-1}$, $D = 5.7 \times 10^{-8} (\text{cm}^2/\text{sec})$, $k_L = 1.2 \times 10^{-4} (\text{cm/sec})$; space-time = 1 hr.

present, substantial advantage may be obtained by constructing the reactor into a large number of zones of uniform composition which would approximate the optimal enzyme ratio profile. As the bulk film resistance increases, the difference in optimal yields between the local optimal control and the uniform distribution also decreases. From Fig. 26, it is clear that as the mass-transfer resistance becomes less limiting ($N_{Sh} \rightarrow \infty$, kinetic control regime), the so-called bang-bang policy (Jackson, 1968) is approached; i.e., the two enzymes are packed in tandem in two linear regimes with a single switch point. At low Sherwood number values (bulk diffusional control), the optimal policy for enzyme arrangement moves closer to one of uniform distribution.

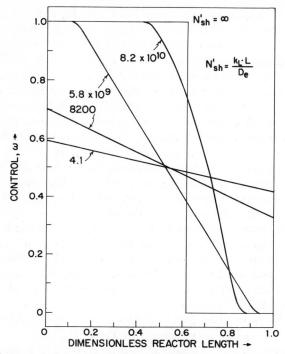

Fig. 26. Influence of Sherwood number on optimal enzyme arrangement in a two-enzyme reactor. Physical system considered is the same as that for the previous figure. Sherwood number values (N'_{Sh}) are indicative of the severity of mass transfer resistances within the reactor. High N'_{Sh} values refer to a regime of kinetic control, i.e., no significant mass-transfer resistances and vice versa. These profiles were generated based on Eqs. (198) through (203). Refer to the discussion on page 311. Parameter values: $k_{f1} = 0.004$ sec^{-1}; $k_{f2} = 0.012$ sec^{-1}; $2l = 0.075$ mm; $D_e = 5.7 \times 10^{-8}$ cm^2/sec; space-time (τ) = 1 hr.

XI. COMPARISON OF RELATIVE EFFICIENCIES OF SEVERAL REACTOR DESIGNS

The overall efficiency of an enzyme reactor design would be determined by myriad factors, as highlighted in the foregoing discussion. Not all these factors have been understood clearly. Although some general analyses and design procedures for enzyme reactors are now known, the dearth of experimental information—particularly on a pilot-plant scale—renders the task of designing and scaling-up an enzyme reactor very difficult. A designer, contemplating the design of a given enzymic reaction system, may be posed with several alternative approaches—perhaps none of them having a clear advantage or design precedence. If comparative information were available on different alternative design schemes, it would facilitate the choice of a particular system. Despite the large body of information available in the literature, a direct comparison of this type is not easy, owing to many variations in experimental and/or theoretical approaches. Presented below is a simplified attempt to compare the relative efficiencies of several reactor configurations, based on research done at our laboratory on free and immobilized enzymes. These data were obtained under conditions that permit their meaningful comparison.

The most important factors governing the overall reactor efficiency include the enzyme loading factor, carrier loading factor, operational stability of enzyme, external and internal diffusional efficiency, and residence time distribution. The efficiency of different reactor configurations with respect to these factors is compared in Table IV. Data on continuous enzyme reactors using free (Bowski *et al.*, 1972), bead-immobilized (Saini *et al.*, 1972), microencapsulated (Mogensen and Vieth, 1973) and collagen membrane-immobilized (Venkatasubramanian *et al.*, 1972; Eskamani, 1972; Wang and Vieth, 1973; Constantinides *et al.*, 1973; Bernath and Vieth, 1974) enzymes and collagen-immobilized whole microbial cells (Saini and Vieth, 1975) formed the basis for the comparison presented in Table IV.

These comparisons are made according to Eqs. (75) and (76), which employ pseudo first-order kinetics. Equation (75) refers to flat sheets of collagen–enzyme membranes, and Eq. (76) relates to spherical beads or microcapsules. The reactor performance equation [Eq. (78)] can be written as

$$\ln(1-X) = -\left[\frac{k_2 E_0 \tau'}{K'_m}\right] \eta \bar{P} \left[\frac{k_L}{k_L + (k_2 E_0/K'_m)\eta l}\right] \quad \text{sheets} \quad (204)$$

$$\ln(1-K) = -\left[\frac{k_2 E_0 \tau'}{K'_m}\right] \eta P \left[\frac{k_L}{k_L + (k_2 E_0/K'_m)\eta(R/3)}\right] \quad \text{beads} \quad (205)$$

where $\bar{P} = la$ for flat sheets and $(R/3)a$ for beads or microcapsules. It is the carrier packing factor expressed in milliliters of carrier per milliliters of fluid.

Equations (204) and (205) refer to flat sheets and beads or microcapsules, respectively. The term $k_2 E_0$ is the catalytic potency of the reactor and has the units International Unit (IU) per milliliter of carrier (1 IU of enzyme activity is defined as the amount of enzyme that would catalyze the conversion of 1 μmole of substrate per minute under specified reaction conditions). An enzyme loading factor of 500 IU per

TABLE IV

ESTIMATE OF RELATIVE EFFICIENCY OF SEVERAL SINGLE-ENZYME REACTOR CONFIGURATIONS[a]

A Enzyme form configuration	B Enzyme load factor (IU/ml)	C $\dfrac{k_2 E_0 \tau}{K'_m}$	D Carrier packing factor	E Max. relative efficiency of C×D	F Microdiffusional capacity factor	G Relative efficiency of F	H Macrodiffusional efficiency ν	I Contact efficiency (residence time factor) relative to plug flow, 90% conversion	J Enzyme stability factor	K Overall maximum relative efficiency: column E×G×H×I×J
I. Free enzyme CSTR/UF[b]	500	1.0	1.0	1.0	∞ (relative)	1.0	1.0	0.25	0.01 (denaturation)	0.0025
II. Microcapsules column	500	1.0	1.0	1.0	~111	0.5	0.5	1.0	0.05 (capsule breakage)	0.0125
III. Beads column	10–250	0.02–0.5	1.0	0.5	30–300 m²/gm	0.25	1.0	1.0	0.05–1.0 (slow leaching)	0.1250
IV. Collagen multipore module	50–500	0.1–1.0	0.5	0.5	100–1000 m²/gm	0.5	1.0	1.0	1.0 (stable limit of several collagen-enzyme modules)	0.2500

[a] Basis: rate expression considered pseudo first-order with respect to substrate.

[b] Continuous-flow stirred-tank reactor/ultrafiltration system.

milliliter of carrier is an effective enzyme loading. For purposes of simplicity, let us examine the case where the dimensionless parameter $k_2E_0\tau'/K'_m$ is assigned the value of unity when k_2E_0 is at this level (see columns B and C in Table IV).

The quantity $kE_0\bar{P}$ can be expressed as the product of the enzyme loading factor expressed in IU per milliliter of carrier, and the reactor loading factor $(\bar{P})$ expressed in milliliters of carrier per milliliter of fluid. The values of the latter factor are shown in column D in Table IV. Values of the effectiveness factor η which defines the microdiffusional efficiency of the system are tabulated in column G.

In Eqs. (204) and (205), the fourth term on the right-hand side represents the macrodiffusional efficiency (ν) of the substrate. Column H (Table IV) shows this effect. Now, Eqs. (204) and (205) can be rearranged in terms of the different efficiency factors as

$$\ln 1/[(1-X)] = [(1.0)P\eta\nu] \tag{206}$$

where (1.0) = unit value of $(k_2E_0\tau'/K'_m)$.

The values of $-\ln(1-X)$ are tabulated in column K (Table IV), which reflects the overall efficiency of the reactor systems, containing also the catalyst stability effect and shown in column J.

For the combined CSTR/ultrafiltration (UF) membrane reactors employing the free enzyme, one further correction has been used; i.e., at the same level of conversion $(X = 0.9)$, the CSTR will require approximately four times the space-time value (τ') required for the packed bed. Therefore, relative to the latter, the CSTR has a space-time factor of 0.25, as shown in column I.

At low conversions, $\ln[1/(1-X)]$ can be approximated as X. Thus at low conversion levels, column K (Table IV) directly gives the relative fractional conversion values.

Examining columns J and K (Table IV), it can be noted that in the case of free enzyme CSTR/UF process, the serious drawback of the system arises from its poor operational stability, the reasons for which were discussed in detail in an earlier section. Enzyme stability could conceivably be increased by attaching the enzyme to soluble high-molecular-weight supports. On the subject of stability it must also be pointed out that the operational life and characteristics of the ultrafiltration membrane itself need to be scrutinized more carefully. Since the enzyme could be adsorbed on the ultrafiltration membrane, the filtration rate and the rejection efficiency could be significantly impaired.

Among the immobilized enzyme reactor systems listed in Table IV, again microcapsule reactors seem to suffer from poor stability (column

J) mainly owing to breakage of capsules. More rigid microcapsules could possibly reduce this problem, but it is also likely to increase the transport impedances. Between packed columns and collagen multipore modules, the main difference arises from the superior microdiffusional capacity of the latter, which is attributable to the very highly open internal structure of collagen matrix—particularly under swollen conditions.

Thus, on the basis of the data appearing in column K (Table IV), it is clear that immobilized enzyme reactor systems are technically advantageous as compared to free or microencapsulated systems for process scale conversion of substrates. Within the former, the biocatalytic modules employing collagen–enzyme complex membranes exhibit definite advantages of overall reactor efficiency.

XII. EPILOG

In the last few years, considerable progress has been made in enzyme reactor engineering. In contrast to earlier approaches to modeling enzyme reactors that lump mass transfer-kinetic effects together implicitly, current mathematical analyses include these effects separately and explicitly as a rational combination of elementary steps. The heterogeneous catalysis literature has rather a corpulent reservoir of information, most of which is applicable to immobilized enzyme systems. Indeed, several such results have successfully been extended to provide a reasonably sound theoretical basis for enzyme reactor design. Theoretical analysis of CSTR and packed-bed reactors are adequate; much remains to be accomplished on modeling fluidized-bed reactors. What is lacking is also the experimental validation of these theoretical results. While theoretical simulations are useful in gaining a qualitative understanding of a given system, their general applicability could be ratified only from proper experimental verification.

Most laboratory data have been collected from small reactors. Even if they agree with theoretical predictions, their use in reactor scale-up is somewhat limited. Therefore, more efforts should be devoted to shift the scene from the laboratory to the pilot plant. Recent efforts made by several university and industrial laboratories in this direction are indeed very encouraging. Pilot-scale experimental data should provide a sound basis to formulate reliable scale-up procedures as well as to make realistic process economic calculations. As of now, little is known on reactor scale-up. Of particular importance is the operational stability of the supported enzyme system. Reliable procedures should be developed so that a priori prediction of catalyst life would be possible. More work is also needed in characterizing the hydrodynamics, including analyses of residence time distributions, of reactor systems of practical importance such as spiral-wound and hollow-fiber membrane reactors. Combined reaction–separation schemes are likely to gain greater importance in the future. Analyses of these systems, particularly with respect to concentration polarization, and related phenomena are important. These efforts should steer the fledgling enzyme technology toward a practical engineering discipline.

APPENDIX

$$g(\tau') = \frac{-\phi_1^2}{2(\phi_1^2 - \phi_2^2)} \left[\frac{N'_{\text{Sh}}}{N'_{\text{Sh}} \cosh \phi_1 + \phi_1 \sinh \phi_1} \right] \exp(-p_1\tau') \tag{A}$$

$$p_1 = \frac{k_{\text{L}}\phi_1 \tanh \phi_1}{X_0(N'_{\text{Sh}} + \phi_1 \tanh \phi_1)} \tag{B}$$

$$p_2 = \frac{k_{\text{L}}\phi_2 \tanh \phi_2}{X_0(N'_{\text{Sh}} + \phi_2 \tanh \phi_2)} \tag{C}$$

$$h = \frac{N'_{\text{Sh}}\phi_1^2(\phi_1 - \phi_2 \cosh \phi_1 \tanh \phi_2)}{\phi_2 \tanh \phi_2 (\phi_1^2 - \phi_2^2)(N'_{\text{Sh}} \cosh \phi_1 + \phi_1 \sinh \phi_1)(1 - N)} \tag{D}$$

where

$$N = \frac{\phi_1 \tanh \phi_1 (N'_{\text{Sh}} + \phi_2 \tanh \phi_2)}{\phi_2 \tanh \phi_2 (N'_{\text{Sh}} + \phi_1 \tanh \phi_1)}$$

$$Y_{\text{Bs}} = \left[\frac{N'_{\text{Sh}}}{N'_{\text{Sh}} + \phi_2 \tanh \phi_2} \right] Y_{\text{BF}}$$
$$- 2 \frac{(\phi_1 \sinh \phi_1 - \phi_2 \cosh \phi_1 \tanh \phi_2)}{N'_{\text{Sh}} + \phi_2 \tanh \phi_2} g(\tau') \tag{E}$$

$$f_1 = \frac{k_{\text{L}}\phi_1 \tanh \phi_1}{x_0(N'_{\text{Sh}} + \phi_1 \tanh \phi_1)} \tag{F}$$

$$f_2 = \frac{k_{\text{L}}\phi_1^2 N'_{\text{Sh}}(\phi_1 \sinh \phi_1 - \phi_2 \cosh \phi_1 \tanh \phi_2)}{x_0(\phi_1^2 - \phi_2^2)(N'_{\text{Sh}} + \phi_2 \tanh \phi_2)(N'_{\text{Sh}} \cosh \phi_1 + \phi_1 \sinh \phi_1)} \tag{G}$$

$$f_3 = \frac{k_{\text{L}}\phi_2 \tanh \phi_2}{x_0(N'_{\text{Sh}} + \phi_2 \tanh \phi_2)} \tag{H}$$

LIST OF SYMBOLS

A	Surface area of a spherical microcapsule/catalyst particle (cm^2)
a	Catalytic surface area per unit reactor fluid volume (cm^2/ml of fluid)
a'	Catalytic activity of supported enzyme (IU/gm catalyst)
$a_{\underline{m}}$	Area for interfacial mass transfer (cm^2)
$\bar{\bar{a}}$	Membrane area for ultrafiltration (cm^2)
b	Ratio of the first-order kinetic constants for the reactions A $\rightarrow$ B and B $\rightarrow$ C, respectively, i.e., k_{f1}/k_{f2}
Bo	Bodenstein number = uR_L/D_a (defined by Eq. 113)
C_R	Reaction capacity of reactor as defined by Eq. (4) (moles/min)
c'	Contact efficiency
D	Diffusion coefficient in bulk phase (cm^2/sec)
D_a	Axial dispersion coefficient (cm^2/sec)
D_e	Effective diffusivity (cm^2/sec)
d_P	Diameter of catalyst particle (cm)
E_0	Enzyme concentration in the reactor (moles/liter)
E_0'	Initial concentration of "unbound" active enzyme (moles/liter)
E_t	Total molar enzyme concentration (moles/liter)
$E_{\text{diff}}, E_{\text{kin}}$	Reactor efficiency for diffusion-controlled and kinetically controlled reactions, respectively
e	Electrostatic charge (coulombs)
F	Faraday constant
G	Superficial mass velocity = $u\rho$ (gm/cm^2 min)
H	Electrostatic partition coefficient as defined by Eq. (47)
h	Partition coefficient
I	Concentration of inactivated enzyme (moles/liter)
J	Mass flux (gm/cm^2 min)
j_D	Factor defined by Eq. (31)
K'	Combined mass transfer-kinetic coefficient as defined by Eq. (75) [ml of reactor fluid/cm^2 (catalyst surface) min]
K_E	k_E/k_E'
K_m, K_m'	Michaelis–Menten constant for free and immobilized enzymes, respectively (moles/liter)
K_P	(Competitive) product inhibition constant (moles/liter)
K_P'	(Noncompetitive) product inhibition constant (moles/liter)
K_S	Substrate inhibition constant (moles/liter)
K_1 through K_8	Dimensionless reaction rate constants
K_1', K_2', K_3'	Rate constants defined by Eqs. (193), (194), and (195), respectively (min^{-1})
k_c	Constant characteristic for the material deposited in the ultrafiltration membranes
k_d', k_d	Substrate-dependent and substrate-independent enzyme decay constants, respectively (hr^{-1})
k_E, k_E'	Forward and backward rate constants for reversible enzyme denaturation (min^{-1})
k_f	Pseudo first-order rate constant (min^{-1})
k_L	Mass-transfer coefficient (cm/sec)
k_{L+M}	Combined fluid and membrane mass-transfer coefficient (cm/sec)
k_{true}	"True" kinetic coefficient without diffusional disguises [min^{-1} (liter/mole)$^{1-m}$ (m = order of reaction)]

k_1, k_2, k_3 Forward reaction rate constants (min^{-1})
k_{-1}, k_{-2}, k_{-3} Backward reaction rate constants (min^{-1})
$\bar{k}$ Boltzmann constant
L Dimensionless membrane thickness $= x/l$
l Half-thickness of enzyme-membrane (cm)
M Electrostatic potential modifier defined by Eq. (53)
M_1 through M_5 Dimensionless composite reaction rate constants
m Reaction order
$\bar{m}$ Proportionality constant in Eq. (117)
m_1 through m_5 Lumped kinetic constants as given by Eq. (97) through Eq. (101)
N_{Pe} Peclet number $= (N_{Re})\,(N_{Sc})$
N_{Re} Reynolds number $= d_p G/\mu$
N_{Sc} Schmidt number $= \mu/\rho D$
N'_{Sh} Modified Sherwood number $= k_L\,l/D_e$
ΔP Pressure drop (kg/cm^2)
P Product concentration (moles/liter)
P' Total substrate permeability coefficient as defined by Eq. (88) (ml/min cm^2)
P'_S Substrate permeability coefficient for enzyme-membrane (ml/min cm^2)
Pr Reactor productivity $= XS_0/\tau'$ (moles/liter min)
$\bar{P}$ Catalyst packing density (ml catalyst/ml reactor fluid)
p Geometric factor, as defined in Eq. (86)
Q Flow rate through the reactor (liter/min)
R Radius of spherical catalyst particle (cm)
R' Diameter of reactor tube (cm)
R_L Reactor length (cm)
R_M (Ultrafiltration) membrane resistance to filtration (liter/min meter^2)
R_0, R'_0 Inner and outer radius of microcapsule, respectively (cm)
R_S, R_E Rejection coefficients for substrate and product, respectively
$\bar{R}$ Gas constant
r Reaction rate (moles/liter min or moles/gm catalyst min)
r_m Mass transfer rate as defined by Eq. (27) (moles/min gm catalyst)
$\imath$ Radial position (cm)
S Substrate concentration (moles/liter)
T Absolute temperature (°K)
t Time (min)
u Superficial fluid velocity in the axial direction, i.e., flow rate divided by cross-sectional area of reactor tube without any packing (cm/min)
u_m Maximal axial velocity in laminar flow (cm/min)
$\bar{V}$ Ratio of bath volume to microcapsule volume [Eq. (123)]
V_f Ultrafiltration rate (liters/min meter^2)
V_m, V'_m Maximal reaction rates for free and immobilized enzymes, respectively (moles/liter min)
v Fluid velocity in the vertical direction in reactor (cm/min)
W Weighting factor as defined in Eq. (195)
X Fractional conversion of substrate
x Distance from the center of plane enzyme-membrane (cm)
x_0 Half-thickness between two consecutive membranes (cm)
Y Dimensionless concentration (i.e., concentration of species divided by inlet substrate concentration)
y Vertical position in the reactor (cm)
y_0 Half-thickness between two flat plates in a spiral-wound reactor (cm)
Z Dimensionless axial position in reactor $= z/R_L$

z Axial position in reactor (cm)

$\bar{z}$ Sign of charge (± 1)

Greek Letters

α $(R_L D)/ul^2$, dimensionless

β K'_m/S_0, dimensionless Michaelis–Menten constant

β' S_0/K'_m, dimensionless

γ Electrostatic surface potential parameter as defined by Eq. (51)

δ Thickness of Nernst boundary layer (cm)

ϵ Void volume fraction of packed-bed reactor

ϵ_d Dielectric constant

ζ Dimensionless reactor length for reactors with porous annular walls

η Effectiveness factor

η_{DE} Effectiveness factor for diffusive and electrostatic effects as defined by Eq. (55)

η_0 Overall effectiveness factor as defined by Eq. (86)

θ Dimensionless time $= (Dt)/R^2$

κ Reciprocal of thickness parameter of diffuse double layer (cm^{-1})

λ Initial fraction of "unbound" active enzyme $= E'_0/E_0$

μ Viscosity of substrate (gm/cm min)

ν Macrodiffusional capacity factor

Ξ Ratio of maximum reaction rate to mass transfer rate as defined by Eq. (56)

ξ Ratio of average channeling length to particle diameter

ρ Density of substrate (gm/ml)

σ Objective function defined by Eq. (89)

τ Reactor space-time $= V_R/Q$ (min)

τ' Reactor space-time based on reactor fluid volume $= V_R\epsilon/Q$, min

ϕ Thiele modulus $= l[(k_{true} S_S{}^{m-1})/D_e]^{0.5}$

ϕ_m Modified Thiele modulus $= l[V'_m/(K'_m D_e)]^{0.5}$

Ψ Electrostatic potential (volts)

ω Molar concentration ratios of enzyme 1 and enzyme 2 in a two-enzyme system

Subscripts

A   Substrate species

A · E   Substrate–enzyme species

B   Product species in monoenzyme system and intermediate species in two-enzyme system

B · E   Product–enzyme species

C   Product species in two-enzyme system

diff   for (bulk) diffusion-controlled reaction

E   Enzyme

e   Reactor exit

F   Fluid phase

I   Denatured enzyme

kin   For kinetically controlled reaction

m   Membrane phase

O   Initial

P   Product

PFR   Plug-flow reactor

S   Substrate

s   Solid phase

t   Time

W   Wall

Suffix

1 First enzyme in a two-enzyme system
2 Second enzyme in a two-enzyme system

ACKNOWLEDGMENT

The authors are grateful to Mr. P. M. Fernandes and Dr. R. Shyam for their assistance. Thanks are also due to Mrs. M. Yuran and Mrs. D. Walker for typing the manuscript.

REFERENCES

Anonymous (1974). *New Sci.* Apr. 4, p. 8.
Aris, R. (1957). *Chem. Eng. Sci.* **6**, 262.
Aris, R. (1969). "Elementary Chemical Reactor Analysis." Prentice-Hall, Englewood Cliffs, New Jersey.
Aris, R. (1972). *Math. Biosci.* **13**, 1.
Bachler, M. J., Strandberg, G. W., and Smiley, K. L. (1970). *Biotechnol. Bioeng.* **12**, 85.
Bar-Eli, A., and Katchalski, E. (1963). *J. Biol. Chem.* **238**, 1690.
Barker, S. A., Emery, A. N., and Novais, J. M. (1971). *Process Biochem.* **6**(10), 11.
Bernath, F. R., and Vieth, W. R. (1974). *In* "Immobilized Enzymes in Food and Microbial Processes" (A. C. Olson and C. L. Cooney, eds.), pp. 157–185. Plenum, New York.
Bischoff, K. B. (1965). *AIChE J.* **11**, 351.
Blanch, H. W., and Dunn, I. J. (1974). *Adv. Biochem. Eng.* **3**, 127.
Bourgeois, P., and Grenier, P. (1968). *Can. J. Chem. Eng.* **46**, 325.
Bowski, L., Shah, P. M., Ryu, D. Y., and Vieth, W. R. (1972). *Biotechnol. Bioeng. Symp.* No. 3, p. 229.
Bransom, S. H., and Pendse, S. (1961). *Ind. Eng. Chem.* **53**, 575.
Broun, G., Selegny, E., Tran-Minh, E., and Thomas, D. (1970). *FEBS Lett.* **7**, 235.
Broun, G., Thomas, D., and Selegny, E. (1972). *J. Membr. Biol.* **8**, 313.
Bruns, D. D., Bailey, J. E., and Luss, D. (1973). *Biotechnol. Bioeng.* **15**, 131.
Bunting, P. S., and Laidler, K. J. (1972). *Biochemistry* **11**, 4477.
Bunting, P. S., and Laidler, K. J. (1974). *Biotechnol. Bioeng.* **16**, 119.
Butterworth, T. A., Wang, D. I. C., and Sinskey, A. J. (1970). *Biotechnol. Bioeng.* **12**, 615.
Carbonell, R. G., and Kostin, M. D. (1972). *AIChE J.* **18**, 1.
Chang, T. M. S. (1972). "Artificial Cells." Thomas, Springfield, Illinois.
Chang, T. M. S., and Malave, N. (1970). *Trans. Am. Soc. Artif. Intern. Organs* **16**, 141.
Chang, T. M. S., Gonda, A., Dirks, J. H., and Malave, N. (1971). *Trans. Am. Soc. Artif. Intern. Organs* **17**, 246.
Charm, J. E., and Lai, C. J. (1971). *Biotechnol. Bioeng.* **13**, 185.
Charm, S. E., and Matteo, C. E. (1970). *In* "Enzyme Purification and Related Techniques" (W. B. Jakoby, ed.), Methods in Enzymology, Vol. 22, p. 476. Academic Press, New York.
Charm, S. E., and Wong, B. L. (1970). *Biotechnol. Bioeng.* **12**, 1103.
Cheftel, C., Ahren, M., Wang, D. I. C., and Tannenbaum, S. R. (1971). *J. Agric. Food Chem.* **19**, 155.
Chibata, I., Tosa, T., Sato, T., Mori, T., and Matsuo, Y. (1972). *Proc. Int. Ferment. Symp., 4th, Ferment. Technol. Today* pp. 383–389.
Choi, P. S. K., and Fan, L. T. (1973). *J. Appl. Chem. Biotechnol.* **23**, 531.

Chu, J. C., Kalil, J., and Wathworth, W. A. (1953). *Chem. Eng. Prog.* **49**, 141.

Churchill, S. W. (1974). "The Interpretation and Use of Rate Data: the Rate Concept." McGraw-Hill, New York.

Cleland, W. W. (1967). *Annu. Rev. Biochem.* **36**, 77.

Closset, G. P., Cobb, J. T., and Shah, Y. T. (1973). *Biotechnol. Bioeng.* **15**, 441.

Closset, G. P., Cobb, J. T., and Shah, Y. T. (1974). *Biotechnol. Bioeng.* **16**, 345.

Constantinides, A., Vieth, W. R., and Fernandes, P. (1973). *Mol. Cell. Biochem.* **1**, 127.

Corno, C., Galli, G., Morisi, F., Bettonte, M., and Stopponi, A. (1972). *Sterke* **24**, 420.

Coughlin, R. W., Charles, M., Allen, B. R., Paruchuri, E. K., and Hasselberger, F. X. (1973). *AIChE Annu. Meet., 66th, Philadelphia* Pap. No. 17b.

Danckwerts, P. V. (1953). *Chem. Eng. Sci.* **2**, 1.

Davidson, B., Vieth, W. R., Wang, S. S., Zwiebel, S., and Gilmore, R. (1974). *AIChE Symp. Ser.* **144**, 70.

Davidson, J. F., and Harrison, D. (1963). "Fluidized Particles." Cambridge Univ. Press, London and New York.

Denbigh, K. G. (1965). "Chemical Reactor Theory." Cambridge Univ. Press, London and New York.

Denbigh, K. G., and Page, F. M. (1954). *Discuss. Faraday Soc.* **17**, 145.

DeSimone, J., and Caplan, S. R. (1973a). *Biochemistry* **12**, 3032.

DeSimone, J. A., and Caplan, S. R. (1973b). *J. Theor. Biol.* **39**, 523.

Dinelli, D. (1972). *Process Biochem.* **7**(8), 14.

Dixon, M., and Webb, E. C. (1964). "Enzymes," 2nd Ed. Longman Press, New York.

Dutta, R., Armiger, W., and Ollis, D. (1973). *Biotechnol. Bioeng.* **15**, 993.

Emery, A. N., and Revel-Chion, L. (1974). *AIChE Natl. Meet., 77th, Pittsburgh* Pap. No. 11b.

Engasser, J. M., and Horvath, C. (1973). *J. Theor. Biol.* **42**, 137.

Engasser, J. M., and Horvath, C. (1974). *Biochemistry* **13**, 3845.

Epton, R., and Thomas, T. H. (1972). *Koch-Light Lab. Bull.* pp. 61–67.

Ergun, S. (1952). *Chem. Eng. Prog.* **48**, 227.

Eskamani, A. (1972). Ph.D. Thesis, Rutgers Univ., New Brunswick, New Jersey.

Fernandes, P. M. (1974). Ph.D. Thesis, Dep. Chem. Biochem. Eng., Rutgers Univ., New Brunswick, New Jersey.

Fernandes, P. M., Constantinides, A., Vieth, W. R., and Venkatasubramanian, K. (1975). *Chemtech.* **5**, 438.

Fink, D. J., Na, T. Y., and Schultz, J. S. (1973). *Biotechnol. Bioeng.* **15**, 879.

Ford, J. R., Lambert, A. H., Cohen, W., and Chambers, R. P. (1972). *Biotechnol. Bioeng. Symp.* No. 3, p. 267.

Ford, J. R., Cohen, W., and Chambers, R. P. (1973). *AIChE Annu. Meet., 75th, Detroit* Pap. No. 23e.

Gellf, G., Thomas, D., and Broun, G. (1974). *Biotechnol. Bioeng.* **16**, 2315.

Ghose, T. K., and Kostick, J. A. (1970). *Biotechnol. Bioeng.* **12**, 921.

Giniger, M. (1973). M.S. Thesis, Dep. Chem. Biochem. Eng., Rutgers Univ., New Brunswick, New Jersey.

Goldberg, S. (1974). M.S. Thesis, Dep. Chem. Biochem. Eng., Rutgers Univ., New Brunswick, New Jersey.

Goldman, R., and Katachalski, E. (1971). *J. Theor. Biol.* **32**, 243.

Goldman, R., Silman, I. H., Kaplan, S. R., Kadem, O., and Katchalski, E. (1965). *Science* **150**, 758.

Goldman, R., Kadem, O., Silman, I. H., Kaplan, S. R., and Katchalski, E. (1968). *Biochemistry* **7**, 486.

Goldman, R., Goldstein, L., and Katchalski, E. (1971). *In* "Biochemical Aspects of Reactions on Solid Supports" (G. R. Stark, ed.), p. 1. Academic Press, New York.

Goldstein, L. (1969). *In* "Fermentation Advances" (D. Perlman, ed.), pp. 391–425. Academic Press, New York.

Goldstein, L., and Katchalski, E. (1968). *Z. Anal. Chem.* **243**, 375.

Goldstein, L., Levin, Y., and Katchalski, E. (1964). *Biochemistry* **3**, 1913.

Gondo, S., Sato, T., and Kusunoki, K. (1973). *Chem. Eng. Sci.* **28**, 1773.

Gregor, H. P., and Rauf, P. W. (1974). *Enzyme Technol. Dig.* **2**(3), 155.

Haas, W. R., Tavlarides, L. L., and Wnek, W. J. (1974). *AIChE J.* **20**, 707.

Hamilton, B. K., Stockmeyer, L. J., and Colton, C. K. (1973). *J. Theor. Biol.* **41**, 547.

Hamilton, B. K., Gardner, C. R., and Colton, C. K. (1974). *AIChE J.* **20**, 503.

Hamilton, W. (1970). *Can. J. Chem. Eng.* **48**, 55.

Hanesian, O., and Opalewski, J. (1967). *J. Chem. Eng. Data* **12**, 357.

Heden, C. G. (1972). *Biotechnol. Bioeng. Symp.* No. 3, 1973.

Ho, L. Y., and Humphrey, A. E. (1970). *Biotechnol. Bioeng.* **12**, 291.

Hornby, W. E., and Filippusson, H. (1970). *Biochim. Biophys. Acta* **220**, 343.

Hornby, W. E., Lilly, M. D., and Crook, E. M. (1966). *Biochem. J.* **98**, 420.

Hornby, W. E., Lilly, M. D., and Crook, E. M. (1968). *Biochem. J.* **107**, 669.

Horvath, C., and Engasser, J. M. (1973). *Ind. Eng. Chem., Fundam.* **12**, 229.

Horvath, C., and Solomon, B. A. (1972). *Biotechnol. Bioeng.* **14**, 885.

Horvath, C., Sardi, A., and Solomon, B. A. (1972). *Physiol. Chem. Phys.* **4**, 125.

Horvath, C., Sardi, A., and Woods, J. S. (1973a). *J. Appl. Physiol.* **34**, 181.

Horvath, C., Solomon, A. B., and Engasser, J. M. (1973b). *Ind. Eng. Chem., Fundam.* **12**, 431.

Horvath, C., Shendelman, L. H., and Light, R. T. (1973c). *Chem. Eng. Sci.* **28**, 375.

Hultin, H. O., Kittrell, J. R., and Laurence, R. L. (1972). *Enzyme Technol. Dig.* **1**, 46.

Hultin, H. O., Laurence, R. L., and Kittrell, J. R. (1974). *Enzyme Technol. Dig.* **2**, 156.

Jackson, R. (1968). *J. Optimization Theory Appl.* **2**(1), 92.

Kallen, R. G., Newirth, T., and Diegelman, M. (1973). *AIChE Annu. Meet., 66th, Philadelphia* Pap. No. 64d.

Karabelos, A. J., Wegner, T. H., and Hanratty, T. J. (1971). *Chem. Eng. Sci.* **26**, 1581.

Katchalski, E., Silman, I., and Goldman, R. (1971). *Adv. Enzymol.* **34**, 445.

Kato, K., Kubota, H., and Wen, C. (1970). *Chem. Eng. Prog., Symp. Ser.* **105**, 87.

Kay, G., and Lilly, M. D. (1970). *Biochim. Biophys. Acta* **198**, 276.

Kay, G., Lilly, M. D., Sharp, A. K., and Wilson, A. J. H. (1968). *Nature (London)* **217**, 642.

Kent, C. A., and Emery, A. N. (1974). *AIChE Natl. Meet., 77th, Pittsburgh* Pap. No. 11e.

Kingma, W. G. (1966). *Process Biochem.* **1**(1), 49.

Kittrell, J. R. (1970). *Adv. Chem. Eng.* **8**, 97.

Kobayashi, T., and Laidler, K. J. (1974a). *Biotechnol. Bioeng.* **16**, 77.

Kobayashi, T., and Laidler, K. J. (1974b). *Biotechnol. Bioeng.* **16**, 99.

Kobayashi, T., and Moo-Young, M. (1971). *Biotechnol. Bioeng.* **13**, 893.

Kobayashi, T., and Moo-Young, M. (1973). *Biotechnol. Bioeng.* **15**, 47.

Kunii, D., and Levenspiel, O. (1969). "Fluidization Engineering." Wiley, New York.

Kunii, D., and Suzuki, M. (1967). *Int. J. Heat Mass Transfer* **10**, 845.

Laidler, K. J., and Bunting, P. S. (1973). "Chemical Kinetics of Enzyme Action," 2nd Ed. Oxford Univ. Press (Clarendon), London and New York.

Lambert, A. H., and Chambers, P. R. (1973). *AIChE Annu. Meet., 66th, Philadelphia* Pap. No. 64c.

Lasch, J. (1973). *Mol. Cell. Biochem.* **2**, 79.

Lasch, J., and Koelsch, R. (1973). *Mol. Cell. Biochem.* **2**, 71.

Laurence, R. L., and Okay, V. (1973). *Biotechnol. Bioeng.* **15**, 217.

Laurence, R. L., Kittrell, J. R., and Hultin, H. O. (1973). *Enzyme Technol. Dig.* **2**, 7.

Leva, M. (1959). "Fluidization." McGraw-Hill, New York.

Levenspiel, O. (1972). "Chemical Reaction Engineering," 2nd Ed. Wiley, New York.

Li, N. N. (1972). *In* "Recent Developments in Separation Science" (N. N. Li, ed.), Vol. 1, p. 202. Chem. Rubber Publ. Co., Cleveland, Ohio.

Lightfoot, E. N., and Scattergood, E. M. (1965). *AIChE J.* **11**, 175.

Lilly, M. D., and Dunnill, P. (1972). *Biotechnol. Bioeng. Symp.* No. 3, p. 221.

Lilly, M. D., and Sharp, A. K. (1968). *Chem. Eng. (London)* **215**, CE12.

Lilly, M. D., Balasingham, K., Warburton, D., and Dunnill, P. (1972). *Proc. Int. Ferment. Symp., 4th, Ferment. Technol. Today* pp. 379–381.

Lilly, M. D., O'Neill, S. P., and Dunnill, P. (1973). *Biochimie* **55**, 985.

Lin, S. H. (1972). *Biophysik* **8**, 302.

McCune, L. K., and Wilhem, R. H. (1949). *Ind. Eng. Chem.* **41**, 1124.

McLaren, D., and Packer, L. (1970). *Adv. Enzymol.* **33**, 245.

Maldonado, O. (1974). Inst. Cent. Am. Invest. Technol. Ind. Personal communication.

Marconi, W., Cecere, F., Morisi, F., Penna, G. D., and Rappuoli, B. (1973). *J. Antibiot.* **26**, 228.

Marconi, W., Gulinelli, S., and Morisi, F. (1974). *Biotechnol. Bioeng.* **16**, 501.

Marsh, D., Lee, Y. K., and Tsao, G. T. (1973). *Biotechnol. Bioeng.* **15**, 483.

Martensson, K. (1974). *Biotechnol. Bioeng.* **16**, 567, 579.

Mattiasson, B., and Mosbach, K. (1971). *Biochim. Biophys. Acta* **235**, 253.

Melrose, G. J. H. (1971). *Rev. Pure Appl. Chem.* **21**, 83.

Michaels, A. S. (1968). *Chem. Eng. Prog.* **64**(12), 31.

Miyamoto, K., Fujii, T., Tamaoki, N., Okazaki, M., and Miura, Y. (1973). *J. Ferment. Technol.* **51**, 566.

Mogensen, A. O., and Vieth, W. R. (1973). *Biotechnol. Bioeng.* **15**, 467.

Mohan, R. R., and Li, N. N. (1974). *Biotechnol. Bioeng.* **16**, 513.

Moo-Young, M., and Kobayashi, T. (1972). *Can. J. Chem. Eng.* **50**, 162.

Mosbach, K. (1971). *Sci. Amer.* **224**(3), 26.

Mosbach, K., and Mattiasson, B. (1970). *Acta Chem. Scand.* **24**, 2093.

Na, H. S., and Na, T. Y. (1970). *Math. Biosci.* **6**, 25.

Newirth, T. L., Diegelman, M. A., Pye, E. K., and Kallen, R. G. (1973). *Biotechnol. Bioeng.* **15**, 1089.

Ollis, D. (1972). *Biotechnol. Bioeng.* **14**, 871.

O'Neill, S. P. (1971). *Biotechnol. Bioeng.* **13**, 493.

O'Neill, S. P. (1972a). *Rev. Pure Appl. Chem.* **22**, 133.

O'Neill, S. P. (1972b). *Biotechnol. Bioeng.* **14**, 675.

O'Neill, S. P. (1972c). *Biotechnol. Bioeng.* **14**, 201.

O'Neill, S. P. (1972d). *Biotechnol. Bioeng.* **14**, 473.

O'Neill, S. P., Wykes, J. R., Dunnill, P., and Lilly, M. D. (1971a). *Biotechnol. Bioeng.* **13**, 319.

O'Neill, S. P., Dunnill, P., and Lilly, M. D. (1971b). *Biotechnol. Bioeng.* **13**, 337.

O'Neill, S. P., Lilly, M. D., and Rowe, P. N. (1971c). *Chem. Eng. Sci.* **26**, 173.

Petersen, E. E. (1965). "Chemical Reaction Analysis." Prentice-Hall, Englewood Cliffs, New Jersey.

Pontryagin, L. S., Boltyanskii, V. G., Garnkrelidge, R. V., and Mischenko, E. F. (1962). "The Mathematical Theory of Optimal Processes" (Engl. transl. by K. N. Trirogoff). Wiley (Interscience), New York.

Porter, M. C. (1972). *Ind. Eng. Chem., Prod. Res. Dev.* **11**, 234.

Quiocho, F., and Richards, F. M. (1966). *Biochemistry* **5**, 4062.

Ramamurthy, K., and Subbaraju, K. (1973). *Ind. Eng. Chem., Process Des. Dev.* **12**, 184.

Regan, D. L., Dunnill, P., and Lilly, M. D. (1974). *Biotechnol. Bioeng.* **16**, 333.

Reynolds, J. H. (1972). U.S. Patent 3,705,084.

Richardson, J. F., and Zaki, W. N. (1954). *Trans. Inst. Chem. Eng.* **32**, 35.

Roberts, G. W., and Satterfield, C. N. (1965). *Ind. Eng. Chem., Fundam.* **4**, 289.

Robertson, C. R., and Waterland, L. R. (1973). *AIChE Annu. Meet., 66th, Philadelphia* Pap. No. 64a.

Robinson, P. J., Dunnill, P., and Lilly, M. D. (1973). *Biotechnol. Bioeng.* **15**, 603.

Rony, P. R. (1971). *Biotechnol. Bioeng.* **13**, 431.

Rony, P. R. (1972). *J. Am. Chem. Soc.* **94**, 8247.

Rovito, B. J., and Kittrell, J. R. (1973). *Biotechnol. Bioeng.* **15**, 143.

Rovito, B. J., and Kittrell, J. R. (1974). *Biotechnol. Bioeng.* **16**, 419.

Ryu, D. Y., Bruno, C. F., Lee, B. K., and Venkatasubramanian, K. (1972). *Proc. Int. Ferment. Symp., 4th, Ferment. Technol. Today* pp. 307–318.

Saini, R., and Vieth, W. R. (1975). *J. Appl. Chem. Biotechnol.* **25**, 115.

Saini, R., Vieth, W. R., and Wang, S. S. (1972). *Trans. N.Y. Acad. Sci.* **34**, 8.

Sampson, D., Hersh, L. S., Cooney, D., and Murphy, G. P. (1972). *Trans. Am. Soc. Artif. Intern. Organs* **18**, 54.

Satterfield, C. N. (1970). "Mass Transfer in Heterogeneous Catalysis." MIT Press, Cambridge, Massachusetts.

Satterfield, C. N., and Sherwood, T. K. (1963). "The Role of Diffusion in Catalysis." Addison-Wesley, Reading, Massachusetts.

Selegny, E., Broun, G., and Thomas, D. (1971). *Physiol. Veg.* **9**, 25.

Self, D. A., Kay, G., and Lilly, M. D. (1969). *Biotechnol. Bioeng.* **11**, 337.

Sharp, A. K., Kay, G., and Lilly, M. D. (1969). *Biotechnol. Bioeng.* **11**, 363.

Shuler, M. L., Aris, R., and Tsuchia, H. M. (1972). *J. Theor. Biol.* **35**, 67.

Shuler, M. L., Tsuchia, H. M., and Aris, R. (1973). *J. Theor. Biol.* **41**, 347.

Shyam, R. (1974). Ph.D. Thesis, Dep. Chem. Biochem. Eng., Rutgers Univ., New Brunswick, New Jersey.

Shyam, R., Davidson, B., and Vieth, W. R. (1975). *Chem. Eng. Sci.* **30**, 669.

Smiley, K. L. (1971). *Biotechnol. Bioeng.* **13**, 309.

Smiley, K. L., and Strandberg, G. W. (1972). *Adv. Appl. Microbiol.* **14**, 13.

Smith, J. M. (1970). "Chemical Engineering Kinetics," 2nd Ed. McGraw-Hill, New York.

Sundaram, P. V., and Hornby, W. E. (1970). *FEBS Lett.* **10**, 325.

Sundaram, P. V., and Laidler, K. J. (1971). *In* "Chemistry of the Cell Interface" (H. D. Brown, ed.), Part A, pp. 255–296. Academic Press, New York.

Sundaram, P. V., Tweedale, A., and Laidler, K. J. (1970). *Can. J. Chem.* **48**, 1498.

Tachauer, E., Cobb, J. T., and Shah, Y. T. (1974). *Biotechnol. Bioeng.* **16**, 545.

Tavlarides, L. L., Haas, W. R., Sroka, J. R., McKay, G. A., and Wnek, W. J. (1974). *Enzyme Technol. Dig.* **3**, 161.

Thomas, D., Tran, M. H., Gellf, G., Domurado, D., Paillot, B., Jacobsen, R., and Broun, G. (1972). *Biotechnol. Bioeng. Symp.* No. 3, p. 299.

Tosa, T., Mori, T., Fuse, N., and Chibata, I. (1967). *Enzymologia* **32**, 153.

Tosa, T., Mori, T., Fuse, N., and Chibata, I. (1969). *Agric. Biol. Chem.* **33**, 1047.

Tosa, T., Sato, T., Mori, T., Matuo, Y., and Chibata, I. (1973). *Biotechnol. Bioeng.* **15**, 69.

Venkatasubramanian, K., and Vieth, W. R. (1973). *Biotechnol. Bioeng.* **15**, 583.

Venkatasubramanian, K., Vieth, W. R., and Wang, S. S. (1972). *J. Ferment. Technol.* **50**, 600.

Venkatasubramanian, K., Saini, R., and Vieth, W. R. (1974a). *J. Ferment. Technol.* **52**, 268.

Venkatasubramanian, K., Vieth, W. R., and Bernath, F. R. (1974b). *In* "Enzyme Engineering" (E. K. Pye and L. B. Wingard, Jr., eds.), Vol. 2, pp. 439–445. Plenum, New York.

Vieth, W. R., and Venkatasubramanian, K. (1973). *Chemtech.* 3(11), 677.

Vieth, W. R., and Venkatasubramanian, K. (1974a). *Chemtech.* 4(1), 47.

Vieth, W. R., and Venkatasubramanian, K. (1974b). *Chemtech.* 4(5), 309.

Vieth, W. R., and Venkatasubramanian, K. (1974c). *Chemtech.* 4(7), 434.

Vieth, W. R., Gilbert, S. G., and Wang, S. S. (1972a). *Trans. N.Y. Acad. Sci.* 34, 454.

Vieth, W. R., Wang, S. S., and Gilbert, S. G. (1972b). *Biotechnol. Bioeng. Symp.* No. 3, p. 285.

Vieth, W. R., Wang, S. S., Bernath, F. R., and Mogensen, A. O. (1972c). *In* "Recent Developments in Separation Science" (N. N. Li, ed.), Vol. 1, pp. 175–201. Chem. Rubber Publ. Co., Cleveland, Ohio.

Vieth, W. R., Mendiratta, A. K., Mogensen, A. O., Saini, R., and Venkatasubramanian, K. (1973). *Chem. Eng. Sci.* 28, 1013.

Vieth, W. R., Gilbert, S. G., Wang, S. S., and Venkatasubramanian, K. (1974). U.S. Patent 3,809,613.

Wang, D. I. C., Sinskey, A. J., and Butterworth, T. A. (1970). *In* "Membrane Science and Technology" (J. E. Flinn, ed.), pp. 112–147. Plenum, New York.

Wang, S. S., and Vieth, W. R. (1973). *Biotechnol. Bioeng.* 15, 93.

Warburton, D., Dunnill, P., and Lilly, M. D. (1973). *Biotechnol. Bioeng.* 15, 13.

Waterland, L. R., Michaels, A. S., and Robertson, C. R. (1974). *AIChE J.* 20, 50.

Weetall, H. H. (1971). *Res. Dev.* 22, 18.

Weetall, H. H. (1973). *Food Prod. Dev.* 1, 46.

Weetall, H. H., and Havewala, N. B. (1972). *Biotechnol. Bioeng. Symp.* No. 3, p. 221.

Weetall, H. H., and Hersh, L. S. (1970). *Biochim. Biophys. Acta* 206, 54.

Weetall, H. H., and Messing, R. A. (1972). *In* "The Chemistry of Biosurfaces" (M. L. Hair, ed.), Vol. 2, pp. 563–595. Dekker, New York.

Weetall, H. H., Havewala, N. B., Pitcher, W. H., Jr., Detar, C. C., Vann, W. P., and Yaverbaum, S. (1974a). *Biotechnol. Bioeng.* 16, 689.

Weetall, H. H., Havewala, N. B., Pitcher, W. H., Jr., Detar, C. C., Vann, W. P., and Yaverbaum, S. (1974b). *Biotechnol. Bioeng.* 16, 295.

Weetall, H. H., Havewala, N. B., Garfinkel, H. M., Buehl, W. M., and Baum, G. (1974c). *Biotechnol. Bioeng.* 16, 169.

Weibel, M. K., and Bright, H. J. (1971). *Biochem. J.* 124, 801.

Wilhem, R., and Kwauk, M. (1948). *Chem. Eng. Prog.* 44, 7.

Wilson, E. J., and Geankoplis, C. J. (1966). *Ind. Eng. Chem., Fundam.* 5, 9.

Wilson, R. J. H., Kay, G., and Lilly, M. D. (1968). *Biochem. J.* 108, 845.

Wingard, L. B., Jr. (1972). *Adv. Biochem. Eng.* 2, 1.

Witt, P. R., Jr., Sair, R. A., Richardson, R. A., and Olson, N. F. (1970). *Brewers Dig.* 45(10), 70.

Worthington, C. C. (1972). Can. Patent 915,979.

Woychik, J. H., and Wondolowski, M. V. (1973). *J. Milk Food Technol.* 36, 31.

Wykes, J. R., Dunnill, P., and Lilly, M. D. (1971). *Biochim. Biophys. Acta* 250, 522.

Zaborsky, O. R. (1973). "Immobilized Enzymes." Chem. Rubber Publ. Co., Cleveland, Ohio.

Zahavi, E. (1971). *Int. J. Heat Mass Transfer* 14, 835.

Zenz, F. A., and Othmer, D. F. (1960). "Fluidization and Fluid-Particle Systems." Reinhold, New York.

Industrial Applications of Immobilized Enzymes and Immobilized Microbial Cells

Ichiro Chibata and Tetsuya Tosa

Research Laboratory of Applied Biochemistry,
Tanabe Seiyaku Co. Ltd., Osaka, Japan

I. INTRODUCTION

Over the past few years, the immobilization of enzymes has been the subject of increased interest, and a number of papers on potential applications of immobilized enzymes have been published. Very recently papers on the immobilization of microbial cells for the purpose of industrial applications also have been published. However, practical industrial systems using immobilized enzymes and immobilized microbial cells have been very limited, and available information on the details of those industrial applications has been extremely sparse.

In 1969, we succeeded in the industrial application of an immobilized enzyme, i.e., immobilized aminoacylase, for continuous production of L-amino acids from acyl-DL-amino acids. This new procedure gave satisfactory results, and is said to be the first industrial application of immobilized enzymes in the world. Since then we also have carried out the industrial application of immobilized microbial cells, applying them in the continuous production of L-aspartic acid from ammonium fumarate using immobilized *Escherichia coli* of higher aspartase activity. In this review these two industrial examples are described along with some other current and potential applications.

II. INDUSTRIAL APPLICATIONS OF IMMOBILIZED ENZYMES

A. Production of L-Amino Acids by Immobilized Aminoacylase

Utilization of L-amino acids for medicine and food has been developing rapidly in recent years. For the industrial production of L-amino acids, fermentative and chemically synthetic methods are considered to have a promising future. However, chemically synthesized amino acids are optically inactive racemic mixtures of L- and D-isomers. To obtain L-amino acid from the chemically synthesized DL-form, optical resolution is necessary.

Generally, optical resolution of racemic amino acids is carried out by physicochemical, chemical, enzymic, and biological methods. Among these methods, the enzymic method using mold aminoacylase (EC 3.5.1.14) is one of the most advantageous procedures, yielding optically pure L-amino acids. The reaction catalyzed by the enzyme is shown as follows:

$$
\underset{\substack{\text{acyl-DL-amino acid}}}{\text{DL–}R\text{–}\underset{\substack{|\\\text{NHCOR}'}}{\text{CHCOOH}}} + H_2O \xrightarrow[\substack{\text{amino-}\\\text{acylase}}]{} \underset{\substack{\text{L-amino acid}}}{\text{L–}R\text{–}\underset{\substack{|\\\text{NH}_2}}{\text{CHCOOH}}} + \underset{\substack{\text{acyl-D-amino acid}}}{\text{D–}R\text{–}\underset{\substack{|\\\text{NHCOR}'}}{\text{CHCOOH}}}
$$

$$\text{---------------- racemization ----------------}$$

A chemically synthesized acyl-DL-amino acid is asymmetrically hydrolyzed by aminoacylase to give a L-amino acid and the unhydrolyzed acyl-D-amino acid. After being concentrated, both products are easily separated by the difference in their solubilities. Acyl-D-amino acid is racemized, and reused for the resolution procedure.

From 1954 to 1969, this enzymic resolution method was employed by Tanabe Seiyaku Co. Ltd., for the industrial production of several L-amino acids. The enzyme reaction was carried out batchwise by incubating a mixture containing substrate and soluble enzyme. However, this procedure had some disadvantages for industrial use. For instance, in order to isolate a L-amino acid from the enzyme reaction mixture, it was necessary to remove enzyme protein by pH and/or heat treatments. If enzyme activity remained, there resulted an uneconomical use of the enzyme. In addition, as a complicated purification procedure was necessary for removal of contaminating proteins and coloring materials, the yield of L-amino acids was lowered. Also much labor was necessary for batch operation. To overcome these disadvantages, we studied extensively the continuous optical resolution of DL-amino acids using a column packed with immobilized aminoacylase.

1. Preparation of Immobilized Aminoacylase

In this section a variety of approaches for immobilization of aminoacylase suitable for industrial purposes are discussed. The results of immobilization of aminoacylase by several methods are summarized in Table I (Chibata *et al.*, 1972).

a. Physical adsorption: An attempt to prepare the immobilized aminoacylase was carried out by physical adsorption of aminoacylase on water-insoluble carriers, such as activated carbon, aluminum oxides (acidic, neutral, and basic), and silica gel. Immobilized aminoacylases having a very low activity were obtained only in the cases of aluminum oxides of acidic and neutral forms (Tosa *et al.*, 1966). The physical adsorption method was not considered suitable for the immobilization of aminoacylase owing to low activity and low yield.

b. Ionic binding: Preparation of immobilized aminoacylase by ionic binding of the enzyme to ion-exchange derivatives was attempted. It was found that weakly basic derivatives of cellulose and Sephadex were suitable carriers for the immobilization of aminoacylase, as the resulting activity and yield of activity were relatively high. However, in the cases of weakly basic derivatives of synthetic resins and cationic derivatives of Sephadex, active preparations were not obtained (Tosa *et al.*, 1966).

c. Covalent binding: The immobilization of aminoacylase by covalent binding to various water-insoluble carriers also was carried out. Among the carriers, the use of diazotized arylaminoglass showed the highest activity; but the immobilized aminoacylase obtained was unstable, indicating that this carrier was not suitable for industrial applications. In the case of CM-cellulose azide, active preparations were not obtained; and in the cases of cyanogen bromide-activated cellulose and Sephadex, the activities and yields were very low. On the other hand, in the case of halogenoacetyl cellulose derivatives, immobilized aminoacylases were obtained in relatively high yield.

Immobilization by attaching the enzyme to water-insoluble carriers with the use of bifunctional reagents, as shown in Table I, was not

TABLE I

VARIOUS IMMOBILIZED AMINOACYLASES AND THEIR ACTIVITIES

Immobilization methods and carriers	Aminoacylase used[a] (unit[b])	Immobilized aminoacylase[a]	
		Activity (unit[b])	Yield of activity (%)
Physical adsorption			
Activated carbon	1210	0	0
Acidic aluminum oxide	1210	13	1.0
Neutral aluminum oxide	1210	10	0.8
Basic aluminum oxide	1210	0	0
Silica gel	1210	0	0
Ionic binding			
PAB-cellulose	1210	0	0
ECTEOLA-cellulose	1210	293	24.2
TEAE-cellulose	1210	623	51.5
DEAE-cellulose	1210	668	55.2
CM-Sephadex C-50	1210	0	0
SE-Sephadex C-50	1210	0	0
DEAE-Sephadex A-25	1210	713	58.9
DEAE-Sephadex A-50	1210	680	56.2
Amberlite IRC-50	1210	0	0
Amberlite IR-4B	1210	0	0
Amberlite IR-45	1210	0	0
Covalent binding			
Diazotized PAB-cellulose	1210	64	5.3
Diazotized arylaminoglass	1210	525	43.4
Diazotized Enzacryl AA	1210	44	3.6
CM-cellulose azide	1210	0	0
BrCN-activated cellulose	1210	12	1.0
BrCN-activated Sephadex	1210	15	1.2
Chloroacetyl cellulose	1210	137	11.3
Bromoacetyl cellulose	1210	339	28.0
Iodoacetyl cellulose	1210	472	39.0
Cross-linking using carrier			
AE-cellulose			
1,4-Butylene dibromide	1440	6	0.4
1,4-Butylene dichloride	1440	6	0.4
Dicyclohexyl carbodiimide	1440	17	1.2
Diiodomethane	1440	5	0.3
Glutaraldehyde	1440	8	0.6
Hexamethylene diisocyanate	1440	23	1.6
Toluene diisocyanate	1440	3	0.2
CM-cellulose			
Dicyclohexyl carbodiimide	1440	1	0.1

(Continued)

TABLE I (*Continued*)

Immobilization methods and carriers	Aminoacylase used[a] (unit[b])	Immobilized aminoacylase[a]	
		Activity (unit[b])	Yield of activity (%)
Cross-linking by bifunctional reagent			
Glutaraldehyde	1440	211	14.7
Toluene diisocyanate	1440	18	1.3
Lattice-entrapping			
Acrylamide	1000	526	52.6
HPMCP-DEAE[c]	1000	190	19.0
Encapsulation			
Nylon	1000	360	36.0
Polyurea	1000	150	15.0
Ethyl cellulose	1000	104	10.4

[a] All immobilized enzyme assays were carried out with the same degree of agitation; the native enzymes were assayed without agitation.

[b] One enzyme unit is defined as that amount of enzyme which hydrolyzes 1 μmole of acetyl-DL-methionine per hour at 37°C.

[c] Diethylaminoethyl derivative of hydroxypropiomethyl cellulose phthalate.

suitable for the immobilization of aminoacylase because of low activity and low yield.

d. Cross-linking by bifunctional reagent: The immobilization of aminoacylase was attempted by the cross-linking methods using bifunctional reagents, such as glutaraldehyde and toluene diisocyanate (Table I). In the case of glutaraldehyde, higher activity and better yield were obtained than with toluene diisocyanate, although neither agent gave activities suitable for industrial application.

e. Lattice-entrapping and encapsulation: Among lattice-entrapping and encapsulation methods, relatively active immobilized aminoacylase was obtained by entrapping the enzyme in polyacrylamide gel lattices (Mori *et al.*, 1972).

2. *Enzymic Properties of Immobilized Aminoacylases*

As described above, relatively active and stable immobilized aminoacylases were obtained by ionic binding to DEAE-Sephadex, covalent binding to iodoacetyl cellulose, and entrapment in polyacrylamide gel lattices. Thus, to select the most suitable preparation for industrial purposes, the enzymic properties of these three immobilized aminoacylases were studied and compared with those of the native enzyme (Tosa *et al.*, 1969a; Sato *et al.*, 1971; Mori *et al.*, 1972). The results are summarized in Table II.

The optimum pH of the immobilized DEAE-Sephadex-aminoacylase for hydrolysis of acetyl-DL-methionine shifted about 0.5 ~ 1.0 pH units more to the acid side than that of the native enzyme (Tosa *et al.*, 1969a). As discussed in detail by Katchalski and co-workers (Levin *et al.*, 1964; Goldstein *et al.*, 1964), this shift may be explained by the redistribution of hydrogen ions between the positively charged enzyme carrier, i.e., DEAE-Sephadex, and the surrounding aqueous medium. This shift was also observed in the case of aminoacylase entrapped in the polyacrylamide gel lattice, but the reason was not clear (Mori *et al.*, 1972).

On the optimum temperature and activation energy, a significant difference was observed between the immobilized enzymes and the native forms. The immobilized DEAE-Sephadex–aminoacylase showed the highest optimum temperature.

On the effects of metal ions, inhibitors, substrate specificity, optical

TABLE II

SUMMARY OF ENZYMIC PROPERTIES OF VARIOUS IMMOBILIZED AMINOACYLASES

Properties	Native amino-acylase[a]	Immobilized aminoacylase[a]		
		Ionic binding to DEAE-Sephadex	Covalent binding to iodoacetyl-cellulose	Entrapping by polyacrylamide
Optimum pH	7.5 ~ 8.0	7.0	7.5 ~ 8.0	7.0
Optimum temp. (°C)	60	72	55	65
Activation energy[b] (kcal/mole)	6.9	7.0	3.9	5.3
Optimum Co^{2+} (mM)[b]	0.5	0.5	0.5	0.5
K_m (mM)[b]	5.7	8.7	6.7	5.0
V_{max} (mole/hr)[b]	1.52	3.33	4.65	2.33
Heat stability (%)[c]				
60°C, 10 min	62.5	100	77.5	78.5
70°C, 10 min	12.5	87.5	62.5	34.5
Operation stability (half-life[d], days)	—	65 Days, 50°C	—	48 Days, 37°C

[a] Data for acetyl-DL-methionine.
[b] All assays done at 37°C and pH 7.0.
[c] Remaining activity.
[d] The time required for 50% of the enzyme activity to be lost.

specificity, and kinetic constants, no marked differences were observed between the immobilized enzymes and the native forms.

Heat stability of the enzyme is an important factor in the industrial application of an immobilized enzyme. Thus, the effect of temperature on the stability of the native and immobilized enzymes was tested at 37°C to 80°C for 10 min. The immobilized DEAE-Sephadex–aminoacylase showed the highest stability. In summary the immobilized aminoacylase entrapped in polyacrylamide gels was most similar to the native enzyme among the three immobilized enzymes tested.

3. Selection of Immobilized Aminoacylase Suitable for Industrial Application

For the industrial application of immobilized aminoacylases, it was necessary to satisfy many conditions. Some of these criteria for the three immobilized aminoacylases are compared in Table III. Generally the enzymes and carriers for immobilization are relatively expensive. Thus, the following two factors were rated most important for industrial purposes: (a) operational stability of the immobilized enzymes, and (b) regenerability of deteriorated immobilized enzyme columns after long periods of operation. As shown in Table III, the covalent binding to iodoacetyl cellulose was difficult, the cost was high, and the regeneration of the deteriorated immobilized enzyme column was impossible. In the case of entrapping in polyacrylamide, the immobilization cost was not so high, but the operational stability

TABLE III
CHARACTERISTICS OF IMMOBILIZED AMINOACYLASES

Characteristics	Immobilized aminoacylases		
	Ionic binding to DEAE-Sephadex	Covalent binding to iodoacetylcellulose	Entrapping by polyacrylamide
Preparation	Easy	Difficult	Medium
Enzyme activity	High	High	High
Cost of immobilization[a]	Low	High	Moderate
Binding force	Medium	Strong	Strong
Operational stability	High	—	Medium
Regeneration[b]	Possible	Impossible	Impossible

[a] Cost of immobilization is compared from the basis for unit production of L-amino acid.

[b] Regeneration of deteriorated immobilized aminoacylase column after operation for long period.

was lower than for ionic binding to DEAE-Sephadex. Thus, the ionically bound DEAE-Sephadex-aminoacylase was chosen as the most advantageous enzyme preparation for the industrial production of L-amino acids.

4. Industrial Applications of DEAE-Sephadex–Aminoacylase Complex

Conditions for the continuous production of L-amino acids using a DEAE-Sephadex-aminoacylase column were investigated in detail (Tosa *et al.*, 1967; Chibata *et al.*, 1972) and are summarized herein.

a. Preparation of immobilized DEAE-Sephadex-aminoacylase: Optimum conditions for preparation of the immobilized aminoacylase having the highest activity are as follows: At 35°C, 1000 liters of DEAE-Sephadex A-25 previously buffered with 0.1 M phosphate buffer (pH 7.0) were stirred with 1100–1700 liters of aqueous solution of native aminoacylase (334,000,000 units) for 10 hr. After filtration the DEAE-Sephadex–aminoacylase complex was washed with water and 0.2 M acetyl-DL-methionine solution. The activity of the resulting immobilized aminoacylase was 167,000–200,400 units per liter of the immobilized enzyme-support preparation. This yield of the activity corresponded to 50–60% of the native aminoacylase used for immobilization.

b. Flow rate of substrate: For the purpose of optical resolution of acyl-DL-amino acids, it was desirable that the asymmetric hydrolysis go to completion. Thus, the relationship between the flow rate of the substrate and the extent of the reaction was investigated (Chibata *et al.*, 1972). The results are shown in Fig. 1. The data show that, at the flow rates of space velocity, SV, of 2.8 hr^{-1} for acetyl-DL-methionine and SV of 2 hr^{-1} for acetyl-DL-phenylalanine, the reactions produced 100% hydrolysis of the L-form of the substrates. This result indicated that the immobilized aminoacylase column was optically specific.

c. Design of enzyme reactor: In order to design the most efficient enzyme column, the following factors were deemed most important: (i) the flow system for the substrate solution, (ii) the effect of column dimensions on the reaction rate, and (iii) the pressure drop through the enzyme column.

The reaction rate was compared for the substrate solution fed to the column either in upward or downward flow and was found to be equal for both flow directions (Tosa *et al.*, 1969b). In practice downward flow was employed so that the air bubbles that evolved from the warmed substrate solution could be easily separated and thus prevent channeling in the column.

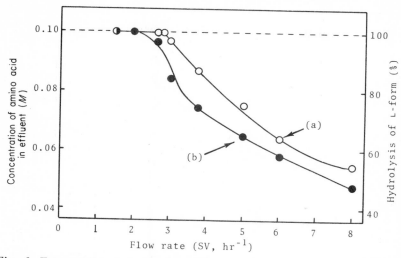

Fig. 1. Extent of hydrolysis of acetyl-DL-amino acids by DEAE-Sephadex–aminoacylase column. A solution of 0.2 M acetyl-DL-methionine (curve a, pH 7.0, containing $5 \times 10^{-4} M$ Co²⁺) or 0.2 M acetyl-DL-phenylalanine (curve b, pH 7.0, containing $5 \times 10^{-4} M$ Co²⁺) was passed through the column at flow rates of SV 1.5–8 hr⁻¹ at 50°C. (From Chibata *et al.*, 1972.)

The effect of column dimensions on the reaction rate was investigated by using several columns of different lengths at a constant column volume. Very little difference was observed in the reaction rate owing to the varied column dimensions (Tosa *et al.*, 1969b). In most cases a uniformly packed column, with substrate solution passed through smoothly, produced the same reaction rate regardless of the column dimensions. However, in some cases, such as the formation of L-aspartic acid from ammonium fumarate by the action of immobilized aspartase (Tosa *et al.*, 1973), the enzyme reaction rate was influenced by the column dimensions.

For calculating the pressure drop through the column, we investigated the application of the Kozeny–Carman equation to the aminoacylase column. The experimentally determined pressure drop was proportional to the flow rate and column length at a specified temperature. Therefore, the pressure drop through the enzyme column could be calculated from the Kozeny–Carman equation (Tosa *et al.*, 1971).

From these results an enzyme reactor system was designed for continuous production of L-amino acids with immobilized aminoacylase (Chibata *et al.*, 1972). The flow diagram is shown in Fig. 2. This system was automatically controlled and operated continuously.

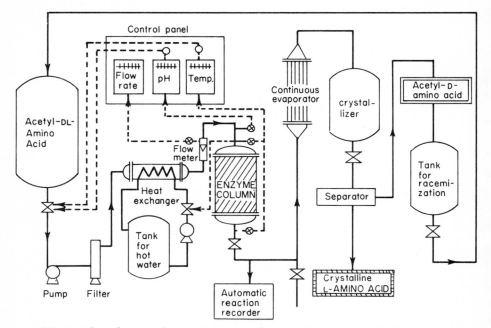

Fig. 2. Flow diagram for continuous production of L-amino acids by using immobilized aminoacylase. (From Chibata *et al.*, 1972.)

d. Stability and regeneration of the immobilized aminoacylase column: The stability of the DEAE-Sephadex-aminoacylase column in industrial operation over a long period of time is shown in Fig. 3 (Chibata *et al.*, 1972). The column maintained more than 60% of the initial activity after more than 30 days of operation, and the half-life (the time required for 50% of the enzyme activity to be lost) of the column was estimated to be about 65 days. This result indicated that the column was very stable and was satisfactory for industrial purposes. As described in Section II,A,3, the capability for regeneration of a deteriorated column was very important. In the case of an aminoacylase column deteriorated because of a long period of operation, it was completely reactivated by the addition of the amount of aminoacylase corresponding to the deteriorated activity, as shown in Fig. 3. This regenerability was one of the merits of the ionic binding method for the immobilization of the enzyme, and this was especially advantageous when the carrier and enzyme were expensive.

Since the water-insoluble carrier DEAE-Sephadex has been much more stable than we had expected, it has been used for over 5 years without significant loss of binding activity or physical decomposition.

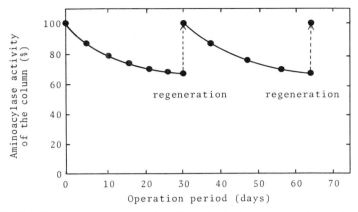

Fig. 3. Stability and regeneration of a DEAE-Sephadex–aminoacylase column. A solution of 0.2 M acetyl-DL-methionine (pH 7.0, containing $5 \times 10^{-4} M$ Co^{2+}) was applied to the column at 50°C at a flow rate of SV 2 hr^{-1}. The activity of the column was determined under standard conditions (Tosa *et al.*, 1969b). Regeneration of the deteriorated column was carried out by a recharge of aminoacylase corresponding to the deteriorated activity. (From Chibata *et al.*, 1972.)

e. Continuous production of L-amino acids: An example of the continuous production of L-methionine using a 1000-liter enzyme column is described. A solution of 0.2 M acetyl-DL-methionine (pH 7.0, containing $5 \times 10^{-4} M$ Co^{2+}) was passed through the aminoacylase column at a flow rate of SV 2 hr^{-1} (2000 liters/hr) at 50°C. From the effluent, L-methionine was isolated as follows. Two thousand liters of the effluent were evaporated, and the separated crude L-methionine was collected by centrifugation and recrystallized from water. The yield was 27 kg (91% of the theoretical); $[\alpha]_D^{25} = +23.4°$ ($c = 3$ in 1N HCl). After separation of the crude L-methionine, the acetyl-D-methionine in the mother liquor was heated at 60°C with acetic anhydride for racemization. The reaction mixture was adjusted to pH 1.8; and the separated acetyl-DL-methionine was collected and reused as substrate. The yield was 36 kg (94% of the theoretical).

Several examples of the production of L-amino acids are summarized in Table IV, which shows the space velocity and the theoretical yield for each amino acid produced in the 1000-liter aminoacylase column.

f. Economic aspects of production of L-amino acids by immobilized aminoacylase: A typical comparison of the cost for production of L-amino acids is shown in Fig. 4 (Chibata *et al.*, 1972). With the immobilized enzyme, the purification procedure for the reaction product became simpler and the yield was higher than in the case of the

TABLE IV
PRODUCTION OF L-AMINO ACIDS BY DEAE-SEPHADEX–AMINOACYLASE
COLUMN OF 1000-LITER VOLUME

L-Amino acids	Space velocity (hr^{-1})	Yield (theory) of L-amino acids	
		In 24 hr (kg)	In 30 days (kg)
L-Alanine	1.0	214	6,420
L-Methionine	2.0	715	21,450
L-Phenylalanine	1.5	594	17,820
L-Tryptophan	0.9	441	13,230
L-Valine	1.8	505	15,150

soluble enzyme. Therefore, less substrate was required for the production of a unit amount of L-amino acid. As shown in Fig. 3, the immobilized aminoacylase was very stable. Thus, the cost of the enzyme was markedly reduced from that of the soluble enzyme. In the case of the immobilized enzyme, the process was automatically controlled.

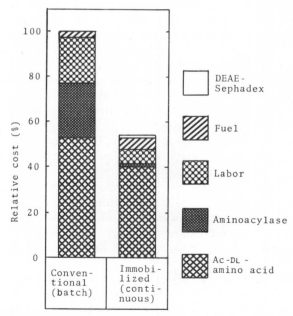

Fig. 4. Comparison of relative cost for industrial production of L-amino acids. (From Chibata *et al.*, 1972.)

Therefore, the labor cost was also dramatically reduced. The overall operating cost of the immobilized enzyme process was about 60% of that of the conventional batch process using the soluble enzyme.

B. Other Current and Potential Applications of Immobilized Enzymes

Current and potential applications of immobilized enzymes in industry for transformations of organic compounds are listed in Table V. Besides immobilized aminoacylase, many other immobilized enzymes have been reported to be under investigation for industrial applications, but accurate information is not available. It is reported (*Chem. Age Int.*, 1974; *Chem. Eng. News*, 1974) that immobilized penicillin amidase is industrially employed in the United States and in Europe for hydrolysis of the side chain of penicillin to produce 6-aminopenicillanic acid used for the production of synthetic penicillins. Furthermore, these journals and Smiley and Strandberg (1972) and Davis (1974) state that in the United States continuous production of fructose from glucose by a column packed with immobilized glucose isomerase is being studied on a pilot-plant scale.

In Japan, glucoamylase has been immobilized by several methods for the purpose of industrial production of glucose from soluble starch (Tominaga *et al.*, 1969; Maeda *et al.*, 1970). Glucose isomerase has been immobilized by ionic binding to DEAE-Sephadex; and the continuous isomerization of glucose to fructose has been reported by Tsumura and Ishikawa (1967). However, these procedures are not yet industrialized in Japan.

An interesting process employing a two-enzyme system has been patented by Toray Inds. Inc. of Japan for the production of L-lysine

TABLE V

CURRENT AND POTENTIAL TRANSFORMATIONS OF ORGANIC COMPOUNDS
IN INDUSTRY BY IMMOBILIZED ENZYMES

Enzymes	Transformation of organic compounds
Aminoacylase	Optical resolution of DL-amino acids
α-Amylase, and glucoamylase	Conversion of starch to glucose
Glucose isomerase	Conversion of glucose to fructose
β-Galactosidase	Hydrolysis of lactose in milk or whey
Penicillin amidase	Production of 6-aminopenicillanic acid from penicillin
Steroid-modifying enzymes	Steroid modification
α-Amino-ε-caprolactam hydrolase and racemase	Production of L-lysine from DL-α-amino-ε-caprolactam

(Fukumura, 1974). According to this system, L-lysine is produced from DL-α-amino-ϵ-caprolactam, which is easily synthesized from cyclohexene, a by-product of nylon synthesis, by the actions of α-amino-ϵ-caprolactam racemase and L-α-amino-ϵ-caprolactam hydrolase as follows:

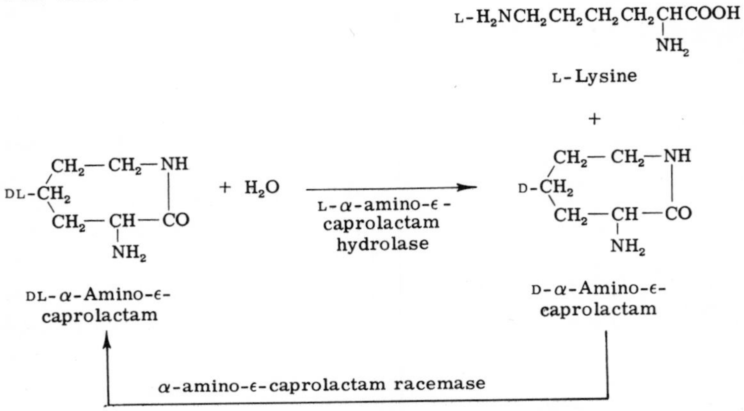

L-$H_2NCH_2CH_2CH_2CH_2CHCOOH$
        |
       NH_2

L - Lysine

+

DL-α-Amino-ϵ-caprolactam $+ H_2O$ $\xrightarrow{\text{L-}\alpha\text{-amino-}\epsilon\text{-caprolactam hydrolase}}$ D-α-Amino-ϵ-caprolactam

α-amino-ϵ-caprolactam racemase

These two enzymes can be immobilized by ionic binding to an anion-exchange polysaccharide and used for production of L-lysine in a column or in a batch system.

III. APPLICATIONS OF IMMOBILIZED MICROBIAL CELLS

A. Production of L-Aspartic Acid by Immobilized *Escherichia coli*

L-Aspartic acid is widely used in medicines and as a food additive and has been industrially produced by fermentative or enzymic methods from ammonium fumarate using the action of aspartase by the following reaction:

$$HOOCCH{=}CHCOOH + NH_3 \underset{\text{aspartase}}{\rightleftharpoons} \text{L-}HOOCCH_2CHCOOH$$
$$NH_2$$

Fumaric acid   L-Aspartic acid

This reaction has been carried out using batch procedures, which have disadvantages for industrial purposes, just as in the case of the native aminoacylase process. Thus, we have studied extensively the continuous production of L-aspartic acid using a column packed with immobilized aspartase (Tosa *et al.*, 1973). As the aspartase is an intracellular enzyme, it was necessary to extract the enzyme from microbial cells before immobilization. Extracted intracellular enzyme is generally unstable, and most of the immobilization methods we tried resulted in low activity and poor yield from unit intact cells. Although entrapment into a polyacrylamide gel lattice gave relatively active immobilized aspartase, its operational stability was not satisfactory, i.e., the half-life was 27 days at 37°C.

Therefore, this immobilized aspartase was considered not satisfactory for the industrial production of L-aspartic acid. If the microbial cells could be immobilized directly, these disadvantages might be overcome. From these points of view, we studied the immobilization of whole microbial cells (Chibata *et al.*, 1974; Tosa *et al.*, 1974), and we succeeded in industrialization of this technique in 1973.

1. Immobilization of Escherichia coli

In this section a variety of approaches for immobilization of *E. coli* for industrial applications are presented. The following methods for immobilization of *Escherichia coli* ATCC 11303 having high aspartase activity were tested:

a. Lattice-entrapping: *E. coli* cells were entrapped in a polyacrylamide gel lattice by employing acrylamide monomer and *N,N'*-methylenebisacrylamide (BIS) for lattice cross-linking.

b. Cross-linking: *E. coli* cells were cross-linked by bifunctional reagents, such as glutaraldehyde or 2,4-toluene diisocyanate.

c. Encapsulation: *E. coli* cells were encapsulated by polyurea produced from 2,4-toluene diisocyanate and hexamethylenediamine.

Among these methods, active immobilized *E. coli* cells were obtained by entrapping the cells in a polyacrylamide gel lattice and by cross-linking the cells with glutaraldehyde. The more active preparation was obtained in the former case. To prepare the most efficient immobilized microbial cells by this polyacrylamide gel method, the type and concentration of bifunctional reagents for lattice cross-linking and the concentration of acrylamide monomer were important factors. Table VI shows the results of immobilization of *E. coli* using various bifunctional reagents for lattice formation. The activity of the immobilized *E. coli* was almost the same except with ethylene urea bisacrylamide and 1,3,5-triacryloyl-*s*-triazine. BIS was chosen because it was commercially available at low cost.

The concentrations of acrylamide monomer and of BIS and the amount of cells to be entrapped were investigated. As a result, a set of optimum conditions for immobilization of *E. coli* were selected.

E. coli cells (10 kg, wet weight) collected from cultured broth were

TABLE VI

IMMOBILIZATION OF *Escherichia coli* CELLS BY THE USE OF VARIOUS
BIFUNCTIONAL REAGENTS FOR ACRYLAMIDE LATTICE CROSS-LINKING[a]

Bifunctional reagents	Aspartase activity (μmoles/hr)	Yield of activity (%)
N,N'-Methylenebisacrylamide (BIS)	1220	67.0
N,N'-Propylenebisacrylamide	1104	60.7
Diacrylamide dimethylether	1048	57.6
1,2-Diacrylamide ethyleneglycol	1136	62.4
N,N'-Diallyl tartardiamide	1320	72.5
Ethylene urea bisacrylamide	128	7.0
1,3,5-Triacryloyl hexahydro-s-triazine	128	7.0

[a] From Chibata *et al*. (1974).

suspended in 40 liters of physiological saline. To this suspension were added 7.5 kg of acrylamide, 0.4 kg of BIS, 5 liters of 5% β-dimethylaminopropionitrile, and 5 liters of 2.5% potassium persulfate. The mixture was allowed to stand at below 40°C for ~10–15 min, and the resulting stiff gel was made into 2–3-mm cubes. The aspartase activity of immobilized *E. coli* obtained under the optimum conditions was 12,000–16,000 μmoles/hr per gram of wet cells.

An interesting phenomenon was observed with these cells. When the immobilized *E. coli* cells were suspended at 37°C for 24–48 hr in substrate solution, i.e., 1 *M* ammonium fumarate solution, pH 8.5, containing 1 m*M* Mg^{2+}, the activity increased 9–10 times. This phenomenon was also recognized when intact cells were incubated in the same solution. This activation was considered to be either adaptive formation of aspartase-protein in the presence of substrate or increase of membrane permeability for substrate and/or product due to autolysis of *E. coli* cells in the gel lattice.

Thus, in order to investigate the adaptive formation of the enzyme, fresh cells or fresh immobilized cells were incubated in 1 *M* substrate solution for 48 hr at 37°C in the absence or in the presence of chloramphenicol at concentrations that completely inhibited protein synthesis. The results shown in Table VII indicate that the enzyme activities increased, even in the presence of chloramphenicol. Therefore, this activation was considered not to be the result of protein synthesis but to be due to increased permeability caused by autolysis of *E. coli* cells in the gel lattice. This was also confirmed by electron micrographs of immobilized *E. coli* cells, after activation, which indicated that lysis of cells had occurred (Chibata *et al*., 1974). Of course, even when lysis of

TABLE VII

EFFECT OF CHLORAMPHENICOL ON ACTIVATION OF INTACT AND
IMMOBILIZED *Escherichia coli* CELLS[a]

Concentration of chloramphenicol (μg/ml)	Aspartase activity (μmoles/hr)			
	Before activation		After activation	
	Intact cells	Immobilized cells	Intact cells	Immobilized cells
0	1700	1310	11,290	12,200
50	1700	1310	11,420	12,340
100	1700	1310	11,510	12,240

[a] From Sato, Tosa, and Chibata (unpublished data).

the cells did occur, the aspartase should not leak out from the gel lattice. But the substrate, ammonium fumarate, and the product, L-aspartate, passed more easily through the gel lattice.

In Fig. 5 comparative results of the aspartase activities of various enzyme preparations per unit of intact cells are summarized. The results show that the immobilized cells were advantageous in comparison with immobilized aspartase for the industrial production of L-aspartic acid. The aspartase activity of autolyzed cell suspensions was as high as that of the activated immobilized cells. However, the latter was much more stable than the former for the industrial produc-

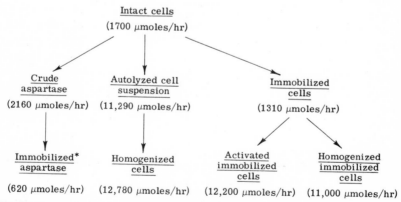

Fig. 5. Comparison of aspartase activity of various enzyme preparations per unit of intact cells. One gram (packed wet weight) of intact cells corresponds to 0.2 gm of dried cells. Numerical values in parentheses are aspartase activities obtained from 1 gm of intact cells. * Crude aspartase was immobilized by entrapping it in a polyacrylamide gel lattice.

tion of L-aspartic acid from ammonium fumarate, as shown in Fig. 6, and continuous operation for long periods of time became possible by using the immobilized cells.

2. Enzymic Properties of Immobilized *Escherichia coli* Cells

The enzymic properties of immobilized *E. coli* cells are summarized for pH, metal ions, and temperature (Chibata *et al.*, 1974). The pH activity profile of immobilized cells on the formation of L-aspartic acid from ammonium fumarate was different from that for intact cells. The immobilized cells showed an optimal activity at pH 8.5 (same as native aspartase), whereas the optimal pH of the intact cells was 10.5.

On the effects of metal ions, although the native and immobilized aspartase were activated by Mn^{2+}, the formation of L-aspartic acid by intact and immobilized cells was not accelerated by this metal ion. On the other hand, investigation of the protective effects of various metal ions against heat inactivation of intact and immobilized cells showed that Ba^{2+}, Ca^{2+}, Mg^{2+}, $Mn,^{2+}$ and Sr^{2+} protected the intact and immobilized cells against thermal inactivation. Furthermore, these protective or stabilizing effects of bivalent metal ions also were investigated using the continuous column process for formation of L-aspartic acid. It was found that Ca^{2+}, Mg^{2+}, and Mn^{2+} had a stabilizing effect on

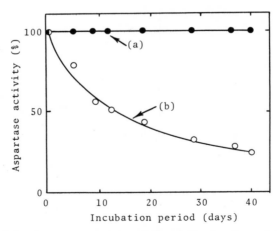

Fig. 6. Stability of aspartase activity of immobilized cells and intact cells. Immobilized cells or intact cells were incubated with a solution of 1 M ammonium fumarate (pH 8.5, containing 1 mM Mg^{2+}) at 37°C for 40 days. As appropriate intervals their remaining activities were determined under standard conditions. Curve a: immobilized cells; curve b: intact cells.

the aspartase activity of immobilized *E. coli* during the continuous enzyme process (Tosa *et al.*, 1974).

The effect of temperature on the formation of L-aspartic acid by the immobilized cells was compared with that of the intact cells, and it was found that the optimal temperature was 50°C in both preparations.

3. Industrial Applications of Immobilized Escherichia coli

Conditions for continuous production of L-aspartic acid from ammonium fumarate by using a column packed with immobilized *E. coli* cells were investigated in detail (Tosa *et al.*, 1974). Unless otherwise noted, the continuous enzyme reaction was carried out by passing 1 M ammonium fumarate containing 1 mM Mg^{2+}, pH 8.5, into the immobilized cell column at 37°C.

a. Flow rate of substrate: A substrate solution was passed through the immobilized cell column at various flow rates, and the rate of formation of L-aspartic acid was measured. The data showed that the flow rate or space velocity of 0.8 hr^{-1} was the maximal flow rate that enabled the complete conversion of ammonium fumarate to L-aspartic acid.

b. Stability of the immobilized cell column: The stability of the immobilized cell column was investigated by continuously passing substrate solution through the column for a long period of time at various temperatures (Fig. 7). Figure 7 shows that the deterioration of

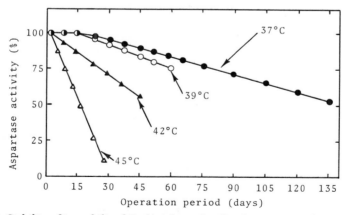

Fig. 7. Stability of immobilized *Escherichia coli* cell column at various temperatures. A solution of 1 M ammonium fumarate (pH 8.5, containing 1 mM Mg^{2+}) was applied to the column at a flow rate of SV 0.6 hr^{-1} at 37°C, 39°C, 42°C, or 45°C for 30–135 days. The activity of the column was determined under standard conditions.

the activity depended on temperature and that the immobilized cell column was very stable. The half-life of the column was estimated to be 120 days at 37°C. If the substrate solution was passed through the column at lower temperature, the half-life of the column was prolonged.

   c. Continuous production of L-aspartic acid. The aspartase reactor system using immobilized *E. coli* cells was essentially the same as that for the immobilized aminoacylase system shown in Fig. 2. A solution of 1 *M* ammonium fumarate containing 1 m*M* MgCl$_2$, pH 8.5, was passed through the immobilized *E. coli* cell column at a flow rate of SV 0.6 hr^{-1} at 37°C. The effluent, 2400 liters, was adjusted to pH 2.8 with 60% H$_2$SO$_4$ at 90°C and then cooled at 15°C for 2 hr. The L-aspartic acid that crystallized out was collected by centrifugation and washed with water. This product was pure without recrystallization; $[\alpha]_D{}^{25} = +25.5$ ($c = 8$ in 6 *N* HCl), and the yield was 3048 kg (95% of theoretical).

   d. Economic aspect of production of L-aspartic acid by immobilized E. coli: A comparison of the costs for production of L-aspartic acid by the conventional batch process using intact cells (Kisumi *et al.*, 1960) and the continuous process using immobilized cells is shown in Fig. 8. As the immobilized cells were very stable, the

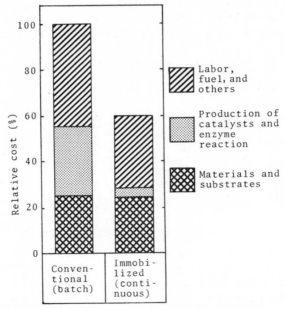

Fig. 8. Comparison of relative cost for industrial production of L-aspartic acid using intact or immobilized *Escherichia coli*.

cost for preparation of the catalyst was greatly reduced, i.e., about one-ninth that of intact cells. In the case of immobilized cells, the process was automatically operated; therefore, the labor cost was also reduced, i.e., about 30% less than that for intact cells. As a result, the overall production cost of the immobilized cell system was about 60% of that of the conventional batch process using intact cells. Furthermore, as stated later, the procedure employing immobilized cells was advantageous from the standpoint of plant waste treatment.

Therefore, it is clear that this new technique is very efficient and superior to the conventional fermentative or enzymic technique. We have been industrially operating this new system in our plant for automatic and continuous production of L-aspartic acid since the autumn of 1973. This is considered to be the first industrial application of immobilized microbial cells in the world.

B. Other Applications of Immobilized Microbial Cells

In addition to the continuous production of L-aspartic acid, we studied efficient continuous methods for the production of useful organic compounds, such as L-citrulline (Yamamoto *et al.*, 1974a), urocanic acid (Yamamoto *et al.*, 1974b), 6-aminopenicillanic acid (Sato *et al.*, 1976), and L-malic acid (Yamamoto *et al.*, 1976) by using immobilized microbial cells.

1. Production of L-Citrulline by Immobilized Pseudomonas putida

L-Citrulline is used for medicines and has been produced from L-arginine by the action of microbial L-arginine deiminase as follows:

$$\text{L-H}_2\text{NCNHCH}_2\text{CH}_2\text{CH}_2\text{CHCOOH} + \text{H}_2\text{O} \xrightarrow[\text{deiminase}]{\text{L-arginine}}$$

$$\underset{\text{L-Arginine}}{\overset{\text{NH} \qquad\qquad\qquad \text{NH}_2}{\|}}$$

$$\text{L-H}_2\text{NCONHCH}_2\text{CH}_2\text{CH}_2\text{CHCOOH} + \text{NH}_3$$

$$\underset{\text{L-Citrulline}}{\overset{}{\text{NH}_2}}$$

In most cases, however, a part of the L-citrulline formed is further converted to L-ornithine by the action of ornithine transcarbamylase. On the other hand, we found that *Pseudomonas putida* has a higher activity of L-arginine deiminase and no activity of ornithine transcarbamylase. The microorganism has been industrially used for the production of L-citrulline (Kakimoto *et al.*, 1971). As these procedures have been carried out in a batch process by incubating a mixture of substrate and fermented broth, they have had some disadvantages for industrial uses.

In order to produce L-citrulline more advantageously, the immobilization of *P. putida* ATCC 4359 having higher L-arginine deiminase activity was investigated, and its immobilization was carried out using the polyacrylamide gel method as in the case of the *E. coli* cells (Yamamoto *et al.*, 1974a). The cells were tightly entrapped in the polyacrylamide gel lattice and did not leak out from the gel lattice after repeated washings with saline and substrate solutions.

Some of the enzymic properties of the immobilized cell L-arginine deiminase were investigated and compared with those of the intact cells. A marked difference was observed between the permeability of substrate or product through the cell wall for the intact and immobilized cells. That is, formation of L-citrulline by the intact cells was scarcely observed in the absence of the surfactant, cetyltrimethylammonium bromide, whereas formation occurred by the immobilized cells without the reagent. This phenomenon indicated that the cell wall of intact cells was a barrier for L-arginine or L-citrulline, and that immobilization of the cells possibly removed this barrier.

No difference was observed between pH activity curves of the intact and immobilized cells on the formation of L-citrulline; and the optimal pH was 5.5–6.0 for both cells. The optimal temperature for the formation of L-citrulline was 37°C for intact cells and 55°C for the immobilized cells. The heat stability of the L-arginine deiminase activity of *P. putida* was considerably increased by immobilization.

The conditions for the continuous production of L-citrulline by using a column packed with the immobilized *P. putida* cells were studied. When an aqueous solution of 0.5 *M* L-arginine hydrochloride (pH 6.0) was passed through the column at 37°C and at flow rates below SV 0.26 hr^{-1}, the reaction was completed. From the column effluent pure L-citrulline was obtained with a total yield of 96% by the concentration and the ion-exchange resin treatments. The stability of the L-arginine deiminase activity of the immobilized cell column was very high, and the half-life of the column was estimated to be about 140 days at 37°C. As described above, this technique was considered to be more efficient for the production of L-citrulline than the batch method using microbial broth.

2. Production of Urocanic Acid by Immobilized Achromobacter liquidum

Urocanic acid is used as a sun-screening agent in the pharmaceutical and cosmetic fields and is produced from L-histidine by the action of microbial L-histidine ammonia-lyase as follows:

$$\underset{\text{L-Histidine}}{\text{L-HC}=\text{C}\overset{\text{CH}_2\text{CHCOOH}}{\underset{\text{NH}_2}{\diagup}} \quad\underset{\text{N}\diagdown\text{C}\diagup\text{NH}}{}\quad\overset{H}{}} \xrightarrow[\text{ammonia-lyase}]{\text{L-histidine}} \underset{\text{Urocanic acid}}{\text{HC}=\text{C}\overset{\text{CH}=\text{CHCOOH}}{\diagup} \quad\text{N}\diagdown\text{C}\diagup\text{NH}\quad\overset{H}{}} \quad + \quad \text{NH}_3$$

urocanase

$$\underset{\text{Imidazolone propionic acid}}{\text{HO}\diagdown\text{C}=\text{C}\overset{\text{CH}_2\text{CH}_2\text{COOH}}{\diagup}\quad\text{N}\diagdown\text{C}\diagup\text{NH}\quad\overset{H}{}}$$

The enzyme is widely distributed in many bacteria, and *Achromobacter liquidum* was found to be one of the most suitable enzyme sources for industrial production of urocanic acid (Shibatani *et al.*, 1974). This enzymic process also has some disadvantages for commercial production of the acid since the procedure is carried out by batch incubation of a mixture of L-histidine and intact cells. To develop a more efficient method, the continuous production of urocanic acid was investigated using immobilized microbial cells having higher L-histidine ammonia-lyase activity (Yamamoto *et al.*, 1974b).

Several microorganisms having high enzyme activity were immobilized in a polyacrylamide gel lattice. *Achromobacter liquidum* IAM 1667 was found to show the highest activity after immobilization. Although the organism had urocanase activity, this undesired activity was removed by a simple heat treatment (70°C, 30 min) before immobilization of the cells (Shibatani *et al.*, 1974).

Enzymic properties of the immobilized *A. liquidum* cells were compared with those of the intact cells. No difference was observed between optimal pH and optimal temperature on the formation of urocanic acid for the intact and immobilized cells. The permeability of substrate or product through the cell wall was increased by immobilization of the cells as in the case of immobilized *P. putida* for the production of L-citrulline from L-arginine.

By using a column packed with the immobilized *A. liquidum* cells, the conditions for the continuous production of urocanic acid were investigated. When an aqueous solution of 0.25 M L-histidine (pH 9.0) containing 1 mM Mg^{2+} was passed through the column at flow rates below SV 0.06 hr^{-1}, L-histidine was completely converted to urocanic acid. From the column effluent pure urocanic acid was crystallized by merely adjusting the pH to 4.7. The yield was 91% of the theoretical.

The enzyme activity of the column was very stable in the presence of Mg^{2+}, and its half-life was about 180 days at 37°C.

This process was also more advantageous for the industrial production of urocanic acid than the batch process using extracted enzyme or microbial broth.

C. Current and Proposed Applications of Immobilized Microbial Cells and Future Prospects

Current and proposed industrial transformations of organic compounds by immobilized microbial cells are listed in Table VIII. Besides our studies previously described, we have been studying the immobilization of *E. coli* having high penicillin amidase activity and *Brevibacterium ammoniagenes* having high fumarase activity for continuous production of 6-aminopenicillanic acid from penicillin and L-malic acid from fumaric acid, respectively. As satisfactory results for industrial production have been obtained by the immobilized microbial cells, the detailed conditions will be published elsewhere.

Further, Mosbach and Larsson (1970) immobilized *Curvularia lunata* belonging to fungus in a polyacrylamide gel lattice, and found that the immobilized cells could convert 11-deoxycortisol to predonisolone by the action of 11-β-hydroxylase contained in the cells.

Takasaki and Kanbayashi (1969) immobilized *Streptomyces* sp. having higher glucose isomerase activity by heat treatment, and investigated the conditions for conversion of glucose to fructose by the immobilized cells. That is, when the cells were heated at 60–85°C for about 10 min, glucose isomerase was fixed inside the cells and did not leak out from the cells, even if the cells were incubated under the conditions for enzyme reaction. Into a column packed with the immobilized *Streptomyces* sp. cells, 40% glucose solution (pH 8.0) containing 5 mM Mg^{2+} and 1 mM Co^{2+} was fed by the method of upward flow at 70°C. The continuous reaction could be carried out for 15 days at the average isomerization rate of 40%. However, it is understood that there are a number of problems yet to be solved for industrial application of this column technique, and the industrial production of fructose from glucose presently is carried out by a batch reaction process with reuse of the immobilized cells two or three times. Therefore, the preparation of better stabilized immobilized cells and the development of a continuous reactor system suitable for industrial uses are expected.

Recently, Vieth of Rutgers University immobilized *Streptomyces venezuelae* cells containing glucose isomerase by attaching them to

TABLE VIII

TRANSFORMATION OF ORGANIC COMPOUNDS BY IMMOBILIZED MICROBIAL CELLS

Microbial cells	Methods of immobilization	Enzyme system	Products	References
Achromobacter liquidum	Entrapping by polyacrylamide	Histidine ammonia-lyase	Urocanic acid	Yamamoto et al. (1974b)
Brevibacterium ammoniagenes	Entrapping by polyacrylamide	Fumarase	L-Malic acid	Yamamoto et al. (1976)
Curvularia lunata	Entrapping by polyacrylamide	11-β-Hydroxylase	Prednisolone	Mosbach and Larsson (1970)
Escherichia coli	Entrapping by polyacrylamide	Aspartase	L-Aspartic acid	Chibata et al. (1974); Tosa et al. (1974)
Escherichia coli	Entrapping by polyacrylamide	Penicillin amidase	6-Aminopenicillanic acid	Sato et al. (1976)
Pseudomonas putida	Entrapping by polyacrylamide	Arginine deiminase	L-Citrulline	Yamamoto et al. (1974a)
Pseudomonas mephitica	Heat treatment	Lipase	Hydrolyzate of tributyrin and triacetin	Kosugi and Suzuki (1973)
Streptomyces sp.	Heat treatment	Glucose isomerase	Fructose	Takasaki and Kambayashi (1969)
Aspergillus oryzae	Entrapping by cellulose nitrate	Aminoacylase	L-Tryptophan	Leuschner (1966)
Fungal spore	Ionic binding to ECTEOLA-cellulose	Invertase	Glucose and fructose	Johnson and Ciegler (1969)
Lichen	Entrapping by polyacrylamide	Depsidase and orsellinic acid decarboxylase	Orcinol and orcinol mono-methyl ether	Mosbach and Mosbach (1966)
Brevibacterium ammoniagenes	Entrapping by polyacrylamide	Multiple	Coenzyme A	Shimizu et al. (1975)
Corynebacterium glutamicum	Entrapping by polyacrylamide	Multiple	L-Glutamic acid	Slowinski and Charm (1973)

collagen by physicochemical bonds to form a whole cell membrane and used this for the isomerization of glucose to fructose (Venkata-subramanian *et al.*, 1974).

Current industrial applications for continuous enzyme-catalyzed reactions using immobilized microbial cells are carried out primarily by the action of single enzymes. However, many chemical substances, especially in fermentative methods, usually are produced by the action of several enzymes. The immobilization of microbial cells for multienzyme reactions has been attempted, as shown at the bottom of Table VIII. Slowinski and Charm (1973) investigated L-glutamic acid formation from glucose, an organic nitrogen source and inorganic ammonium salts by using *Corynebacterium glutamicus* (a glutamic acid-producing bacteria) entrapped in a polyacrylamide gel lattice. From this work they estimated that the immobilized cells probably could be used in a column with continuous processing. However, there were a number of difficulties to be overcome, especially the supply of air or oxygen into the column. In addition Shimizu *et al.* (1975) have investigated immobilized *Brevibacterium ammoniagenes* for continuous production of coenzyme A from pantothenic acid, L-cysteine, and ATP.

As described above, several kinds of microbial cells having an enzyme of higher activity can be easily immobilized and stabilized by the entrapping method using polyacrylamide gel. Continuous enzyme processing by the immobilized microbial cells will be advantageous in the following cases: (1) when the enzymes are intracellular, (2) when the enzymes extracted from microbial cells are unstable, (3) when the enzymes are unstable during and after immobilization, (4) when the microorganism contains no other enzymes that catalyze interfering side reactions or when those interfering enzymes can be readily inactivated or removed, and (5) when the substrates and products are not high molecular weight compounds and can easily pass through the gel lattice.

Another aspect to be considered is the volume of liquid to be processed. For the unit production of a desired compound, the volume of fermentation broth is much smaller in the case of the continuous method using immobilized cells than in the case of conventional batch fermentative methods. Thus, the continuous process using immobilized cells is very advantageous from the point of reducing plant water pollution. Also, in the case of batch fermentative methods feedback inhibition sometimes occurs even at low concentrations of an accumulated compound; whereas this inhibition does not occur in the continuous process.

In the future, studies on immobilized microbial cells will be developed extensively as will those on immobilized enzymes. As this new technique is very efficient and superior to the conventional fermentative and enzymic methods in certain cases, it will be the subject of increased interest in the fermentative industry.

IV. CONCLUSIONS

From our experiences with industrial applications of immobilized enzymes and immobilized microbial cells, the following factors are to be considered important: (1) cost of carriers or reagents for immobilization of enzymes or microbial cells; (2) activity of immobilized enzyme and yield from native enzyme or intact cells; (3) stability of immobilized enzyme or immobilized cells during operation; (4) regenerability of the deteriorated immobilized enzyme or microbial cells after long periods of operation.

Besides these conditions, a number of factors should be considered for industrial applications of immobilized enzymes, including immobilized cells, as shown in Table IX. A continuous column system employing an immobilized enzyme is suitable in those cases where the cost of enzyme and/or the enzyme reaction rate are high. In the case of the column system, the enzyme reaction can be easily con-

TABLE IX

FACTORS TO BE CONSIDERED FOR INDUSTRIAL APPLICATIONS OF IMMOBILIZED ENZYMES

	Factors		Soluble enzyme Batch system	Immobilized enzymes	
				Batch system	Column system
Enzyme	Cost:	High	—	Suitable	Suitable
		Low	Suitable	—	—
	Reuse		Impossible	Possible	Possible
	Stability		Low	Moderate to high	High
Enzyme reaction	Control		Difficult	Difficult	Easy
	Rate:	High	—	—	Suitable
		Low	Suitable	Suitable	—
Product	Purity		Low	High	High
	Yield		Low	High	High
Equipment	Initial cost		Low	Moderate	High
	Automation		Difficult	Difficult	Easy
	Applicability		High	High	Moderate
Running cost	Labor cost		High	Moderate	Low
Scale merit			Low	Low	High

trolled, automation of the process is readily performed, and the running cost is reduced. By employing an immobilized enzyme, a product of higher purity is obtained in higher yield. However, for the column system the initial cost of equipment is relatively high and the applicability for use with a wide variety of enzyme–substrate combinations is lower than with a batch system; but the possibility for merit due to scale-up of equipment can be anticipated.

In the future, if the immobilization of multienzyme systems can be developed, then energy generation and oxidation–reduction reactions can be efficiently and easily carried out. Immobilized enzyme systems and immobilized microbial systems are expected to become highly advantageous bioreactors or catalysts for industrial production of many useful chemical compounds such as steroids, antibiotics, peptides, nucleic acids, and coenzymes.

REFERENCES

Chem. Age Int. (1974). **108**(Jan. 4/11), p. 19.
Chem. Eng. News (1974). Feb. 4, p. 14.
Chibata, I., Tosa, T., Sato, T., Mori, T., and Matuo, Y. (1972). *Proc. Int. Ferment. Symp., 4th, Ferment. Technol. Today* pp. 383–389.
Chibata, I., Tosa, T., and Sato, T. (1974). *Appl. Microbiol.* **27**, 878–885.
Davis, J. C. (1974). *Chem. Eng.* Aug. 19, 52–54.
Fukumura, T. (1974). Jpn. Patent 753,051.
Goldstein, L., Levin, Y., and Katchalski, E. (1964). *Biochemistry* **3**, 1913–1919.
Johnson, D. E., and Ciegler, A. (1969). *Arch. Biochem. Biophys.* **130**, 384–388.
Kakimoto, T., Shibatani, T., Nishimura, N., and Chibata, I. (1971). *Appl. Microbiol.* **22**, 992–999.
Kisumi, M., Ashikaga, Y., and Chibata, I. (1960). *Bull. Agric. Chem. Soc. Jpn.* **24**, 296–305.
Kosugi, Y., and Suzuki, H. (1973). *J. Ferment. Technol.* **51**, 895–903.
Leuschner, F. (1966). Ger. Patent 1,227,855.
Levin, Y., Pecht, M., Goldstein, L., and Katchalski, E. (1964). *Biochemistry* **3**, 1905–1913.
Maeda, H., Miyado, S., and Suzuki, H. (1970). *Hakko Kyokaishi* **28**, 391–397 [*Chem. Abstr.* **75**, 62167].
Mori, T., Sato, T., Tosa, T., and Chibata, I. (1972). *Enzymologia* **43**, 213–226.
Mosbach, K., and Larsson, P. (1970). *Biotechnol. Bioeng.* **12**, 19–27.
Mosbach, K., and Mosbach, R. (1966). *Acta Chem. Scand.* **20**, 2807–2810.
Sato, T., Mori, T., Tosa, T., and Chibata, I. (1971). *Arch. Biochem. Biophys.* **147**, 788–796.
Sato, T., Tosa, T., and Chibata, I. (1976). *Eur. J. Appl. Microbiol.* **2**, 153–160.
Shibatani, T., Nishimura, N., Nabe, K., Kakimoto, T., and Chibata, I. (1974), *Appl. Microbiol.* **27**, 688–694.
Shimizu, S., Morioka, H., Tani, Y., and Ogata, K. (1975). *J. Ferment. Technol.* **53**, 77–83.
Slowinski, W., and Charm, S. E. (1973). *Biotechnol. Bioeng.* **15**, 973–979.
Smiley, K. L., and Strandberg, G. W. (1972). *Adv. Appl. Microbiol.* **15**, 13–38.

Takasaki, Y., and Kanbayashi, A. (1969). *Kogyo Gijutsuin Biseibutsu Kogyo Gijutsu Kenkyusho Kenkyu Hokoku* No. 37, 31–37 [*Chem. Abstr* **74**, 139538].

Tominaga, T., Nimi, M., and Sugihara, H. (1969). Jpn. Patent 69-1,360 [*Chem. Abstr.* **71**, 2152].

Tosa, T., Mori, T., Fuse, N., and Chibata, I. (1966). *Enzymologia* **31**, 214–224.

Tosa, T., Mori, T., Fuse, N., and Chibata, I. (1967). *Biotechnol. Bioeng.* **9**, 603–615.

Tosa, T., Mori, T., and Chibata, I. (1969a). *Agric. Biol. Chem.* **33**, 1053–1059.

Tosa, T., Mori, T., Fuse, N., and Chibata, I. (1969b). *Agric. Biol. Chem.* **33**, 1047–1052.

Tosa, T., Mori, T., and Chibata, I. (1971). *J. Ferment. Technol.* **49**, 522–528.

Tosa, T., Sato, T., Mori, T., Matuo, Y., and Chibata, I. (1973). *Biotechnol. Bioeng.* **15**, 69–84.

Tosa, T., Sato, T., Mori, T., and Chibata, I. (1974). *Appl. Microbiol.* **27**, 886–889.

Tsumura, N., and Ishikawa, M. (1967). *Nippon Shokuhin Kogyo Gakkai-Shi* **14**, 539–540 [*Chem. Abstr.* **69**, 64824].

Venkatasubramanian, K., Saini, R., and Vieth, W. R. (1974). *J. Ferment. Technol.* **52**, 268–278.

Yamamoto, K., Sato, T., Tosa, T., and Chibata, I. (1974a). *Biotechnol. Bioeng.* **16**, 1589–1599.

Yamamoto, K., Sato, T., Tosa, T., and Chibata, I. (1974b). *Biotechnol. Bioeng.* **16**, 1601–1610.

Yamamoto, K., Tosa, T., Yamashita, K., and Chibata, I. (1976). *Eur. J. Appl. Microbiol.* in press.

Subject Index

A

Achromobacter liquidum, immobilized, urocanic acid production by, 350–353

Acid anhydride polymers, protein coupling to, 60

Acid phosphatase, immobilization of, 26

Acrylic polymers, as supporters for enzyme immobilization, 50–54, 82–83, 85–87

Acyl azide polymers, protein coupling to, 60

Acylation reactions, for enzyme covalent coupling, 43–62

Adsorption methods
  for enzyme immobilization, 24, 25–30
  list of adsorbents, 28–29

Affinity chromatography, of enzymes, 5

Affinity supports, enzyme adsorption by, 26

Agarose,
  structure of, 90
  as support for enzyme immobilization, 47–48, 90, 92

Agarose derivatives, as enzyme adsorbents, 29

Alanine, copolymers of, as supports for enzyme immobilization, 56

Aldehydes, polymeric, in enzyme covalent coupling, 69–70

Alginic acid, as support for enzyme immobilization, 49

Alkaline phosphatase
  immobilization of, 6, 26, 29
  diffusion, 184

Alkylation, for enzyme covalent coupling, 62–64

Alumina, as enzyme adsorbent, 25, 28

Amidination, for enzyme covalent coupling, 68–69

Amino acid polymers, as supports for enzyme immobilization, 56

L-Amino acids, production of, using immobilized enzymes, 11, 330–342

Aminoacylase (immobilized)
  enzymic properties, 333–341
  preparation, 11, 27, 331–333
  use of, 329

α-Amino-ε-caprolactam hydrolase, immobilized, use of, 341–342

α-Amino-ε-caprolactam racemase, immobilized, use of, 341–342

6-Aminopenicillanic acid, production of, using immobilized microbial cells, 352

Amylase
  immobilized, 2, 25
  uses, 341

α-Amylase(s)
  immobilized, 27
  use in sugar production, 12

β-Amylase, immobilization of, 25

Amyloglucosidases
  immobilized, 27
  in multienzyme system, 206
  use in sugar production, 12

Analytical applications of immobilized enzymes, 10

Arylation, for enzyme covalent coupling, 62–64

Aspartase, use in aspartic acid production, 12, 329–330, 342–343

L-Aspartic acid, production of, 342–349, 353

Aspergillus oryzae, immobilized, L-tryptophan production by, 353

ATP creatine phosphotransferase, immobilized kinetics, 198

B

Bead-polymerization procedure, for enzyme entrapment, 31–37

Bentonite, as enzyme adsorbent, 25, 28

A 6
B 7
C 8
D 9
E 0
F 1
G 2
H 3
I 4
J 5